GENETIC BASIS FOR RESPIRATORY CONTROL DISORDERS

GENETIC BASIS FOR RESPIRATORY CONTROL DISORDERS

Edited by

Claude Gaultier

Robert Debré Hospital, Paris, France

Claude Gaultier
Hospital Robert Debré
University Denis Diderot Paris 7
Paris, France

ISBN-13: 978-0-387-70764-8   e-ISBN-13: 978-0-387-70765-5

Library of Congress Control Number: 2007933152

Printed on acid-free paper.

9 8 7 6 5 4 3 2 1

springer.com

To all children living with genetic respiratory control disorders, especially those who are members of the Association Française du Syndrome d'Ondine.

To Agnès Cazorla who proved a limitless source of expert assistance, Antoinette Wolfe for her faithful support, and Jorge Gallego for his keen and constructive comments.

To all those who are in my heart.

Introduction

Developmental respiratory control disorders constitute a broad spectrum of conditions ranging from the fairly common (apnoea of prematurity) to the very rare congenital central hypoventilation syndrome (CCHS). Investigations of respiratory control in newborns and infants are challenging to perform and remain far outside the scope of routine practice. Yet, early diagnosis and treatment is crucial, since respiratory control impairments at early developmental stages compromise brain oxygenation and may generate irreversible motor and cognitive disorders.

Early clinical observations showed family clustering of some of the respiratory control disorders, such as CCHS, suggesting a role for genetic factors. The description of the genome in mice and subsequently in humans opened up the field of research into the genetic basis for respiratory control disorders. The data of greatest clinical relevance were obtained in CCHS, a condition whose link to a *PHOX2B* gene mutation allows parental counselling and prenatal diagnosis. However, despite this advance in knowledge, many questions regarding the pathogenic mechanisms of CCHS remain to be resolved. Among other respiratory control disorders, Prader-Willi and Rett syndromes have been the focus of major advances regarding the pathogenesis of the respiratory control deficit. Even the complex and multifactorial pathogenesis of sudden infant death syndrome and obstructive apnoea syndrome is coming to light. These advances in research are the fruit of combined studies in mutant newborn mice and in humans.

This book is the first comprehensive collection of recent data on the genetic control of respiratory control development in mice from early to late stages of maturation. Respiratory phenotypes of mutant newborn mice lacking genes involved in the respiratory control phenotype are described, with emphasis on their relevance to unravelling the pathogenesis of respiratory control disorders.

Many scientists from different horizons, including paediatricians, geneticists, neurobiologists, and respiratory physiologists, accepted to contribute to this book. I wish to thank them for their outstanding work.

We have advanced in our research. However, we need to move even farther. Genetics will play a major role in the future. Genetic variability may govern not only the normal programming of respiratory control, but also the processing — via adaptation and plasticity — that occurs when the infant's homeostasis is disturbed by prenatal and/or postnatal insults. Further research should look into genetic and environmental interactions. Data obtained so far have opened up a new field of research into therapeutic strategies aimed at correcting respiratory deficits in newborn mice, with the goal of designing treatments for genetic respiratory control disorders in humans.

Claude Gaultier

Preface

If you want to know how something works, build it from scratch. With biological systems that self-assemble, this may be out of reach. A powerful alternative is to sketch out the blueprint and/or interfere with the building plan; watch carefully what happens and interpret the outcome with even more care.

Mammals must breathe at birth. The nervous system must generate a rhythmic motor output driving respiratory muscles and then modulate its frequency and patterns of muscle contraction and relaxation to assure an adequate supply of oxygen and removal of carbon dioxide; reflexes that control and protect the lung and upper airway, as well as coordination with swallowing must also work to (near) perfection. Then, as the infant mammal matures, with alternations in body mass, metabolism, lung and muscle mechanics, sleep and wake patterns, the nervous system must adapt, all on the fly with no scheduled down time for changes.

Breathing was one of the first behaviors studied by the earliest bona fide neuroscientists, late in the 18th century. However, the interests of neuroscientists in breathing faded, and studies of breathing became the domain of physiologists mostly interested in the lung and gas exchange. The brain, the engine of breathing, was essentially delegated the role of a black box transforming information about blood gases into a rhythmic pattern of motor activity that pumped the lungs. Starting in the 1950s, neurophysiologists rediscovered breathing as a worthy problem, their technique of choice was nerve and single neuron recordings, mostly from adult cats, then later rats. In 1986, the neonatal rat became a powerful experimental model, as its brainstem and spinal cord could be explanted to a recording chamber, allowing newly developed *in vitro* techniques to be applied to understanding basic mechanisms for breathing. The *in vitro* preparations from newborn rodents piqued interest in the developmental neurobiology of breathing, further fueled by the increasing realization of health problems associated with serious developmental disorders of breathing in humans in the early stages of postnatal life. Understanding how the nervous system reliably wired itself *in utero* to work at birth and matured hand in hand with changes in the lung, respiratory muscles and body size became in interesting and increasing pressing problem.

The present book summarizes the significant, and in many cases, landmark advances in the past decade in exploiting genetics and molecular biology to understand the neural control of breathing. Developmental disorders that show significant abnormalities in breathing include congenital central hypoventilation syndrome (CCHS), Rett syndrome, Prader-Willi syndrome, and sudden infant death syndrome (SIDS). Remarkably, many of these diseases appear associated with mutations in a single gene, CCHS-*Phox2B*; Rett syndrome-*MECP2*, or multiple genes, each of which may result in a phenotype leading to a similar diagnosis (SIDS). Mouse models with mutations/deletion of these genes result in phenotypes that in many cases show remarkable resemblance of the human breathing disorders. This allows systematic investigation of the sequelae from gene defect to breathing dysfunction. These disease-related genes are far from being the only ones of interest. The correct wiring and prenatal development of neural circuits generating respiratory rhythm, transforming rhythm into motor pattern, and providing signals related to blood gases and mechanical state of the lung and respiratory muscles requires a complex and delicate orchestration of precisely timed expression of key molecules, particularly transcription factors, guidance molecules, and receptors. Both serendipity and systematic investigation have unveiled numerous genes whose mutation/deletion produce significant changes in breathing in transgenic mice. These genes include those coding for: Krox20, MafB, Bdnf, GAD67, Mash1, Rnx, Nurr1, PACAP, Phox2b. Significant advances in understanding the particular features of the development of the brainstem (such as rhombomeric specification) where the key circuits for rhythm generation, central chemoreception and peripheral chemo- and mechano-receptor afferent processing are located, provide an essential basis for delineating the effects of mutated/deleted genes.

We all breathe differently: patterns of breathing in different states (sleep/wake, exercise) vary, as do responses to hypoxia and hypercapnia. Remarkable progress is being made in understanding the relationship between genetic and phenotypic variance in mice, e.g., GENENETWORK (www.genenetwork.org), and we should be optimistic that before long we will delineate the gene networks responsible for this variability, and the associated mechanisms. In some cases, breathing variability includes serious disorders, such as obstructive sleep apneas (OSA). Widely recognized as a significant health problem in the adult population, OSA, once thought rare in early life, is quite prevalent in infants and older children, and may underlie developmental disorders such as attention deficit hyperactivity disorder (ADHD). To the degree that there is a genetic component, it is likely to involve multiple genes, and tools to get at these genes are becoming increasingly powerful.

While the chapters in this volume represent remarkable progress, we still have a long way to go before we obtain a clear picture of the genetic and developmental neurobiology of breathing and associated diseases, but based on the work here, optimism is appropriate.

Jack L. Feldman

Contributors

Michael J. Ackerman, M.D., Ph.D.

Department of Pediatrics, Medicine, Molecular Pharmacology & Experimental Therapeutics, Mayo Clinic College of Medicine
200 First St. S.W., Rochester, Minnesota 55905
U.S.A.
Email: ackerman.michael@mayo.edu

Tiziana Bachetti, Ph.D.

Laboratorio di Genetica Molecolare, Istituto Giannina Gaslini
16148 Genova
ITALY
Email: tiziana.bachetti@tin.it

Elizabeth M. Berry-Kravis, M.D., Ph.D.

Departments of Neurology, Pediatrics, and Biochemistry, Rush University Medical Center
1653 W. Congress Parkway, Chicago Illinois 60612
U.S.A.
Email: elizabeth_m_berry-kravis@rush.edu

John Bissonnette, M.D.

Oregon Health and Science University, 3181 SW Sam Jackson Park Road, Portland, Oregon, 97239
U.S.A.
Email: bissonne@ohsu.edu

Bruno C. Blanchi, Ph.D.

Semel Institute for Neuroscience and Human Behavior, Mental Retardation Research Center, Neuroscience Research Building, Suite 355,
University of California-Los Angeles, Los Angeles, CA, 90095-7332
U.S.A.
Email: bblanchi@mednet.ucla.edu

Jean-François Brunet, Ph.D.

CNRS UMR 8542, Ecole Normale Supérieure,
46 rue d'Ulm, 75005 Paris
FRANCE
Email: jfbrunet@biologie.ens.fr

Isabella Ceccherini, Ph.D.

Laboratorio di Genetica Molecolare, Istituto Giannina Gaslini
16148 Genova
ITALY
Email: isa.c@unige.it

Jean Champagnat, Ph.D.

UPR 2216 Neurobiologie Génétique et Intégrative, Institut de Neurobiologie
Alfred-Fessard, C.N.R.S.
1, avenue de la terrasse, Gif sur Yvette 91198 Cedex,
FRANCE
Email: jean.champagnat@iaf.cnrs-gif.fr

Jack L. Feldman, Ph.D.
Department of Neurobiology, University of California Los Angeles, Los Angeles, CA 90095-1763
U.S.A.
Email: feldman@ucla.edu

Gilles Fortin, Ph.D.

UPR 2216 Neurobiologie Génétique et Intégrative, Institut de Neurobiologie
Alfred-Fessard, C.N.R.S.
1, avenue de la terrasse, Gif sur Yvette 91198 Cedex,
FRANCE
Email : gilles.fortin@inaf.cnrs-gif.fr

Jorge Gallego, Ph.D.
Inserm U676, Hôpital Robert Debré, Université Paris 7,
48 Bd Sérurier, 75019, Paris
FRANCE
Email: gallego@rdebre.inserm.fr

Claude Gaultier, M.D., Ph.D.

Service de Physiologie, Hôpital Robert Debré
48 Bd Sérurier 75019 Paris
FRANCE
Email: claude.gaultier@rdb.aphp.fr

Christo Goridis, Ph.D.

CNRS UMR 8542, Ecole Normale Supérieure,
46 rue d'Ulm, 75005 Paris
FRANCE
Email: goridis@wotan.ens.fr

David Gozal, M.D.

Kosair Children's Hospital Research Institute, University of Louisville, Baxter
Biomedical Research Building, Suite 204, 570 South Preston St.
U.S.A.
Email: david.gozal@louisville.edu

John J. Greer, M.D., Ph.D.

Department of Physiology, University of Alberta, 513 HMRC, Edmonton, AB,
Canada T6G 2S2
CANADA
Email: john.greer@ualberta.ca

Fang Han, M.D.

Department of Medicine, Louis Stokes DVA Medical Center, Case Western Re-
serve University, Cleveland OH
U.S.A.
Email: hanfang1@hotmail.com

Ronald M. Harper, Ph.D.

Department of Neurobiology, David Geffen School of Medicine, University of
California Los Angeles, Los Angeles, CA 90095-1763
U.S.A.
Email: rharper@ucla.edu

Gerard Hilaire, Ph.D.

CNRS, Formation de Recherche en Fermeture, FRE 2722,
280 Boulevard Sainte Marguerite, 13009 Marseille
FRANCE
Email: hilaire@marseille.inserm.fr

Rajesh Kumar, Ph.D.

Department of Neurobiology, David Geffen School of Medicine, University of
California Los Angeles, Los Angeles, CA 90095-1763
U.S.A.
Email: rkumar@mednet.ucla.edu

Paul M. Macey, Ph.D.

Department of Neurobiology, David Geffen School of Medicine, University of
California Los Angeles, Los Angeles, CA 90095-1763
U.S.A.
Email: pmacey@ucla.edu

Mary L. Marazita, Ph.D.
Center for Craniofacial and Dental Genetics, Division of Oral Biology, Oral and
Maxillofacial Surgery, Human Genetics, Graduate School of Public Health,
University of Pittsburgh/School of Dental Medicine, 100 Technology Drive,
Suite 500, Cellomics Building, Pittsburgh, Pennsylvania 55905
U.S.A.
Email: marazita@sdmgenetics.pitt.edu

Gordon S. Mitchell, Ph.D.
Department of Comparative Biosciences, University of Wisconsin, 2015 Linden
Drive, Madison, WI, 53706
U.S.A.
Email: mitchell@svm.vetmed.wisc.edu

Sanjay R Patel, M.D., M.S.

Case Western Reserve University, Cleveland OH 44106
U.S.A.
Email: Srp20@case.edu

Nélina Ramanantsoa, M.S.

Inserm U676, Hôpital Robert Debré, Université Paris 7
48 Bd Sérurier, 75019, Paris
FRANCE
Email: nelina.ramanantsoa@rdebre.inserm.fr

Susan Redline, M.D., M.P.H.

Case Western Reserve University, Cleveland OH 44106-6006
U.S.A.
Email: susan.redline@case.edu

Michael H. Sieweke, Ph.D.

Centre d'Immunologie de Marseille Luminy, CNRS-INSERM-Univ. Med.,
Campus de Luminy, Case 906, 13288 Marseille Cedex 09
FRANCE
Email: sieweke@ciml.univ-mrs.fr

Kingman P. Strohl, M.D.

Department of Medicine, Louis Stokes DVA Medical Center, Case Western Re-
serve University, Cleveland OH
U.S.A.
Email: kpstrohl@aol.com

Clarke G. Tankersley, Ph.D.

Division of Physiology, Bloomberg School of Public Health, The Johns Hop-
kins University, 615 N. Wolfe Street, Baltimore, MD 21205
U.S.A.
Email: ctankers@jhsph.edu

Muriel Thoby-Brisson, Ph.D.

UPR 2216 Neurobiologie Génétique et Intégrative, Institut de Neurobiologie
Alfred-Fessard, C.N.R.S., 1, avenue de la terrasse, Gif sur Yvette 91198 Cedex
FRANCE
Email : muriel.thoby-brisson@inaf.cnrs-gif.fr

Ha Trang, M.D., Ph.D.

Service de Physiologie, Hôpital Robert Debré
48 Bd Sérurier 75019 Paris
FRANCE
Email : ha.trang@rdb.aphp.fr

Vanessa Vaubourg, M.S.

Inserm U676, Hôpital Robert Debré, Université Paris 7
48 Bd Sérurier, 75019, Paris
FRANCE
Email: vanessa.vaubourg@rdebre.inserm.fr

Debra E. Weese-Mayer, M.D.

Pediatric Respiratory Medicine, Rush Children's Hospital, Rush University
Medical Center, 1653 West Congress Parkway, Chicago, IL 60612, U.S.A.
Email: Debra_E_Weese-Mayer@rush.net

John V. Weil, M.D.

University of Colorado Health Sciences Center, 4200 E. Ninth Ave, Denver,
Colorado, 80220
U.S.A.
Email: john.weil@uchsc.edu

Rachel Wevrick, Ph.D.

Department of Medical Genetics, University of Alberta, 513 HMRC, Edmon-
ton, AB, Canada T6G 2S2
CANADA
Email: rwevrick@ualberta.ca

Mary A. Woo, D.N.Sc.
School of Nursing, University of California Los Angeles, Los Angeles, CA
90095-1702 U.S.A.
U.S.A.
Email: mwoo@sonnet.ucla.edu

Motoo Yamauchi, M.D.
Department of Medicine, Louis Stokes DVA Medical Center, Case Western Reserve University, Cleveland OH
U.S.A.
Email: mountain@pastel.ocn.ne.jp

Contents

Abbreviations

aCSF	Artificial cerebrospinal fluid
A2A	Adenosine type 2A receptor
ACE	Angiotensin converting enzyme
ADHD	Attention-deficit hyperactivity disorder
AHI	Apnea hypopnea index
AIH	Acute intermittent hypoxia
Akt	Protein kinase B
ALS	Amyotrophic lateral sclerosis
AMPA	Alpha-amino-3-hydroxy-5-methyl-4-isoxazolepropionic acid
ANKB	Human gene encoding ankyrin B
ANS	Autonomic nervous system
AP	Area postrema
APOE	Apolipoprotein E
AS	Active sleep
ASC	Achaete-scute proneural complex
ATP	Adenosine tri-phosphate
B6	C57B1/GJ mice
BDNF	Brain derived neurotrophic factor
bHLH	Basic helix-loop-helix
BM	Branchiomotor
BMI	Body mass index
BMP	Bone morphogenic protein
BMP2	Human gene encoding bone morphogenic protein-2
BrS	Brugada syndrome
BSA	Body surface area
Beta-2n AchR	Beta2 subunit of the nicotinic acetylcholine receptor
C	Controls
C3	C3H/HEJ mice
C_3	Cervical spinal segment 3
CACNA1C	Calcium channel gene A1C
CAV3	Human gene encoding caveolin 3
CB	Carotid body
CCHS	Congenital central hypoventilation syndrome
CFEOMII	Congenital fibrosis of extra-oculomotor muscles type II

Cftr	Mouse gene encoding cystic fibrosis conductance regulator
CIH	Chronic intermittent hypoxia
CNS	Central nervous system
CO_2	Carbon dioxide
COPD	Chronic obstructive pulmonary disease
CORS	Cerebello-oculo-renal syndrome
CPAP	Continuous positive airway pressure
CPVT	Catecholaminergic polymorphic ventricular-tachycardia
CSA	Central sleep apnoea
CSAS	Central sleep apnoea syndrome
CSN	Carotid sinus nerve
CSSs	Chromosomal substitution strains
CV	Coefficient of variation
CYP1A1	Human gene encoding cytochrome P-450 1A1
DA	Dopamine
DAMGO	[D-Ala2,methyl-Phe4, Gly-ol5]enkephalin
DAS	Differential ability scales
Dbh, DBH, DßH	Dopamine-ß-hydroxylase
dmnX	Dorsal motor nucleus of the vagus nerve
DOI	2, 5-dimethoxy-4-iodoamphetamine
E	Embryonic day
Ece1	Endothelin-converting enzyme 1
ECE1	Human gene encoding endothelin converting enzyme 1
Edn1	Endothelin 1
EDN1	Human gene encoding endothelin 1
Edn3	Endothelin 3
Ednra	Endothelin receptor a
Eln	Gene for elastin
EMSA	Electrophoresis mobility shift assay
EN1	Human gene encoding engrailed 1
Eph	Ephrin
ERK	Extracellular regulated kinases 1 and 2
F1	First-generation offspring
5HIAA	5-hydroxylindole acetic acid
5HT	5-hydroxytryptamine (serotonin)
5HT2	Serotonin type 2 receptor
5HT2A	Serotonin type 2A receptor
5HT7	Serotonin type 7 receptor
5HTT	Serotonin transporter
fMRI	Functional magnetic resonance imaging

G	Tissue damping
GA	Geldanamycin
GABA	Gamma-aminobutyric acid
GAD	GABA-synthesizing enzyme
Galphaq	Guanine nucleotide binding protein, alpha q subunit
Gdnf	Glial cell line-derived neurotrophic factor
GFP	Green fluorescent protein
GFRA1	GDNF Family receptor alpha1
GH	Growth hormone
Gs protein	Stimulatory guanine nucleotide binding protein
GSTT1	Human gene encoding glutathione S-transferase
H	Tissue elastance
HASH1	Human homologue of *Mash1*
HCVR	Hypercapnic ventilarory response
HF	High frequency embryonic activity
HH	Hamburger and Hamilton
HPLC	High pressure liquid chromatography
HSCR	Hirschsprung disease
HSP	Heat shock protein
HTN	Hypertension
HVR	Hypoxic ventilatory response
I	Enteric interneuron
Ih	Hyperpolarization-activated cationic current
IVth	Trochlear nucleus
IH	Intermittent hypoxia
IIIrd	Oculomotor nucleus
IQ	Intellectual quotient
IXth	Petrosal ganglion
JBTS(1/2/3)	Joubert syndrome
KCNE1	Human gene encoding potassium channel E1
KCNE2	Human gene encoding potassium channel E2
KCNEJ2	Human gene encoding potassium channel EJ2
KCNH2	Human gene encoding potassium channel H2
KCNQ1	Human gene encoding potassium channel Q1
Kir2.2	Potassium channel Kir2.2
Kir6.2	Potassium channel Kir6.2
Krox20	Mouse Krüppel box gene
Kv1.1	Potassium channel Kv1.1

LC	Locus coeruleus
LF	Low frequency embryonic rhythm
LG	Loop gain
LQTS	Long QT syndrome
LTF	Long-term facilitation
M	Enteric motoneuron
MAOA	Monoamine oxidase A
MAP kinase	Mitogen-activated protein kinase
Mash1	Mouse gene encoding mammalian achaete-scute
MD	Mean diffusivity
MECP2	Human gene encoding methyl-CpG-binding protein 2
MeCP2	Human methyl-CpG-binding protein 2
Mecp2	Mouse gene encoding methyl-CpG-binding protein 2
Mecp2	Mouse methyl-CpG-binding protein 2
MHD	MAGE homology domain
MR	Medullary raphe
MT	Motor twitches
N_2	Nitrogen
NA	Noradrenaline, (nor)adrenergic
nA	Nucleus ambiguous
NADPH oxidase	Nicotinamide adenine dinucleotide phosphate-oxidase
Ndn	Mouse gene encoding necdin
NGF	Nerve growth factor
NK1	Neurokinin-1
NK1R	Neurokinase1 receptor
Nkx	Natural killer homologue-related homeobox genes
NMDA	*N*-methyl-D-aspartate
NREM	Non-rapid eye movement sleep
nTS	Nucleus of the solitary tract
O_2	Oxygen
OR	Odds Ratio
ORX	Orexin
OSA	Obstructive sleep apnea
P	Partial pressure
PABPN1	Poly(A) binding protein nuclear 1
PACAP	Pituitary adenylate cyclase-activating peptide
PAHs	Polycyclic aromatic hydrocarbons

pC	Paracardiac ganglia
PCR	Polymerase chain reaction
PEEP	Positive end-expiratory pressure
pFRG	Parafacial respiratory group
PHFD	Post-hypoxic frequency decline
Phox	Paired-like homeobox transcription factor
PHOX2A	Human gene encoding paired-like homeobox 2a
Phox2a	Mouse gene encoding paired-like homeobox 2a
PHOX2B	Human gene encoding paired-like homeobox 2b
Phox2b	Mouse gene encoding paired-like homeobox 2b
PHVD	Post-hypoxic ventilatory decline
PKC	Protein kinase C
pLTF	Phrenic long-term facilitation
pMNv	Neuroepithelial domain of the hindbrain
PMOC	Proopiomelanocortin
PNS	Peripheral nervous system
Pre-BötC	Pre-Bötzinger Complex
PTT-A	Preprotachykinin-A
PV curves	Pressure-volume curves
PWS	Prader-Willi Syndrome
QS	Quiet sleep
QTc	Corrected QT interval on ECG
QTL	Quantitative trait loci
r	Rhombomere regulator
REM	Rapid eye movement sleep
RET	Human gene encoding rearranged during transfection factor
Ret	Mouse gene encoding rearranged during transfection factor
RNAi	RNA interference
RNX	Human gene encoding respiratory neuron homeobox
Rnx	Mouse gene encoding respiratory neuron homeobox
ROS	Reactive oxygen species
RRG	Respiratory rhythm generator
RTN	Retrotrapezoid nucleus
RTT	Rett syndrome
RyR2	Human gene encoding ryanodine receptor 2

S	Enteric sensory neuron
SA	Sympatho-adrenergic
SaO$_2$	Arterial oxygen saturation
SHR	Spontaneously hypertensive
SAP	Saporin
SCI	Spinal cord injury
SCN4B	Human gene encoding sodium channel 4B
SCN5A	Human gene encoding sodium channel 5A
SERT	Serotonin transporter protein
7-NI	7-nitroindazole
SH	Sustained hypoxia
SHR	Spontaneously hypertensive rat
SIDS	Sudden infant death syndrome
siRNA	Small, interfering RNA
Slc	Solute carrier
SNP	Single nucleotide polymorphisms
SOD-1	Superoxide dismutase 1
SOD1^{G93A}	Human superoxide dismutase 1 with G93A mutation
SST	Somatostatin
STP	Short-term potentiation
SubP	Substance P
SWS	Slow-wave sleep
TAC1	Tachykinin-1
Th, TH	Tyrosine hydroxylase
T_I/T_{TOT}	Ratio of inspiratory-to-total time
TLC	Total lung capacity
TLX2	Human gene encoding T-cell leukemia 2 homeobox
TLX3	Human gene encoding T-cell leukemia 3 homeobox
TRH	Thyrotropin releasing hormone
Trk	Tyrosine kinase
Trk A	Tyrosine kinase A
Trk B	Tyrosine kinase B
TrkB	Tropomycin related kinase B
Tubb3	ßIII-tubulin
UAR	Upper airways resistance
UDP	Uridine diphosphate
V	Trigeminal nerve
VC	Vital capacity
V$_{30}$	Volume at 30 cmH$_2$0 of airway pressure
V$_{CO2}$	Carbon dioxide production

V_D	Dead-space volume
V_E, VE	Minute ventilation
VEGF	Vascular endothelial growth factor
V_{EQ}	V_E normalized for V_{O2} or V_{CO2}
VGLUT	Vesicular transporter of glutamate
VII	Facial nerve
VLM	Ventro-lateral medulla
VM	General visceral motor
VM/BM	Column of branchial and visceral motoneurons
VNTR	Variable number tandem repeat
V_{O2}	Oxygen consumption
VRG	Ventral respiratory group
V_T, VT	Tidal volume
V_T/T_I	Mean inspiratory flow
WISC-III	Wechsler intelligence scale for children
WKR	Wistar-Kyoto rat
WT	Wild type
XI	Glossopharyngeal nerve
XII	Hypoglossal nerve
Xth	Vagal ganglion
Z_{rs}	Impedance of the respiratory system

1. Respiratory control disorders: from genes to patients and back

David GOZAL

Kosair Children's Hospital Research Institute, and Departments of Pediatrics, Pharmacology and Toxicology, University of Louisville, Louisville, Kentucky, U.S.A.

1.1 Introduction

Our understanding of respiratory control has undergone multiple evolutionary waves since the initial steps pointing to the role of the brain as the source of breathing. In fact, the discipline of neurosciences in general, and more particularly the field of neural cardiorespiratory control, has changed so radically that classic principles that at one time became dogmatic in medicine, have now undergone multiple iterative modifications, such as to account for the flurry of discoveries challenging these original and traditionally accepted concepts.

There is no denial to the evidence that breathing is a critically important and vital function. Yet, there is a paradoxical attitude assumed by most clinicians in their reluctance to become intimately familiar with this important subject. The reasons leading to such aversion of the topic are unclear. May be it is because of the complexity of the interactions between the brain and the lungs. It is indeed conceptually difficult for any clinician to think in terms of several organ interactions, particularly as the specialties become more and more sub-specialized. Alternatively, teaching of the anatomical and physiological elements of respiratory control during medical school is generally fraught with very complicated and unclear anatomical maps and functional descriptions, most of which are void of any quickly memorizable algorithms, such that the location and functional roles of the various nuclei within the brainstem becomes more of a guessing exercise rather than a truly integrated and logical process. If the teacher is unclear, how could we expect the student clinicians to become enthusiastic and avid learners of our field? The uncertainty of the precise elements associated with specific neural respiratory pathways, the complexities of neuronal firing activities, and the multitude of neurotransmitters each with brainstem nucleus-dependent opposing roles on the same respiratory function are undoubtedly tangible obstacles for the medical student or even for the seasoned practitioner. No wonder then that when a patient presents with some kind of mysterious alterations in their control of respiration there is an almost automatic detachment by the residents and fellows involved in the care of these patients and absolute dependency on the opinions and decisions formulated by the one physician who seems to be less intimidated by the situation. Clearly, this has to change, especially when we consider that we are on the brink of an explosion of knowledge on all areas of biomedical disciplines. The field of respiratory control has evolved tremendously in recent years, and we are now witnessing the initial discovery of several of the genes that control the development and maturation of multiple neurally-controlled respiratory functions. These are indeed exciting times, and the contributions to the field now permit a much more coherent and therefore much more intellectually attractive scientific discourse that should be and can be applied to our daily activities in the care of our patients. I will not focus on these exciting discoveries, since these are reviewed in great detail within multiple chapters in this book. Instead, I will delineate the importance of early identification and treatment of the "rare" patient who presents with a disorder of respiratory control.

1.2 Effect of sleep on breathing

Because most of the alterations in gas exchange will occur during sleep in the context of a patient with a respiratory control disorder, it is important to review, if only briefly, the effects of sleep states on breathing. In addition to substantial frequency and tidal volume changes during and between sleep states, normal breathing during sleep is characterized by frequent short pauses in breathing that are associated with transitions in state, body movements, or transient elevations in blood pressure. Disordered breathing, as expressed by hypoxia, hypoventilation, or total cessation of airflow can emerge during sleep, with such marked declines in oxygen saturation as to compromise survival. Sleep state effects on breathing interact with a number of other structural respiratory and brain disorders, often accentuating the primary disorder to intolerable levels for survival. On the other hand, sleep states can assist breathing in other disease conditions, and can provide relief in certain circumstances that interfere with ventilation during waking.

The cyclic activity of the respiratory central pattern generator is further modulated by suprapontine sites that include important efferent projections to areas mediating the sleep-wake cycle, thermoregulation, and circadian rhythmicity. Respiratory control areas also receive afferent inputs from central and peripheral chemoreceptors and from other receptors within the respiratory pump (upper airways, chest wall, and lungs). During fetal life, breathing is discontinuous and coincides with rapid eye movement (REM)-like sleep. After birth, respiratory rhythm is established as a continuous activity to maintain cellular oxygen and carbon dioxide homeostasis. Respiratory pattern instability during sleep is typically present during early life, such that apneic episodes lasting for only a few seconds are extremely common in preterm infants, and their frequency is reduced in full-term infants [16]. These apneic episodes are usually of a central nature in the full-term infant and primarily occur during REM sleep. The greater respiratory instability during REM sleep, as compared to NREM sleep, may be a consequence of phasic inhibitory-excitatory phenomena that characterize REM sleep. An important aspect of REM sleep is the muscular paralysis, or atonia, that is a principal characteristic of the state in addition to bursts of phasic eye movements. The atonia is widespread, and includes virtually all of the striated musculature, except for the diaphragm, in which components are alternately activated during REM sleep [41]. That paralysis is of significant importance in for example, infants with compliant thoracic walls and weak intercostal muscles or in patients with increased upper airway collapsibility such as those with enlarged tonsils and adenoids. Indeed, the cranial motor pools innervating muscles of the upper airway, which include the hypoglossal (XII), XI via X, IXth, VII, and Vth motor pools, and which supply the muscles of the tongue, larynx, styloglossus, dilator nares, tensor palati and masseter muscles, respectively, will lose tone during REM sleep, and therefore set up the stage for upper airway collapse.

1.3 Diagnostic approaches to the patient with suspected abnormalities in respiratory control

Apnea of prematurity, apparent life-threatening events, sudden infant death syndrome, obstructive sleep apnea, and central apnea and hypoventilation are relatively frequent conditions in the pediatric age range and are associated with substantial morbidity and mortality. While most of our diagnostic approaches to such conditions are usually passive, i.e., we observe and record cardiorespiratory patterns during wakefulness and sleep, it is sometimes necessary to complement such recordings with more specific testing. For example, exclusion of underlying lung, cardiac or neuromuscular diseases is usually a pre-requisite for the accurate diagnosis. To this effect, imaging of the brain and spinal cord, echocardiogram, radiological assessment of lung structure and diaphragmatic excursion are routinely conducted. Unfortunately, we have not yet reached the stage, whereby functional imaging studies are routinely used for improved assessments of patients presenting with altered respiratory control [24; 29; 30; 32; 48]. In some cases, metabolic studies aiming to determine abnormalities of fatty or organic acids or other selected enzymes may be useful in conjunction with an evaluation by a geneticist. In other situations, it may be worthwhile to determine the functional status of peripheral and central chemoreceptors. Multiple tests have been developed over the years, but as rule of thumb, most will require a well equipped physiological laboratory, preferably with extensive experience in such tests. For example, sudden hypoxic and hypercapnic transients, as well as hyperoxic and alternate breath tests have all been successfully used in the assessment of infants and children with suspected dysfunction of peripheral chemoreceptors [5-9; 19-21;

23; 27; 33; 39; 42]. Furthermore, the use of rebreathing techniques such as hyperoxic hypercapnia and isocapnic hypoxia have allowed for determination of hypercapnic and hypoxic drives both during waking and sleep in children with and without disorders of respiratory control [11; 18; 26; 34; 35; 40; 43]. Finally, the recent discovery of genes associated with specific conditions has prompted wider utilization of such tests when such disorders are suspected. For example, it is now quite routine to assess for the presence of *PHOX2B* mutations among children presenting with central hypoventilation [13; 40], and such tests represent a formidable advance in our ability to separate across different disease presenting with a common phenotype.

1.4 Potential consequences of delayed diagnosis and treatment

Certain disorders of respiratory control will interfere with ventilation to the extent that development of neural structures may be affected. For example, we have gained substantial incremental knowledge in recent years on the different consequences of intermittent and sustained hypoxia during sleep, in the context of sleep apnea. For example, several studies in children with obstructive sleep apnea have documented significantly reduced IQ scores (obtained from the Wechsler Intelligence Scale for Children – WISC-III) compared with control children [4]. In these studies, the probability for lower normal or borderline range performance was much higher in the presence of sleep-disordered breathing. More recently, we have documented significantly impaired General Conceptual Ability scores (a measure of IQ obtained from the Differential Ability Scales; DAS) in school-age [37] and preschool-age [36] children with sleep apnea when compared with control children. Academic performance is affected [17], and nocturnal hypoxia, particularly when of intermittent nature, seems particularly deleterious to normal cognitive development in children [1; 2; 28; 46]. These principles apply to children with disorders of control of breathing, who may be even more susceptible to the injurious brain processes activated by the occurrence of repeated hypoxemia or asphyxic episodes during sleep [47]. As such, the impact of such events affects not only cognition and behavior, but also quality of life [10; 12; 14; 25].

1.5 Where do we go from here?

While substantial advances have occurred in the process of evaluating and treating children with respiratory control disorders, particularly as it relates to invasive and non-invasive home mechanical ventilation, we are still in the early phases of understanding the variance in the phenotype of such diseases. Several chapters in this book are dedicated to the exploration of genes that underlie specific conditions, and such overviews clearly recapitulate the linkage between a gene and a cluster of specific symptoms and signs. For example, a recent study among *PHOX2B* carriers indicated that those with nonpolyalanine repeat mutations would be more likely to display more severe disruption of *PHOX2B* function [3], while the presence of either missense or frameshift heterozygous mutations appeared to be linked to a higher risk for development of tumors of the autonomic nervous system [45]. However, we still do not know, how genes involved in the pathophysiology of certain diseases interact with other genes, and how environmental conditions modify the phenotypic presentation of a specific disease. For example, we have previously shown that intermittent hypoxia during gestation or during the early days after birth can elicit substantial and lifelong alterations of respiratory patterning in rats [22; 38]. Similar alterations in the regulation of respiratory rhythm have been observed in offspring of cigarette smoking mothers [31]. As discussed in yet another chapter in this volume, these environmental exposures can not only modify the phenotype, but could potentially be used to restore function [15].

1.6 Conclusion

Improvements in our understanding of the genes and their functions in the regulation of the complexity of respiratory control mechanisms in the human will undoubtedly permit the busy clinician to apply such concepts to medical practice. Such accrued knowledge will undoubtedly ameliorate the quality of the care that physicians currently provide to their patients, and further exploration of gene-gene interactions, gene-environment interactions through generation of animal models should permit not only improved diagnostics but also formulation of novel therapeutic strategies. To the statement of Rollo May, a noted psychologist in the 20th century "*It is an ironic habit of human beings to run faster when we have lost our way*", I would add that we need to run even faster, now that we are finally finding our way.

Acknowledgements

DG is supported by National Institutes of Health grants SCOR 2P50-HL-60296 (Project 2), RO1 HL-65270, and RO1-HL-69932, The Children's Foundation Endowment for Sleep Research, and by the Commonwealth of Kentucky Challenge for Excellence Trust Fund.

References

1.   Bass JL, Corwin M, Gozal D, Moore C, Nishida H, Parker S, Schonwald A, Wilker RE, Stehle S, Kinane TB (2004) The effect of chronic or intermittent hypoxia on cognition in childhood: a review of the evidence. Pediatrics 114: 805-816

2.   Beebe DW (2006) Neurobehavioral morbidity associated with disordered breathing during sleep in children: a comprehensive review. Sleep 29: 1115-1134

3.   Berry-Kravis EM, Zhou L, Rand CM, Weese-Mayer DE (2006) Congenital central hypoventilation syndrome: PHOX2B mutations and phenotype. Am J Respir Crit Care Med 174: 1139-1344

4.   Blunden SL, Beebe DW (2006) The contribution of intermittent hypoxia, sleep debt and sleep disruption to daytime performance deficits in children: consideration of respiratory and non-respiratory sleep disorders. Sleep Med Rev 10: 109-118

5.   Bouferrache B, Filtchev S, Leke A, Freville M, Gallego J, Gaultier C (2002) Comparison of the hyperoxic test and the alternate breath test in infants. Am J Respir Crit Care Med 165: 206-210

6.   Bouferrache B, Filtchev S, Leke A, Marbaix-Li Q, Freville M, Gaultier C (2000) The hyperoxic test in infants reinvestigated. Am J Respir Crit Care Med 161: 160-165

7.   Bouferrache B, Krim G, Marbaix-Li Q, Freville M, Gaultier C (1998) Reproducibility of the alternating breath test of fractional inspired O_2 in infants. Pediatr Res 44: 239-246

8.   Chardon K, Bach V, Telliez F, Cardot V, Tourneux P, Leke A, Libert JP (2004) Effect of caffeine on peripheral chemoreceptor activity in premature neonates: interaction with sleep stages. J Appl Physiol 96: 2161-2166

9.   Chua TP, Coats AJ (1995) The reproducibility and comparability of tests of the peripheral chemoreflex: comparing the transient hypoxic ventilatory drive test and the single-breath carbon dioxide response test in healthy subjects. Eur J Clin Invest 25: 887-892

10.   Crabtree VM, Varni JW, Gozal D (2004) Health-related quality of life and depressive symptoms in children with suspected sleep-disordered breathing. Sleep 27: 1131-1138

11.   Dahan A, DeGoede J, Berkenbosch A, Olievier IC (1990) The influence of oxygen on the ventilatory response to carbon dioxide in man. J Physiol 428: 485-499

12.   Dellborg C, Olofson J, Midgren B, Caro O, Skoogh BE, Sullivan M (2002) Quality of life in patients with chronic alveolar hypoventilation. Eur Respir J 19: 113-120

13.   Doherty LS, Kiely JL, Deegan PC, Nolan G, McCabe S, Green AJ, Ennis S, McNicholas WT (2007) Late-onset central hypoventilation syndrome: a family genetic study. Eur Respir J 29: 312-316

14.   Doran SM, Harvey MT, Horner RH (2006) Sleep and developmental disabilities: assessment, treatment, and outcome measures. Ment Retard 44: 13-27

15.   Fuller DD, Golder FJ, Olson EB, Mitchell GS (2006) Recovery of phrenic activity and ventilation after cervical spinal hemisection in rats. J Appl Physiol 100: 800-806

16.   Gaultier C (1995) Cardiorespiratory adaptation during sleep in infants and children. Pediatr Pulmonol 19: 105-117

17.   Gozal D (1998) Sleep-disordered breathing and school performance in children. Pediatrics 102: 616-620
18.   Gozal D (2001) Central chemoreceptor function in children. Pediatr Pulmonol Suppl 23: 110-113
19.   Gozal D, Arens R, Omlin KJ, Jacobs RA, Keens TG (1995) Peripheral chemoreceptor function in children with myelomeningocele and Arnold-Chiari malformation type 2. Chest 108: 425-431
20.   Gozal D, Arens R, Omlin KJ, Ward SL, Keens TG (1994) Absent peripheral chemosensitivity in Prader-Willi syndrome. J Appl Physiol 77: 2231-2236
21.   Gozal D, Marcus CL, Shoseyov D, Keens TG (1993) Peripheral chemoreceptor function in children with the congenital central hypoventilation syndrome. J Appl Physiol 74: 379-387
22.   Gozal D, Reeves SR, Row BW, Neville JJ, Guo SZ, Lipton AJ (2003) Respiratory effects of gestational intermittent hypoxia in the developing rat. Am J Respir Crit Care Med 167: 1540-1547
23.   Haider AZ, Rehan V, Al-Saedi S, Alvaro R, Kwiatkowski K, Cates D, Rigatto H (1995) Effect of baseline oxygenation on the ventilatory response to inhaled 100% oxygen in preterm infants. J Appl Physiol 79: 2101-2105
24.   Harper RM, Macey PM, Woo MA, Macey KE, Keens TG, Gozal D, Alger JR (2005) Hypercapnic exposure in congenital central hypoventilation syndrome reveals CNS respiratory control mechanisms. J Neurophysiol 93: 1647-1658
25.   Hsiao KH, Nixon GM (2007) The effect of treatment of obstructive sleep apnea on quality of life in children with cerebral palsy. Res Dev Disabil, in press
26.   Katz-Salamon M (2004) Delayed chemoreceptor responses in infants with apnoea. Arch Dis Child 89: 261-266
27.   Katz-Salamon M, Eriksson M, Jonsson B (1996) Development of peripheral chemoreceptor function in infants with chronic lung disease and initially lacking hyperoxic response. Arch Dis Child Fetal Neonatal 75: F4-F9
28.   Kheirandish L, Gozal D (2006) Neurocognitive dysfunction in children with sleep disorders. Dev Sci 9: 388-399.
29.   Kumar R, Macey PM, Woo MA, Alger JR, Harper RM (2006) Elevated mean diffusivity in widespread brain regions in congenital central hypoventilation syndrome. J Magn Reson Imaging 24: 1252-1258
30.   Kumar R, Macey PM, Woo MA, Alger JR, Keens TG, Harper RM (2005) Neuro-anatomic deficits in congenital central hypoventilation syndrome. J Comp Neurol 487: 361-371
31.   Luo Z, McMullen NT, Costy-Bennett S, Fregosi RF (2007) Prenatal nicotine exposure alters glycinergic and GABAergic control of respiratory frequency in the neonatal rat brainstem-spinal cord preparation. Respir Physiol Neurobiol, in press
32.   Macey KE, Macey PM, Woo MA, Harper RK, Alger JR, Keens TG, Harper RM (2004) fMRI signal changes in response to forced expiratory loading in congenital central hypoventilation syndrome. J Appl Physiol 97: 1897-1907
33.   McClean PA, Phillipson EA, Martinez D, Zamel N (1998) Single breath of CO_2 as a clinical test of the peripheral chemoreflex. J Appl Physiol 64: 84-89
34.   Modarreszadeh M, Bruce EN, Hamilton H, Hudgel DW (1995). Ventilatory stability to CO_2 disturbances in wakefulness and quiet sleep. J Appl Physiol 79: 1071-1081
35.   Mohan RM, Amara CE, Cunningham DA, Duffin J (1999) Measuring central-chemoreflex sensitivity in man: rebreathing and steady-state methods compared. Respir Physiol 115: 23-33

36.   Montgomery-Downs HE, Crabtree VM, Gozal D (2005) Cognition sleep and respiration in at-risk children treated for obstructive sleep apnea. Eur Resp J 25: 336-342

37.   O'Brien LM, Mervis CB, Holbrook CR, Bruner JL, Smith NH, McNally N, McClimment MC, Gozal D. (2004) Neurobehavioral correlates of sleep disordered breathing in children. J Sleep Res 13: 165-172

38.   Reeves SR, Mitchell GS, Gozal D (2006) Early postnatal chronic intermittent hypoxia modifies hypoxic respiratory responses and long-term phrenic facilitation in adult rats. Am J Physiol Regul Integr Comp Physiol 290: R1664-R1671

39.   Rigatto H, Brady JP, de la Torre Verduzco R (1975) Chemoreceptor reflexes in preterm infants: I. The effect of gestational and postnatal age on the ventilatory response to inhalation of 100% and 15% oxygen. Pediatrics 55: 604-613

40.   Riley DJ, Santiago TV, Daniele RP, Schall B, Edelman NH (1977) Blunted respiratory drive in congenital myopathy. Am J Med 63: 459-466

41.   Sieck GC, Trelease RB, Harper RM (1984) Sleep influences on diaphragmatic motor unit discharge. Experimental Neurology 85: 316-335

42.   Sohrab S, Yamashiro SM (1980) Pseudorandom testing of ventilatory response to inspired carbon dioxide in man. J Appl Physiol 49: 1000-1009

43.   Swaminathan S, Paton JY, Ward SL, Jacobs RA, Sargent CW, Keens TG (1989) Abnormal control of ventilation in adolescents with myelodysplasia. J Pediatr 115: 898-903

44.   Trang H, Dehan M, Beaufils F, Zaccaria I, Amiel J, Gaultier C (2005) French CCHS Working Group. The French Congenital Central Hypoventilation Syndrome Registry: general data, phenotype, and genotype. Chest 127: 72-79

45.   Trochet D, O'Brien LM, Gozal D, Trang H, Nordenskjold A, Laudier B, Svensson PJ, Uhrig S, Cole T, Niemann S, Munnich A, Gaultier C, Lyonnet S, Amiel J (2005) PHOX2B genotype allows for prediction of tumor risk in congenital central hypoventilation syndrome. Am J Hum Genet 76: 421-426

46.   Urschitz MS, Wolff J, Sokollik C, Eggebrecht E, Urschitz-Duprat PM, Schlaud M, Poets CF (2005) Nocturnal arterial oxygen saturation and academic performance in a community sample of children. Pediatrics 115: 204-209

47.   Vanderlaan M, Holbrook CR, Wang M, Tuell A, Gozal D (2004) Epidemiologic survey of 196 patients with congenital central hypoventilation syndrome. Pediatr Pulmonol 37: 217-229

48.   Woo MA, Macey PM, Macey KE, Keens TG, Woo MS, Harper RK, Harper RM (2005) fMRI responses to hyperoxia in congenital central hypoventilation syndrome. Pediatr Res 57: 510-518

2. Hereditary aspects of respiratory control in health and disease in humans

John V. WEIL

University of Colorado Health Sciences Center, 4200 E. Ninth Ave, Denver, Colorado, 80220, U.S.A.

2.1 Introduction

An early step toward discovery of heritable influences on ventilatory control was the finding of differences in the strength of ventilatory responses among human subjects. Here we review observations that these responses tended to be similar among family members and identical twins suggesting a role for heredity and that these effects were most evident for the ventilatory response to hypoxia than to hypercapnia. Findings in humans and animals suggest that the effect is largely on the strength of peripheral chemoreception.

2.2 Inter-individual variation in human ventilatory control

Among the earliest indications of variability in ventilatory response were the findings by Schaeffer [43] of inter-individual differences in the hypercapnic ventilatory response and the later observation by Beral and Read that hypercapnic ventilatory response was decreased in Enga tribesmen of New Guinea compared to Caucasians [6].

Because of the greater dependence of oxygenation on ventilation at high altitude than at sea level, some of the earliest differences in hypoxic ventilatory response were found among individuals with varied breathing and oxygen levels at altitude. Chiodi had found lesser ventilation among natives than sojourners at high altitude in the Andes [13], which lead to the observation by Severinghaus of profound decreases in hypoxic ventilatory response in Andean high altitude natives compared to those in newcomers [47]. The role of chronic hypoxia was underscored by the finding of similarly decreased hypoxic ventilatory responses in patients with life-long hypoxemia due to cyanotic congenital heart disease [48]. These findings suggested that hypoxia from birth might be required, but later studies found depressed hypoxic ventilatory response associated with chronic altitude exposure beginning later in life [11; 53].

Early measurements of the hypoxic ventilatory response were calculated from the increase in slope of hypercapnic ventilatory response at a single level of hypoxia compared to that in hyperoxia. The hypoxic response was thus often assessed from only two levels of oxygenation. Subsequently the development of oxygen tension sensors and improved oximeters with real-time control of end-tidal PCO_2 levels during progressive induction of hypoxia permitted a continuous assessment of the response over a range of oxygen levels. The resulting improved precision afforded the opportunity to detect minor differences in hypoxic ventilatory response both within and among individuals [54].

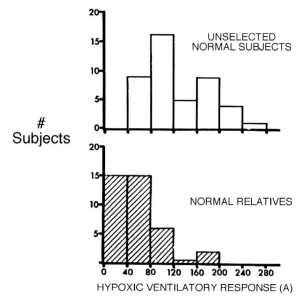

Fig. 1. Upper panel: broad distribution of hypoxic ventilatory responses among normal unselected subjects. Responses were measured during progressive isocapnic hypoxia and expressed as the shape parameter A, an index of steepness of the hyperbolic response. Redrawn from [19]. Lower panel: responses in first degree relatives of endurance athletes, and healthy relatives of patients with idiopathic hypoventilation or chronically hypercapnic chronic obstructive pulmonary disease. The responses in relatives of subjects with low responses or of patients with hypoventilation are shifted to the low end of the spectrum. Drawn from [21]; [39]; [46]; [36]. Figure reproduced with permission from [56].

This lead to studies, which showed wide variation of hypoxic ventilatory response among normal subjects at low altitude (Fig. 1, upper panel) [19; 36]. The responses, which measured the relationship of declining end-tidal oxygen tension to rising ventilation under isocapnic conditions, were assessed as the shape parameter, A, which indicates the steepness of the response. Responses spanned a range broad of seven-fold and were distributed in a non-normal fashion with suggestion of a possibly bimodal configuration.

In addition, hypoxic ventilatory responses at the high and low ends of the range seemed to be found in persons with particular attributes. Low responses were seen in patients with primary hypoventilation (Fig.1 lower panel) [21; 38].

Athletes, most notably those with success in endurance events were also found to have decreased hypoxic ventilatory response compared to non-athletic control subjects [7; 12; 31; 45] although one study found no decrease in marathoners [31]. It is not entirely clear whether decreased hypoxic ventilatory response in endurance athletes in endowed or acquired. Studies indicate that training of unconditioned subjects fails to lower hypoxic ventilatory response [30; 32] although other studies found a decrease [1; 23]. However, no studies have replicated the very long term conditioning of the typical athlete. As indicated below, studies of athlete's families suggest a preexistent contribution independent of training.

The strength of hypoxic ventilatory response has also been linked to performance at high altitude, with high hypoxic responses found in mountain climbers capable of unusually high altitude climbs [33; 44] and low responses in subjects with poor adaptation to altitude manifested as acute mountain sickness [4; 10; 34; 38] or high altitude pulmonary edema [4; 18; 20; 34].

2.3 Population and species differences

2.3.1 Studies in humans

Several reports describe differences in ventilatory control among geographically diverse populations. These include the early report of decreased hypercapnic ventilatory response among Enga tribesmen mentioned earlier. Much of the focus has been on the role of potential differences in hypoxic ventilatory response in relation to variation in adaptation to high altitude. Tibetans seem possessed of superior altitude adaptation and have higher hypoxic ventilatory response than Han Chinese and Andean Aymara [5; 56; 57]. Further, individuals of mixed Han-Tibetan ancestry have hypoxic ventilatory response greater than those of pure Han lineage pointing to a dominant effect of Tibetan ancestry [15]. The increased hypoxic ventilatory response of Tibetans reflects in part a resistance to the blunting effect of long-term hypoxic exposure on hypoxic ventilatory response, mentioned earlier, which is a common feature of other populations [37].

These population differences have commonly been considered to suggest genetic effects on ventilatory control, but a recent analysis suggested that they may be unrelated to genetic distance and may instead reflect differential adaptation to hypoxia [50].

2.3.2 Studies in animals

Differences in hypoxic ventilatory response have been found, both among, and within animal species. Decreased responses are found in species with excellent adaptation to high altitude. Bar-headed geese, which fly at exceptionally high altitude, have lower hypoxic ventilatory responses than the low altitude pekin duck [8]. This was seen in birds raised at low altitude and thus the lower response in the geese could not be ascribed to chronic hypoxic exposure. Variation has also been evident among strains in rats (see Chapter 9) and mice (see Chapter 10).

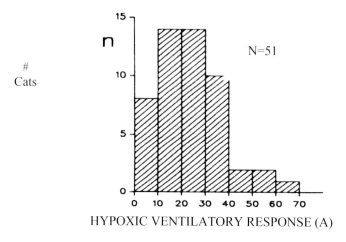

Fig. 2. Interindividual variation in ventilatory response to isocapnic hypoxia among cats. Hypoxic responses are expressed as the shape parameter A, an index of steepness of the hyperbolic response of ventilation to decrease in oxygen tension. Responses were assessed during wakefulness with plethysmographic techniques and were found to be reproducible on repeat testing on several days indicating stable differences among cats. Data from [52].

A study of hypoxic ventilatory response in awake cats showed a large range of responses similar to that seen in humans (Fig. 2) [52]. Repeat measurements on separate days were similar for individual cats indicating stable interindividual differences.

2.4 Familial clusters

In the early 1970's Hudgel observed a young asthmatic patient with frequent episodes of hypoventilation and severe hypoxemia, which seemed disproportionate to his mild airway obstruction [21]. Studies of the patient's ventilatory responses during asthmatic remission showed profoundly decreased hypoxic response with a normal response to hypercapnia indicating that the low hypoxic response was not attributable to ventilatory limitation. Studies of the patient's parents and siblings who were in good health, found a family cluster of low hypoxic, but normal hypercapnic responses (Fig. 3). Similar findings were seen in studies of a second case of unexplained hypoventilation in which first degree relatives had low hypoxic ventilatory responses with no depression of the hypercapnic ventilatory response [36].

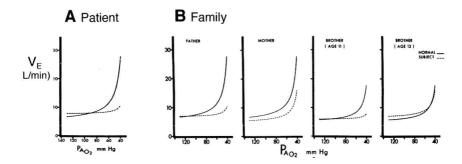

Fig. 3. Panel A: decreased hypoxic ventilatory response to isocapnic hypoxia in a patient with hypoventilation (dashed line), compared to average control value (solid line). Panel B: decreased responses in the patient's healthy parents and siblings (dashed line) compared to average control values of similar age. Data from [21], reproduced with permission from [56].

As mentioned earlier, low values of hypoxic ventilatory response are found in endurance athletes. A study of families of runners who had won events of a mile or longer at the state or higher levels showed clusters of low hypoxic responses in the runner's nonathletic parents and siblings (Fig. 4) [46]. The findings suggest that decreased hypoxic ventilatory response may be a pre-existent attribute of individuals capable of endurance exercise. Reasons for such linkage are unclear. It might be that this reflects a general cellular ability to maintain normal metabolic function with less metabolic error signal at lower oxygen tensions manifested as lesser ventilatory stimulation and better skeletal muscle function at lower oxygen tensions in blood and skeletal muscle.

Differences in hypoxic ventilatory response have also been found among families of patients with chronic obstructive pulmonary disease (COPD). Observations were stimulated by the variation in chronic stable ventilatory status among patients with COPD. It has long been apparent that such patients span a ventilatory spectrum ranging from individuals who chronically maintain nearly normal ventilation (pink puffers or fighters) to those with chronic hypoventilation (blue bloaters or non-fighters) [41; 42]. Early studies had shown that $PaCO_2$ in stable COPD patients was not clearly explained by the severity of airway obstruction, but was associated with decreased ventilatory effort response to hypercapnia (measured as respiratory work or occlusion pressure) [28; 35]. Hypoxic ventilatory responses measured as occlusion pressure were found to be lower in COPD patients with hypoxemia than in those who remained well oxygenated with similar degree of airways obstruction [9].

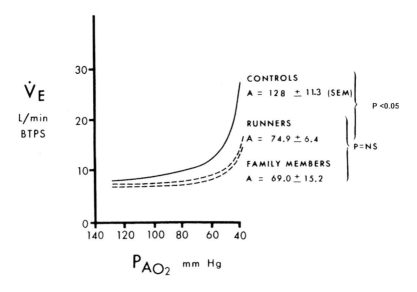

Fig. 4. Depressed values for hypoxic ventilatory response in endurance runners and their healthy, non-athletic parents and siblings. Reproduced with permission from [46].

The possibility that familial factors might contribute to differences in stable ventilation in patients with COPD was explored in studies of families of hypercapnic and normocapnic patients. Healthy offspring of patients with chronic hypoventilation showed lower hypoxic ventilatory response than those of patients with similar airway obstruction who maintained normal ventilation (Fig. 5) [16; 24; 39].

Further, the extent of hypoventilation (measured as $PaCO_2$) during acute exacerbations of airways obstruction in COPD patients was inversely related to the strength of the hypoxic response of their offspring (Fig. 6) [24]. In these studies the linkage of ventilatory responses of offspring to the ventilatory status of the patients was mainly, or exclusively, to the hypoxic response with little or no relationship evident for the response to hypercapnia. Overall, these findings pointed to familial determinants of the hypoxic ventilatory response and of ventilation in the face of chronic airway obstruction.

Familial clustering of decreased hypoxic ventilatory responses has also been found in relation to patients with sleep apnea (see Chapter 8), but not with patients with the obesity hypoventilation syndrome [22].

When combined, hypoxic responses in first degree relatives of endurance athletes and patients with hypoventilation show a distribution is strongly shifted to lower end of the broad spectrum seen in the general population (Fig. 1, lower panel). These familial effects on the hypoxic response seemed to be quite strong given that they were evident with small groups of subjects.

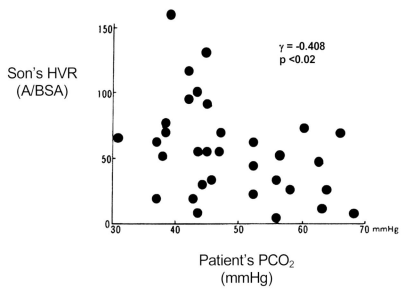

Fig. 5. Familial influence on ventilation in COPD. Ventilation in the COPD patients is correlated with the hypoxic response of their healthy offspring. Hypoxic responses are plotted as the shape parameter A, an index of steepness of the hyperbolic response normalized for body surface area (BSA). Drawn from [24] and reproduced with permission from [56].

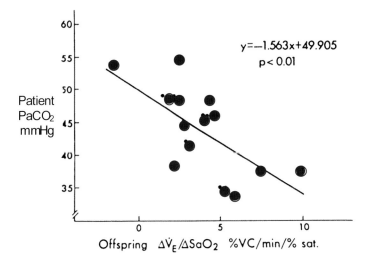

Fig. 6. Familial influence on ventilation in patients with COPD. Alveolar ventilation measured as $PaCO_2$ in patients during acute exacerbation of COPD is correlated with hypoxic ventilatory response of their offspring. Ventilatory responses were measured as the slope of the linear response of increasing ventilation (V_E) to decreasing arterial oxygen saturation (SaO_2) normalised for vital capacity (VC). (Drawn from data of [16] and reproduced from [56].

2.5 Genetics *vs.* environment

Population and familial differences in ventilatory control could reflect either environmental or genetic mechanisms. To explore this issue studies were undertaken to compare responses among monozygotic and dizygotic twins. The approach was to compare similarity of responses between the two members of a twin pair (within-pair variance) for the two classes of twins. The extent to which there is greater similarity of responses within pairs of monozygotic (identical) twins, measured as within-pair variance, compared to that of dizygotic (fraternal) twins is taken as an indicator of genetic contribution. Studies by the Denver, Hokkaido and Edinburgh groups found greater concordance within adult monozygotic than in dyzgotic twins suggesting a genetic contribution to the hypoxic response [14; 24; 29] (Fig. 7). Similar concordance was found in infant monozygotic twins [51]. Findings for the hypercapnic response among twins varied. No genetic effect was evident in work by the Denver investigators [14] and by Arkinstall [3], while two studies by the Hokaido group indicated a genetic contribution [24; 25]. However, these latter findings were complicated by measurement of some of the hypercapnic responses at euoxic oxygen tensions of 90 mmHg, rather than under the usual hyperoxia conditions, which might have added a "hypoxic" contribution to the hypercapnic response. Ultimately the investigators did a direct comparison of hypercapnic responses during euoxia and hyperoxia and found a genetic effect for the hypercapnic response only in hypoxia and agreed that a genetic influence on the response to "pure", hyperoxic, hypercapnia was likely small or absent [27].

In studies of families and of twins, coherence of hypoxic ventilatory responses were evident with very small group sizes subjects suggesting a strong heritable effect.

2.6 Locus of hereditary effects

Hereditary influences on the hypoxic ventilatory response could reflect an influence on chemosensitivity, on central nervous system processing of chemoreceptor signals, or on respiratory mechanics.

Studies in humans and animals indicate a dominant influence on responses to hypoxia rather than to hypercapnia suggesting an effect on peripheral chemosensitivity. The lack of clear effect on the hypercapnic response limits the likelihood that respiratory mechanics account for the findings. The role of an effect on peripheral chemosensitivity is further supported by a study in adult female twins, which showed a genetic influence on a rapid test, consisting of the administration of two breaths of oxygen during steady state hypoxia, pointing to an effect on fast-responding peripheral chemosensitivity [2]. Similar findings were evident in infant twins, in whom a single breath oxygen test showed greater concordance among monozygotic twins [51].

These effects could reflect influences either the peripheral chemoreceptor (carotid body) *per se* or on the central translation of chemoreceptor input into ventilatory output. This was explored in cats, which as mentioned show a broad range of interindividual differences in hypoxic response. Simultaneous measurement of carotid sinus nerve and ventilatory responses to hypoxia showed that the responses were correlated, suggesting that differences in carotid body hypoxic sensitivity were contributors to variation of the hypoxic ventilatory response (Fig. 8) [52].

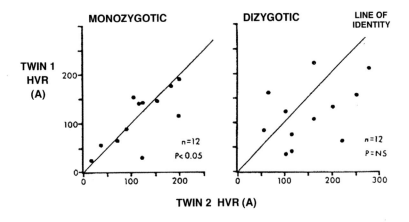

Fig. 7. Hypoxic ventilatory responses (HVR), plotted as the shape parameter A, measured within pairs of monozygotic (left panel) and dizygotic (right panel) twins. The data are plotted in relation to the line of identity (solid line). The findings show greater similarity of hypoxic ventilatory response within pairs of monozygotic than is dizygotic twins suggesting a genetic influence on the response. Reproduced from [14].

Further, the ratio of ventilatory to carotid sinus nerve response was unchanged over the broad range of hypoxic ventilatory responses indicating that differences in central nervous system translation of peripheral chemoreceptor activity to ventilation were unlikely factors. Thus it appears that among cats the variation in hypoxic ventilatory response is most likely a reflection of variable peripheral chemosensitivity. To explore the potential heritability of this effect, we began a breeding colony to study the offspring of high and low responder cats. Unfortunately the effort was aborted by high costs, but studies were completed in 12 offspring of low responders. It was found that both the ventilatory and carotid sinus nerve responses of the offspring tended to cluster at the low end of the range of hypoxic responses with values similar to those of their parents (J. Weil, unpublished data), (Fig. 8).

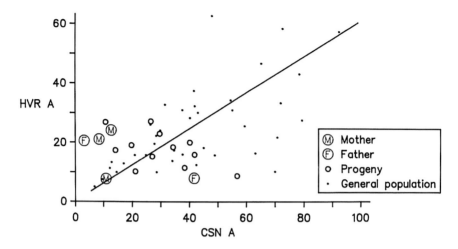

Fig. 8. Ventilatory and carotid sinus nerve responses to hypoxia in cats (filled circles). The wide distribution of ventilatory response shows correlation with carotid sinus nerve responses suggesting a role of varied hypoxic sensitivity of the carotid body in the variation in ventilatory response. Similarity of responses in low responder parents (open circles M & F) and their offspring (open circles) suggests an hereditary effect. From [52] and unpublished data.

These effects on the response of the carotid body could reflect variation in intrinsic chemosensitivity or the influence of neural or humoral modulation. This was addressed in a comparison of strains of rats which found that carotid sinus nerve responses to hypoxia were greater in the spontaneously hypertensive (SHR) than in the Fischer 344 (F344) strain (Fig. 9, panel A) [55]. Hypoxic responses measured in isolated carotid bodies by fluorometric measures of carotid body cytosolic free calcium in response to hypoxic superfusion showed greater response in the SHR strain (Fig. 9, panel B). The findings suggest the existence of inter-strain differences in intrinsic hypoxic chemosensitivity. Thus, collectively the findings are consistent with a possible role of genetically directed effects on intrinsic chemosensitivity in variation of hypoxic ventilatory response. However, the extent to which these findings in animals apply to human variation in hypoxic ventilatory response remains uncertain.

There is little information concerning the question of which specific genes might be involved in variation of hypoxic ventilatory response. Studies in rats suggest possible roles for genes on chromosomes 9 and 18 in differences among Dahl Salt Sensitive and Brown Norway strains [17] and in mice a role for genes on chromosome 9 possibly reflecting effects on dopamine D2 receptor or acetylcholine nicotinic receptor expression have been suggested [49].

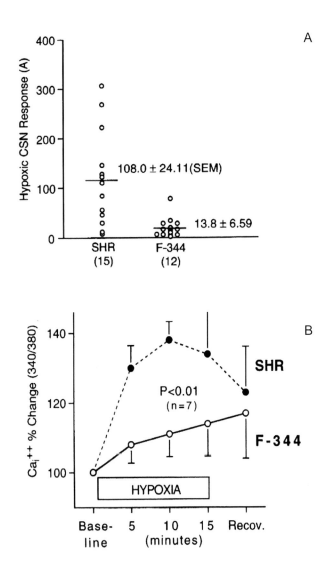

Fig. 9. Panel A: carotid sinus nerve responses to hypoxia are greater in SHR than in F344 rats. Panel B: isolated carotid bodies show greater responses to superfusion hypoxia measured as the increase in cytosolic free calcium. Reproduced [55].

2.7 Conclusion

Ventilatory control shows considerable variation among individual humans. Familial influences are evident in the similarity of ventilatory responses among first degree relatives and a genetic effect is indicated by concordance within identical twins. These factors have their predominant effects on the ventilatory response to hypoxia rather than to the response to hypercapnia. This suggests that heritable influences may act on peripheral chemosensitivity. Studies of interindividual variation in hypoxic ventilatory responses in animals show correlation with carotid body responses, which appear to reflect intrinsic differences in hypoxic sensitivity. Finally, variation in the hypoxic ventilatory response seems related to ventilatory adaptation and function in exposure to high altitude, in endurance athletics and in chronic obstructive pulmonary disease.

References

1. Adamczyk W, Tafil-Klawe M, Chesy G, Klawe JJ, Szeliga-Wczysla M, Zlomanczuk P (2006) Effects of training on the ventilatory response to hypoxia. J Physiol Pharmacol 57 Suppl 4: 7-14
2. Akiyama Y, Nishimura M, Suzuki A, Yamamoto M, Kawakami Y (1991) Peripheral chemosensitivity assessed by the modified transient O_2 test in female twins. Chest 100: 102-105
3. Arkinstall WW, Nirmel K, Klissouras V, Milic-Emili J (1974) Genetic differences in the ventilatory response to inhaled CO_2. J Appl Physiol 36: 6-11
4. Bandopadhyay P, Selvamurthy W (2000) Suggested predictive indices for high altitude pulmonary oedema. J Assoc Physicians India 48: 290-294
5. Beall CM (2000) Tibetan and Andean patterns of adaptation to high-altitude hypoxia. Hum Biol 72: 201-208
6. Beral V, Read DJ (1971) Insensitivity of respiratory centre to carbon dioxide in the Enga people of New Guinea. Lancet 2: 1290-1294
7. Bjurstrom RL, Schoene RB (1987) Control of ventilation in elite synchronized swimmers. J Appl Physiol 63: 1019-1024
8. Black CP, Tenney SM (1980) Oxygen transport during progressive hypoxia in high-altitude and sea-level waterfowl. Respir Physiol 39: 217-239
9. Bradley CA, Fleetham JA, Anthonisen NR (1979) Ventilatory control in patients with hypoxemia due to obstructive lung disease. Am Rev Respir Dis 120: 21-30
10. Burtscher M, Flatz M, Faulhaber M (2004) Prediction of susceptibility to acute mountain sickness by SaO_2 values during short-term exposure to hypoxia. High Alt Med Biol 5: 335-340
11. Byrne-Quinn E, Sodal IE, Weil JV (1972) Hypoxic and hypercapnic ventilatory drives in children native to high altitude. J Appl Physiol 32: 44-46
12. Byrne-Quinn E, Weil JV, Sodal IE, Filley GF, Grover RF (1971) Ventilatory control in the athlete. J Appl Physiol 30: 91-98
13. Chiodi H (1957) Respiratory adaptations to chronic high altitude hypoxia. J Appl Physiol 10: 81-87
14. Collins DD, Scoggin CH, Zwillich CW, Weil JV (1978) Hereditary aspects of decreased hypoxic response. J Clin Invest 62: 105-110

15. Curran LS, Zhuang J, Sun SF, Moore LG (1997) Ventilation and hypoxic ventilatory responsiveness in Chinese-Tibetan residents at 3,658 m. J Appl Physiol 83: 2098-2104

16. Fleetham JA, Arnup ME, Anthonisen NR (1984) Familial aspects of ventilatory control in patients with chronic obstructive pulmonary disease. Am Rev Respir Dis 129: 3-7

17. Forster HV, Dwinell MR, Hodges MR, Brozoski D, Hogan GE (2003) Do genes on rat chromosomes 9, 13, 16, 18, and 20 contribute to regulation of breathing? Respir Physiol Neurobiol 135: 247-261

18. Hackett PH, Roach RC, Schoene RB, Harrison GL, Mills W (1988) Abnormal control of ventilation in high-altitude pulmonary edema. J Appl Physiol 64: 1268-1272

19. Hirshman CA, McCullough RE, Weil JV (1975) Normal values for hypoxic and hypercapnic ventilatory drives in man. J Appl Physiol 38: 1095-1098

20. Hohenhaus E, Paul A, McCullough RE, Kucherer H, Bartsch P (1995) Ventilatory and pulmonary vascular response to hypoxia and susceptibility to high altitude pulmonary oedema. Eur Respir J 8: 1825-1833

21. Hudgel DW, Weil JV (1974) Asthma associated with decreased hypoxic ventilatory drive. A family. Ann Intern Med 80: 623-625

22. Jokic R, Zintel T, Sridhar G, Gallagher CG, Fitzpatrick MF (2000) Ventilatory responses to hypercapnia and hypoxia in relatives of patients with the obesity hypoventilation syndrome. Thorax 55: 940-945

23. Katayama K, Sato Y, Morotome Y, Shima N, Ishida K, Mori S, Miyamura M (2000) Cardiovascular response to hypoxia after endurance training at altitude and sea level and after detraining. J Appl Physiol 88: 1221-1227

24. Kawakami Y, Irie T, Kishi F, Asanuma Y, Shida A, Yoshikawa T, Kamishima K, Hasegawa H, Murao M (1981) Familial aggregation of abnormal ventilatory control and pulmonary function in chronic obstructive pulmonary disease. Eur J Respir Dis 62: 56-64

25. Kawakami Y, Yamamoto H, Yoshikawa T, Shida A (1984) Chemical and behavioral control of breathing in adult twins. Am Rev Respir Dis 129: 703-707

26. Kawakami Y, Yoshikawa T, Shida A, Asanuma Y, Murao M (1982) Control of breathing in young twins. J Appl Physiol 52: 537-542

27. Kobayashi S, Nishimura M, Yamamoto M, Akiyama Y, Kishi F, Kawakami Y (1993) Dyspnea sensation and chemical control of breathing in adult twins. Am Rev Respir Dis 147: 1192-1198

28. Lane DJ, Howell JB (1970) Relationship between sensitivity to carbon dioxide and clinical features in patients with chronic airways obstruction. Thorax 25: 150-159

29. Leitch AG (1976) Chemical control of breathing in identical twin athletes. Ann Hum Biol 3: 447-454

30. Levine BD, Friedman DB, Engfred K, Hanel B, Kjaer M, Clifford PS, Secher NH (1992) The effect of normoxic or hypobaric hypoxic endurance training on the hypoxic ventilatory response. Med Sci Sports Exerc 24: 769-775

31. Mahler DA, Moritz ED, Loke J (1982) Ventilatory responses at rest and during exercise in marathon runners. J Appl Physiol 52: 388-392

32. Markov G, Orler R, Boutellier U (1996) Respiratory training, hypoxic ventilatory response and acute mountain sickness. Respir Physiol 105: 179-186

33. Masuyama S, Kimura H, Sugita T, Kuriyama T, Tatsumi K, Kunitomo F, Okita S, Tojima H, Yuguchi Y, Watanabe S, Honda Y (1986) Control of ventilation in extreme-altitude climbers. J Appl Physiol 61: 500-506

34.   Matsuzawa Y, Fujimoto K, Kobayashi T, Namushi NR, Harada K, Kohno H, Fu-kushima M, Kusama S (1989) Blunted hypoxic ventilatory drive in subjects sus-ceptible to high-altitude pulmonary edema. J Appl Physiol 66: 1152-1157
35.   Matthews AW, Howell JB (1976) Assessment of responsiveness to carbon diox-ide in patients with chronic airways obstruction by rate of isometric inspiratory pressure development. Clin Sci Mol Med 50: 199-205
36.   Moore GC, Zwillich CW, Battaglia JD, Cotton EK, Weil JV (1976) Respiratory failure associated with familial depression of ventilatory response to hypoxia and hypercapnia. N Engl J Med 295: 861-865
37.   Moore LG (2000) Comparative human ventilatory adaptation to high altitude. Respir Physiol 121: 257-276
38.   Moore LG, Harrison GL, McCullough RE, McCullough RG, Micco AJ, Tucker A, Weil JV, Reeves JT (1986) Low acute hypoxic ventilatory response and hy-poxic depression in acute altitude sickness. J Appl Physiol 60: 1407-1412
39.   Mountain R, Zwillich C, Weil J (1978) Hypoventilation in obstructive lung dis-ease. The role of familial factors. N Engl J Med 298: 521-525
40.   Ohyabu Y, Usami A, Ohyabu I, Ishida Y, Miyagawa C, Arai T, Honda Y (1990) Ventilatory and heart rate chemosensitivity in track-and-field athletes. Eur J Appl Physiol Occup Physiol 59: 460-464
41.   Robin ED, O'Neill RP (1963) The fighter versus the non-fighter. Control of venti-lation in chronic obstructive pulmonary disease. Arch Environ Hlth 7: 125-129
42.   Scadding JG (1963) Meaning of diagnostic terms in bronchopulmonary disease. Brit Med J 2: 1435-1430
43.   Schaeffer KE (1958) Respiratory pattern and respiratory response to CO_2. J Appl. Physiol. 13: 1-14
44.   Schoene RB (1982) Control of ventilation in climbers to extreme altitude. J Appl Physiol 53: 886-890.
45.   Schoene RB, Robertson HT, Pierson DJ, Peterson AP (1981) Respiratory drives and exercise in menstrual cycles of athletic and nonathletic women. J Appl Physiol 50: 1300-1305
46.   Scoggin CH, Doekel RD, Kryger MH, Zwillich CW, Weil JV (1978) Familial as-pects of decreased hypoxic drive in endurance athletes. J Appl Physiol 44: 464-468
47.   Severinghaus JW, Bainton CR, Carcelen A (1966) Respiratory insensitivity to hypoxia in chronically hypoxic man. Respir Physiol 1: 308-334.
48.   Sorensen SC, Severinghaus JW (1968) Respiratory insensitivity to acute hypoxia persisting after correction of tetralogy of Fallot. J Appl Physiol 25: 221-223
49.   Tankersley CG (2001) Selected contribution: variation in acute hypoxic ventila-tory response is linked to mouse chromosome 9. J Appl Physiol 90: 1615-1622
50.   Terblanche JS, Tolley KA, Fahlman A, Myburgh KH, Jackson S (2005) The acute hypoxic ventilatory response: testing the adaptive significance in human populations. Comp Biochem Physiol A Mol Integr Physiol 140: 349-362
51.   Thomas DA, Swaminathan S, Beardsmore CS, McArdle EK, MacFadyen UM, Goodenough PC, Carpenter R, Simpson H (1993) Comparison of peripheral chemoreceptor responses in monozygotic and dizygotic twin infants. Am Rev Respir Dis 148: 1605-1609
52.   Vizek M, Pickett CK, Weil JV (1987) Interindividual variation in hypoxic venti-latory response: potential role of carotid body. J Appl Physiol 63: 1884-1889
53.   Weil JV, Byrne-Quinn E, Sodal IE, Filley GF, Grover RF (1971) Acquired at-tenuation of chemoreceptor function in chronically hypoxic man at high altitude. J Clin Invest 50: 186-195

54.   Weil JV, Byrne-Quinn E, Sodal IE, Friesen WO, Underhill B, Filley GF, Grover RF (1970) Hypoxic ventilatory drive in normal man. J Clin Invest 49: 1061-1072
55.   Weil JV, Stevens T, Pickett CK, Tatsumi K, Dickinson MG, Jacoby CR, Rodman DM (1998) Strain-associated differences in hypoxic chemosensitivity of the carotid body in rats. Am J Physiol 274: L767-774
56.   Weil JV (2003) Variation in human ventilatory control-genetic influence on the hypoxic ventilatory response. Respir Physiol Neurobiol 135: 239-346
57.   Wu T, Li S, Ward MP (2005) Tibetans at extreme altitude. Wilderness Environ Med 16: 47-54
58.   Zhuang J, Droma T, Sun S, Janes C, McCullough RE, McCullough RG, Cymerman A, Huang SY, Reeves JT, Moore LG (1993) Hypoxic ventilatory responsiveness in Tibetan compared with Han residents of 3,658 m. J Appl Physiol 74: 303-311

3. Phox2b and the homeostatic brain

Jean-François BRUNET and Christo GORIDIS

CNRS UMR 8542, Ecole Normale Supérieure, 46 rue d'Ulm, 75005 Paris, France

3.1 Introduction

The nervous system, like any other organ, forms under the control of developmental transcription factors. These transcription factors, expressed in varied combinations, commonly called "transcriptional codes", specify regions of the brain and, subsequently, the multitude of neuronal types arising from them. This extremely combinatorial system entails that the expression pattern of any individual transcription factor in the forming brain has little predictive value for the final wiring and function of its constituent neurons. One transcription factor, however, enigmatically stands out: the paired-like homeobox gene *Phox2b*, specifically expressed and required in neurons which go on to form the visceral reflex circuits that control the digestive, cardiovascular and respiratory systems. We will describe the system-wide implication of *Phox2b* in the ontogeny of the visceral nervous system and discuss its embryological, physiopathological and evolutionary ramifications.

3.2 Expression pattern of *Phox2* genes

3.2.1 Phox2b

The expression pattern of *Phox2b* has been examined most extensively in mouse [18; 58; 78] but less comprehensive studies in other vertebrates (Xenopus) [75], man [3], zebrafish [30], and chicken (J. Gay and J.-F. Brunet, unpublished data) are congruent. *Phox2b* expression is strictly restricted to the nervous system,

Table 1. Structures expressing *Phox2* genes

List of structures expressing *Phox2* genes. Structures depending only on *Phox2b* are in plain text. Structures depending only on *Phox2a* are in italics; Structures depending on both genes are in bold. Structures for which no data is available yet are in small font. The spinal interneurons expressing *Phox2a* but not *Phox2b* have been omitted for clarity. VLM: ventro-lateral medulla.

PNS	CNS
Geniculate ganglion	*Oculomotor nucleus* (nIII)
Petrosal ganglion	*Trochlear nucleus* (nIV)
Nodose ganglion	Trigeminal nucleus (nV)
Carotid body	Facial nucleus (nVII)
Sympathetic ganglia	Nucleus ambiguus (nA)
Parasympathetic ganglia	Dorsal motor n. of the vagus (dmnX)
Ciliary	Salivatory nuclei
Sphenopalatine	Cochlear and vestibular efferent nuclei
Otic	Nucleus of the solitary tract (nTS)
Submandibular and sublingual	Area postrema (AP)
Paracardiac	(Nor)adrenergic centers: A1-5,
All others	**A6 (locus cœruleus)**, A7, C1-3
Enteric neurons	Retrotrapezoid nucleus
	Unidentified interneurons of the VLM

both central and peripheral [58], and within the nervous system, to a handful of neuronal types that are discussed below. In every neural structure examined to date, expression starts very early: in the central nervous system (CNS), *Phox2b* is detected either in dividing progenitors of the neuroepithelium or right after cell cycle exit of post-mitotic precursors; and in the peripheral nervous system (PNS) at the very inception of ganglion formation.

Neuronal types expressing *Phox2b* in the CNS and PNS are listed in Table 1. This seemingly random set of neurons, on closer inspection, turns out to match with uncanny accuracy the anatomical concept proposed by Blessing [9]: "visceral neurons, afferent and efferent" to replace the historically muddled notion of "autonomic nervous system", that is, those neurons that maintain bodily homeostasis through the reflex control of the digestive, cardiovascular and respiratory functions. On the afferent path of visceral reflexes, *Phox2b* is expressed in primary visceral sensory neurons, forming the three distal ganglia of cranial nerves VII, IX and X, the geniculate, petrosal and nodose ganglia, as well as in the carotid body, a chemosensory organ presynaptic to the petrosal ganglion. It is also expressed in their central targets, the second-order visceral sensory neurons of the hindbrain (forming the nucleus of the solitary tract (nTS)) and their associated chemosensory center, the area postrema (AP). On the efferent path, *Phox2b* is expressed in all autonomic ganglia (sympathetic, parasympathetic and enteric) as well as their presynaptic partners, the "general visceral motor" (VM) neurons of the hindbrain — to the notable exclusion of preganglionic sympathetic neurons, located in the spinal cord. Sympathetic premotor neurons, or at least the majority of them (the C1 adrenergic center) [70] express *Phox2b*. *Phox2b* is also expressed in the branchiomotor (BM) neurons located in cranial motor nuclei V (trigeminal),

VII (facial), IX (nucleus ambiguus (nA) pars compacta) and XI (spinal accessory nucleus), which innervate the muscles that motorize the face and neck. Those are included neither in the concept of "autonomic nervous system", nor in the set of visceral motoneurons recognized by Blessing (who, rather idiosyncratically, classifies them as "somatic" [9]). However, a global zoological perspective reveals the bias in this exclusion: in aquatic vertebrates, BM neurons control breathing, one of the three cardinal visceral functions. Moreover, their targets (branchial arch-derived muscles) are ontogenetically quite distinct from the somite-derived muscles innervated by somatic motoneurons [52]. Finally, their transcriptional code ($Phox2a^+$, $Phox2b^+$, $Tbx20^+$, $Hb9$, $Lhx3/4^-$) and their ontogenetic origin (the pMNv neuroepithelial domain of the hindbrain [23]) is identical to that of VM neurons. They should thus be considered as *bona fide* visceral motoneurons (a vindication of their old-fashioned name of "special visceral motoneurons") that have been recruited, together with their target muscles, for non visceral functions in terrestrial vertebrates.

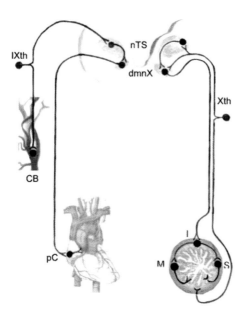

Fig. 1. Two examples of visceral reflex circuits made entirely of *Phox2b*-positive neurons. (Left) Pathway that slows down the heart in response to hyperoxia. (Right) Digestive reflexes utilizing either intrinsic enteric neurons or passing through the CNS via the vagus nerve. Note that some sensori-motor (vago-vagal) synapses bypass the nucleus of the solitary tract. IXth: petrosal ganglion; Xth: vagal ganglion, CB: carotid body; dmnX: dorsal motor nucleus of the vagus nerve; I: enteric interneuron; M: enteric motoneuron; nTS: nucleus of the solitary tract; pC: paracardiac ganglia; S: enteric sensory neuron.

Thus, many well characterized reflex pathways controlling heart rate, blood pressure, or digestive function pass through three, four our five relays, all of which express *Phox2b* (except sympathetic preganglionics). Examples of such pathways are represented in Fig. 1. This represents, by far, the most extensive correlation between the expression of a transcription factor and a neuronal circuit, or set thereof. Two other examples, however limited to two or three relays on an afferent pathway, deserve mention: the expression of *DrgII* in both primary and secondary somatic afferents (reminiscent of the expression of *Phox2b* in primary and secondary visceral afferents) [15; 64] and the expression of *Math1* along the proprioceptor pathway [7].

From a developmental viewpoint, it is striking that *Phox2b*-positive neurons share little besides their visceral function. They share neither position nor lineage: *Phox2b*-expressing neurons arise from the neural crest, epibranchial placodes or the neural tube and, in the latter, both from the ventral (e.g. VM neurons) and dorsal aspects (e.g. the nTS). The only anatomical demarcation is that, within the CNS, *Phox2b*-positive cells are confined to the hindbrain and mid-hindbrain boundary. *Phox2b*-expressing neurons do not share conspicuous phenotypic traits either: they can be sensory, motor or interneurons and they display a large variety of neurotransmitter phenotypes: cholinergic (e.g. VM neurons), noradrenergic (e.g. sympathetic ganglionic neurons), dopaminergic (e.g. some petrosal ganglionic neurons [12]), glutamatergic (e.g. some nTS neurons), serotonergic (e.g. some enteric neurons) and many more.

For the sake of exhaustiveness, one should finally list the few mismatches between the expression pattern of *Phox2b* and the visceral nervous system. The most conspicuous one, already mentioned, consists in preganglionic sympathetic neurons of the spinal cord, which do not express *Phox2b* and, based on their transcriptional code and differentiation pathway, look very much like spinal somatic motoneurons [76]. Conversely, the three *Phox2b*-expressing nuclei situated at the mid-hindbrain boundary are either marginally related to the visceral nervous system (the locus cœruleus (LC)) or not at all (the oculomotor (IIIrd) and trochlear (IVth) nuclei). Finally, the ventro-lateral medulla (VLM) contains a large class of *Phox2b*-positive interneurons whose status remains to be assessed.

3.2.2 Phox2a, the paralogue of Phox2b

Like most homeobox genes in vertebrates, *Phox2b* has a closely related paralogue. The homeodomains of Phox2a and Phox2b are identical in human, rat and mouse and diverge at one residue in chicken, xenopus, and zebrafish. Their N-terminal sequences have extensive similarity, while their C-terminus is very divergent. As its name betrays, *Phox2a* was characterized before *Phox2b*, and it was in fact the expression pattern of the former, very similar to the latter, that drew attention to the correlation of *Phox2* genes with the visceral reflex circuits [77]. However, a closer look at the dynamic of expression of both genes soon revealed a sharp divide between the temporal sequences of *Phox2a* and *Phox2b* expression [58]. At every site where the matter was examined, one of the two genes is expressed first and the other one slightly later (and under the control of the first), leading to pervasive co-expression, at least transiently. This is evidence that the promoters of *Phox2a* and *Phox2b* have radically diverged after the duplication of the ancestral *Phox2* gene, in line with the lack of extensive sequence homologies in the promoter regions. The dividing line between the respective patterns of expression has a striking topographic neatness: in the CNS it passes at the border between the second and first rhombomeres and in the PNS, at the border between head and neck (Fig. 2). Rostral to this line, *Phox2a* precedes *Phox2b* in every *Phox2*-positive neuronal group examined, while caudal to it, *Phox2b* precedes *Phox2a*. Finally, *Phox2a* is expressed alone in a few interneurons in the spinal cord [77].

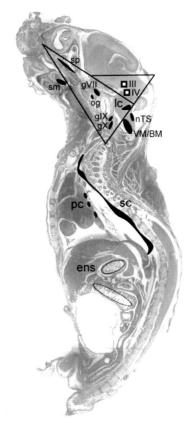

Fig. 2. Topographical relationship of *Phox2*-dependent structures. A representative sample of *Phox2*-dependent ganglia and nuclei are schematized on a parasagittal section of a P0 mouse pup. The structures contained in the large triangle depend on both *Phox2a* and *Phox2b*, those below, only on *Phox2b*. The most rostral ones, enclosed in the small triangle, depend on *Phox2a* only. ens: enteric nervous system; g: ganglion; lc: locus cœruleus; nTS: nucleus of the solitary tract; pc: paracardiac ganglia; og: otic ganglion; sc: sympathetic chain; sm: submandibular ganglion; sp: sphenopalatine ganglion; VM/BM: column of branchial and visceral motoneurons.

3.3 Gross phenotype of mouse mutants for *Phox2* genes

3.3.1 Phox2b knock-outs

Phox2b knock-outs die at around embryonic day (E) 12 with signs of congestive heart failure [59], but they can be rescued until prenatal stages by administration of noradrenergic agonists [55]. Examination of E10.5 to E18.5 mutant embryos shows that the vast majority of *Phox2b*-expressing neural structures are missing (listed in plain and bold text in Table 1 and see Fig. 2), precluding postnatal survival. In the PNS, sympathetic, parasympathetic, enteric neurons [59] as well as adrenal medullary and carotid body cells [40; 59] are absent and the geniculate, petrosal and nodose ganglia are massively atrophic [18; 59]. In the hindbrain, BM and VM nuclei [57], the nTS, the AP [18] and all (nor)adrenergic centers [55] are absent. The fate of the unidentified interneurons in the VLM has not been examined. Differentiation of *Phox2b*-dependent neurons is stalled very early in all cases examined, followed by cell death or fate switches. For example, enteric neuronal precursors invade the esophagus, but fail to express *Ret*, never migrate past the rostral stomach, and then degenerate [59]. Sympathetic precursors aggregate close to the dorsal aorta but never differentiate and soon disappear by apoptosis [59] (and see below). In the CNS, BM/VM neurons are never born [57] while nTS precursors are born but their migration route is changed and few survive, with a presumably altered fate [18]. Thus, in most cases, the loss of *Phox2b* expression disrupts every aspect of subsequent neuronal differentiation. Whether this pleïotropic action reflects a large array of transcriptional targets or a high position in a transcriptional cascade, or both, is not known.

3.3.2. Phox2a knock-outs

Phox2a knock-outs die soon after birth, with an empty stomach and a collapsed esophagus evocative of an impaired swallowing reflex [48]. Histological examination at birth reveals that all *Phox2*-positive neural structures located in the head (e.g. the sphenopalatine and otic parasympathetic ganglia) and mid-hindbrain junction (the LC and the IIIrd and IVth motor nuclei) are missing, while the cranial sensory ganglia are atrophic. Thus, several *Phox2*-dependent structures depend on both *Phox2a* and *Phox2b* (Table 1 and Fig. 2). It is notable that in all these structures *Phox2a* lies upstream of *Phox2b*, raising the possibility that *Phox2a* merely serves as a surrogate promoter for *Phox2b*, which would control differentiation. This hypothesis is supported by reciprocal knock-ins [16] (and see below).

A mutation in human *PHOX2A*, one of which is a clear null, causes, in the homozygous state, Congenital Fibrosis of Extra-Oculomotor Muscles Type II (CFEOMII), whose clinical tableau is essentially explained by the agenesis of the IIIrd and IVth motor nuclei [11; 51; 88]. In remarkable contrast with the mouse mutation, the condition is viable, perhaps reflecting a lesser role for human *PHOX2A* than murine *Phox2a* in cranial ganglion formation.

3.3.3 Mutations in Phox2b and the drive to breathe

As we have seen so far, *Phox2b* is expressed and required in most core relays of the neural control of digestion and cardiovascular function (barring sympathetic preganglionics). Concerning the neural control of breathing, assessment of the role of *Phox2b* has been hampered by the intrinsic complexity of this function in terrestrial vertebrates, a somewhat patchier knowledge of the anatomy and the embryonic death of *Phox2b* null mutants, which precludes a physiological analysis.

At the advent of terrestrial life, the muscle control of breathing has been for the most part discharged on skeletal muscles, innervated by spinal somatic motoneurons (which do not express *Phox2b*). As argued by Blessing [10], this is paralleled by the evolution of breathing, from a purely homeostatic function to a partially voluntary and relational one (including vocalization, sniffing, etc.). However, as discussed earlier, the ancestral, i.e. aquatic form of breathing, is controlled by *Phox2b*-dependent neurons, the "visceral special" or branchial motoneurons, which innervate gill-associated muscles. On the afferent side of respiratory reflexes, *Phox2b* is required in such paramount providers of sensory feedback as the nodose ganglion (which contains the cell bodies of pulmonary stretch receptors), the carotid body (which contains oxygen and carbon dioxide sensors), the petrosal ganglion (which innervates the carotid body) and the nTS (which integrates chemosensory and barosensory information). *Phox2b* is also required for the generation of all noradrenergic centers, among which A5 and the LC function as modulators of the respiratory rhythm (reviewed in [31]). Incidentally, the disappearance of the LC, but not of A5 (which have opposite effects on respiratory rhythm [31] has been incriminated in the slowed-down breathing of perinatal *Phox2a* null fetuses [85].

Recently, human genetics has unveiled a new dimension of the involvement of *Phox2b* in the neurophysiology of breathing: a range of mutations in *PHOX2B* was discovered that are associated, in a heterozygous state, with a complex dysautonomia: Central Congenital Hypoventilation Syndrome (CCHS) [3]. Thereafter, the rate of detection of *PHOX2B* mutations in CCHS steadily rose to reach 96% and mutations in *PHOX2B* can now be considered as necessary and sufficient for this condition [86]. CCHS comprises a whole list of partially penetrant symptoms that read staightforwardly like a list of *Phox2b* expression sites: partial agenesis of the enteric nervous system (or Hirschprung disease), sympathoadrenal tumors, dismotility of the intrinsic muscles of the eye (that receive sympathetic and parasympathetic innervation), etc. (a more complete description will be found elsewhere in this volume, see Chapter 4). However, the defining, fully penetrant symptom of CCHS consists in respiratory arrests typically occurring during sleep, which have been attributed to a reduced, in the worst cases abolished sensitivity to CO_2. This implies that the CO_2 response passes through an obligatory *Phox2b*-positive dependent neuronal relay. The main locus of CO_2 sensitivity (to which the carotid body marginally contributes) is controversial but widely held to be at the ventral medullary surface [10]. Two of the better documented candidates are raphe serotonergic nuclei [66] and the Retro-Trapezoid Nucleus (RTN), a group of glutamatergic neurons found at the medullary surface, just beneath the facial nucleus [50]. The latter turn out to express *Phox2b* [73]. Moreover, *Phox2b*-expressing CO_2-sensitive neurons of the RTN receive direct projections from nTS neurons which relay O_2 responsiveness by virtue of an input from the carotid body, via the petrosal ganglion [73]. Thus, subject to the confirmation of the RTN as key to central CO_2 sensitivity, a major pathway for monitoring blood gazes would consist of an entirely *Phox2b*-positive four-relay circuit. Intriguingly, the RTN is anatomically coextensive with the parafacial respiratory group (pFRG), a recently identified respiratory oscillator [53] proposed to couple to the pre-Bötzinger complex (pre-BötC) [41], the best documented respiratory rhythm generator in terrestrial vertebrates [65]. Thus, even though the pre-BötC itself does not express *Phox2b* [8] it is possible that *Phox2b* also controls some aspects of the circuitry responsible for respiratory rhythm generation. This putative role could underlie the observation that in the most severe cases of CCHS, affected infants will not spontaneously breathe at birth [28].

The mutations that cause CCHS, mostly expansions of a 20 residue-long alanine stretch in the C-terminal part of the protein (but also frame shift and missense mutations) [86] are probably not null mutations (and thus do not cause disease by a simple haploinsufficiency) on the grounds that, in mouse, *Phox2b* heterozygotes display a much subtler phenotype [17; 18; 22], that patients with heterozygous deletions of the *PHOX2B* region do not have CCHS (reviewed in [86]) and that different *PHOX2B* mutations are associated with different combinations and frequencies of symptoms [25; 86]. Rather, the mutant protein may cause CCHS by a dominant-negative mechanism or by toxic gain-of-function [5; 79]. Moreover, mice in which the polyalanine expansion has been introduced into the *Phox2b* locus die at birth due to respiratory failure (V. Dubreuil, J. Amiel, C. G. and J-F. B., unpublished observation), in striking contrast to the heterozygous *Phox2b* mutants, definitely ruling out pure haploinsufficiency as the pathogenic mechanism for this mutation. In man, an argument for a dominant-negative mechanism or cellular toxicity is that a fraction of CCHS patients have strabismus [26], implying a dysfunction of the *Phox2a*-dependent IIIrd and IVth motor nuclei, where *Phox2b* is expressed but not required.

CCHS patients have a greatly increased risk of developing sympathoadrenal tumors, which may also occur in the absence of CCHS in patients with frameshift mutations [49; 62; 80; 81; 84], and *PHOX2B* is the first identified gene in which germline mutations predispose to neuroblastoma [49; 80]. Abrogation of its proneural activity, demonstrated in motoneuron precursors [60], may underlie the occurrence of these tumors.

3.4 Cellular functions of *Phox2* genes

The striking match between the territory of *Phox2b* action and the core circuits of the visceral nervous system makes this complex anatomical entity, *de facto*, a developmental unit. A novel, unexpected "superclass" of neurons (that includes many subclasses) is thus defined: the visceral neuron. It seems likely that downstream of their common transcriptional determinant — *Phox2b* — and upstream of their common physiological role-controlling the viscera- visceral neurons follow some common differentiation pathway. One hypothesis is that this pathway could determine connectivity: neurons are "visceral" primarily by virtue of projecting to the viscera (and/or to other visceral neurons). Tantalizingly, a role in axonal guidance was found for the C. elegans orthologue of *Phox2b* [63], but the connectivity of the five neurons expressing it is uncertain and their function unknown, precluding a parallel with the vertebrate condition. While this putative *Phox2b*-dependent common genetic program has yet to be uncovered, candidate gene approaches combined with loss- and gain-of-function experiments have started to reveal various pleïotropic roles for *Phox2b* in the early differentiation of specific neuronal types.

3.4.1 (Nor)adrenergic differentiation

Historically, the first functional correlate proposed for *Phox2*-expressing neurons was the (nor)adrenergic phenotype (expressed by sympathoadrenal (SA) cells, noradrenergic centres A1-7, adrenergic centres C1-3, and transiently by parasympathetic, enteric and cranial sensory neurons) although it represents only a subset of the expression pattern [83]. Its role in noradrenergic differentiation is still a much studied molecular aspect of *Phox2* function.

Sympathetic neurons are the *Phox2b*-dependent cell-type, whose transcriptional control is best understood. The transcription factors Phox2a, Phox2b, Gata2, Gata3, dHAND, Mash1 and Tfap2 have been assessed for their role in regulating sympatho-adrenergic (SA) differentiation traits, including these factors themselves, *tyrosine hydroxylase* (*Th*), and *dopamine-β-hydroxylase* (*Dbh*), and pan-neuronal genes such as *SCG10*, or *βIII-tubulin*. The combined data are a challenge to schematic representations [27; 47]. Fig. 3 aims at emphasizing three aspects of these complex regulatory interactions that are discussed below.

Phox2b is absolutely required for SA differentiation: all other markers depend on it — except *Mash1* (whose maintenance, though, is *Phox2b*-dependent) [59]. In contrast, SA differentiation is not completely abrogated in the absence of *Mash1*, *Gata2/3* or *dHAND*: only some SA markers are affected by their inactivation and their expression, rather than abolished, is delayed [56], not maintained [46] or attenuated [45; 82]. Similarly, at the organ level, the loss of *Phox2b* leads to agenesis of the sympathetic chains [59] whereas the loss of *Mash1* or *Gata3* only leads to atrophy [29; 32; 56; 82]. One interpretation of this dichotomy is that *Phox2b* is a direct and obligate regulator of all other markers (except *Mash1*) while *Mash1*, *dHand* and *Gata2/3* are cofactors enhancing *Phox2b* action (for such

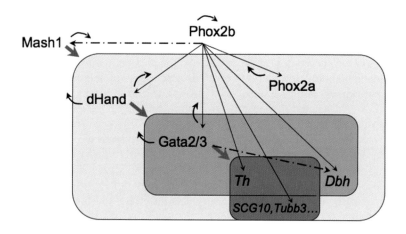

Fig. 3. Transcriptional network controling sympathoadrenal differentiation, including *Th*, *Dbh*, and generic neuronal differentiation traits. Partial dependence is represented by a thick arrow pointing at a frame enclosing the set of downstream markers, complete dependence by a thin arrow pointing at an individual downstream gene. A curvy arrow represents a putative feedback regulation revealed by gain-of-function experiments. A broken line represents a demonstrated role in maintenance rather than activation. Tfap2 is omitted for simplicity, since its epistatic relationships have not been explored in any detail, apart from its independence from *Phox2* genes [34]. *Dbh*: dopamine β-hydroxylase. *Th*: tyrosine hydroxylase. *Tubb3*: βIII-tubulin.

a role for dHand, see [68]). An alternative interpretation is that the loss of function of *Phox2b* leads to cell death so abruptly that other factors do not have the time to partially compensate.

Another complexity is that the transcriptional "cascade" of SA differentiation is rife with countercurrents and whirlpools, as suggested by gain-of-function experiments in which genes switched on in parallel can activate each other, and downstream genes prove capable of inducing upstream ones [37; 71; 72; 82]. The cross-regulations among the members of this transcriptional network might insure the maintenance of upstream genes, thus the robustness and stability of the pathway. A maintenance role is indeed likely for *dHand* [45] and demonstrated for *Gata3* [46]. Transactivation of *Phox2b* by *Mash1*, *Gata2/3* and *dHand* may underlie their capacity to specify the SA phenotype, to various extents, when forced-expressed.

Finally, the case of *Phox2a* begs for clarification: the gene was isolated and studied before *Phox2b* [83; 90] and this serendipitous antecedence, combined with the fact that *Phox2a* lies downstream of *Phox2b* in sympathoblasts [58; 59] and can induce a noradrenergic phenotype by gain of function [43; 72] has sometimes led to consider it as a key effector of *Phox2b* in SA differentiation (e.g. [14; 38]). However, SA differentiation proceeds normally in the absence of *Phox2a* [48], implying that *Phox2b* can fully compensate for any role of *Phox2a* in SA differentiation. To unmask this putative role (which cannot be assessed in the *Phox2b* knock-out, since *Phox2a* is no longer expressed) a Phox2a-only situation was created by placing the *Phox2a* coding sequence under the control of the *Phox2b* locus [16]. This *Phox2b*^{KIPhox2a} allele failed to rescue the *Phox2b* knock out phenotype in the sympathetic chain, demonstrating that, despite its identical DNA binding domain, Phox2a is lacking properties found in Phox2b that are required to trigger *Dbh* expression and SA differentiation *in vivo*. All in all, if one cannot exclude that *Phox2a* has yet undetected roles in the late maturation or postnatal maintenance of the SA phenotype, it can be ruled out as a determinant of this lineage. The SA differentiation observed after forced expression of *Phox2a* [44; 72] most likely results from the up-regulation of *Phox2b*, by virtue of the feedback mechanisms outlined above. It is thus unfortunate that most of the molecular work on the role of Phox2 promoters and proteins during SA differentiation, summarized below, has so far focused almost exclusively on *Phox2a*.

At the molecular level, sympathetic differentiation has been shown to require signaling by BMPs (reviewed in [27]), and, at least *in vitro*, elevated cAMP levels (reviewed in [38]). While the connection between BMP signaling and *Phox2* genes has not been explored, cAMP may impinge on *Phox2a* at two levels: on its expression via increased CREB-mediated transcription [6] and on its activity by stimulating, paradoxically, the dephosphorylation of the protein [1; 14]. Conversely, the ERK1/2 kinase has been shown to inhibit Phox2a [39]. As made clear earlier, if these pathways turn out to be relevant *in vivo*, it is likely that they impinge also on *Phox2b*, which was not considered in these studies. A preliminary characterization of the human *PHOX2B* promoter has appeared in [36].

Downstream of *Phox2b*, the only directly regulated promoters that have been studied are that of *PHOX2B* itself and of *PHOX2A*, both shown to contain sites for activation by Phox2b [13; 24 ; 35] and, most extensively, that of *Dbh*. Schemes for the mechanism of *Dbh* promoter activation, based on the work of several labs [1; 2; 42; 68; 74; 87; 89] and featuring AP1, CREB, CBP, dHand and Phox2a can be found in [74] and [38]. There again, any physiological relevance implies that Phox2a is replaced by Phox2b, at least at the onset of *Dbh* transcription.

Not only peripheral noradrenergic and adrenergic cell types depend on *Phox2b*, but also central ones. Among known factors required for noradrenergic differentiation in the PNS, only Mash1 and Tfap2a, along with Phox2b, are also required in the CNS [32; 34].

(Nor)adrenergic centers are loose neuronal aggregates called A1-7 and C1-3, with the exception of A6 or locus cœruleus (LC), which forms a compact nucleus. The LC is born and located in rhombomere one (r1) [4], all other centers are caudal to it, in the myelencephalon (r2-7). Thus, according to the rule outlined above (see § 3.2.2), the LC switches on *Phox2a* before *Phox2b* and depends on both genes [48; 55], whereas all other noradrenergic centers depend on *Phox2b*, not *Phox2a* [48; 55] (Fig. 2). Does the dependence of the LC on *Phox2a* (in addition to *Phox2b*) indicate a direct role of the former in noradrenergic differentiation? In both *Phox2b* null [55] and *Phox2b*^{KIPhox2a} embryos [16], the lack of *Dbh* expression in Phox2a-positive LC precursors shows that *Phox2a* alone cannot trigger central *Dbh* expression. Conversely, the normal LC of *Phox2a*^{KIPhox2b} embryos shows that the Phox2a protein is dispensable. Altogether these data suggest that in the LC, the only *Phox2a*-dependent noradrenergic cell group, *Phox2a* acts merely as a surrogate promoter for *Phox2b* which, like elsewhere in the PNS and CNS, controls noradrenergic differentiation per se. It remains possible that *Phox2a* plays a role in the maintenance of noradrenergic differentiation of LC cells, in which *Phox2b* expression, at least in mouse, is transient [55].

3.4.2 Motoneuronal differentiation

Phox2b is required for the development of two main types of cranial motoneurons: VM neurons (presynaptic to parasympathetic and enteric ganglia) located in the dorsal motor nucleus of the vagus nerve (dmnX), the nA and the salivatory nuclei and BM neurons (innervating branchial arch-derived muscles) located in the Vth, VIIth, nA and XIth nuclei. VM/BM neurons, although they eventually settle in a latero-dorsal region of the hindbrain, are born from the ventral-most region of the proliferative neuroepithelium (the so-called pMNv domain) [23]. Their birth occurs from E9 to E10.5 (as late as E12.5 for facial motoneurons). Subsequently, the same neuroepithelial domain gives rise to serotonergic neurons, except in the region where facial BM neurons are born [61]. During motoneuron generation, *Phox2b* is expressed in the dividing progenitors of pMNv where it acts as a proneural gene, promoting cell cycle exit and neuronal differentiation [19; 20; 60]. In *Phox2b* knock outs, no BM/VM neuron is ever born, and the neuroepithelium adopts prematurely its later mode of neurogenesis: it gives rise to much fewer post-mitotic precursors that differentiate into serotonergic neurons [60; 61]. When ectopically expressed in the spinal cord, *Phox2b* promotes the differentiation of neurons whose molecular identity and axonal trajectory are similar to those of BM/VM neurons, more specifically the caudal-most ones, found in the XIth nucleus [19; 33]. Thus, in pMNv, in addition to its proneural role, *Phox2b* serves as a phenotypic switch: it specifies the motoneuronal fate while it suppresses the serotonergic one — which takes over, once *Phox2b* is down-regulated. A third and minor class of *Phox2b*-dependent efferent neurons ("motoneurons" in the broad sense) in the medulla are the cochlear and vestibular efferent neurons [78], which modulate the sensitivity of inner ear hair cells.

Finally, a fourth category of Phox2-expressing motoneurons arises at the mesencephalic-metencephalic junction: the oculomotor (IIIrd) and trochlear (IVth) nuclei which are most often classified as somatic. However, their transcription factor code sets them sharply apart from *bona fide* somatic motoneurons: they do not express *HB9*, but instead express both *Phox2* genes as well as low levels of *Tbx20* (H. Dufour and J-F. B., unpublished data) and depend on *Phox2a* [58]. They should thus be put in a class of their own, albeit more related to BM/VM motoneurons than to somatic ones.

At the molecular level, the role of *Phox2b* in motoneuronal generation has been little studied. The activation of *Phox2b* in the pMNv domain has been proposed to require the integration of rostro-caudal and dorso-ventral patterning information, via *Hoxb1/b2* and *Nkx2.2*, respectively, through a proximal enhancer of the gene [69]. Downstream of *Phox2b*, the proneural activity uncovered in BM/VM neurons may involve the downregulation of *Hes* and *Id* antineurogenic genes [20], and is most likely not mediated by classical proneural genes of the bHLH class [60]. It could conceivably involve a direct up-regulation of CDK inhibitors, as has been shown for *Phox2a* in SA cells [54].

In conclusion to this section, it should be emphasized that the expression of both *Phox2b* and *Phox2a* not only starts very early, but lasts for very long, throughout adult life in some neuronal classes. It is therefore possible that these transcription factors have late developmental roles in addition to those uncovered so far in simple knock-outs, or even post-developmental roles in the adult, that can be revealed only by inducible knock-outs.

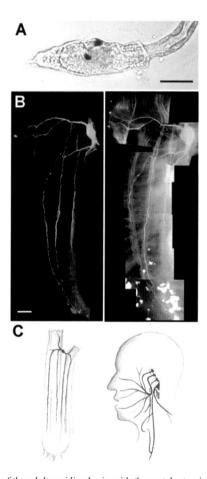

Fig. 4. Homology of the adult ascidian brain with the vertebrate visceral nervous system. A, Expression of a *CiPhox2::YFP* transgene in the neck of the swimming larva, which is homologous to the vertebrate hindbrain [21]. B, The *CiPhox2::YFP* transgene highlights the adult cerebral ganglion, and its nerves coursing along the "body wall muscles". C, Schematic of the anatomy of ciona's motoneurons (left) and human branchial motoneurons (right), in blue. Scale bars represent 100 μm in A, 1mm in B. (Modified from [21]. Copyright 2006, National Academy of Sciences, U.S.A.).

3.5 Ancestry of the homeostatic brain

The extreme specificity of *Phox2b* for the visceral nervous system has been used to explore its evolution. In the adult urochordate *Ciona intestinalis*, in striking accordance with the arguably "visceral" nature of this animal [67], which does little, but breathe and speed through a huge perforated pharynx, the orthologue of *Phox2b/a*, *CiPhox2* is widely expressed in the brain [21] (Fig. 4). At least some of the *CiPhox2*$^+$ neurons are *CiTbx20*$^+$/*CiMnr2*$^-$/ *CiChAT*$^+$, a molecular code that makes them homologous to the BM/VM neurons of vertebrates. Moreover, like the latter, they derive from a hindbrain-like region of the embryonic (and larval) CNS (Fig. 4A). And like BM neurons, their role is to innervate muscles, that despite their appellation ("body wall muscle"), could be just as well considered, on account of the anatomy of this animal, as "branchial" (Fig. 4B,C).

Thus, cranial motoneurons of the branchial class, used for feeding and breathing purposes, are candidates at representing the ancestral form of the visceral nervous system. In such a scenario, this *Phox2b*-dependent primitive "homeostatic brain" would have been elaborated in the vertebrate lineage, notably by the addition of peripheral structures derived from epibranchial placodes and neural crest, under the control, remarkably, of the same gene.

Conclusion

Phox2b is a transcription factor specifically required for the ontogeny of the visceral nervous system, more precisely for proper development of the control of the three cardinal homeostatic functions: blood circulation, digestion and respiration. This unusual type of specificity makes *Phox2b* a bona fide "master gene" for visceral reflex circuits. It explains that, in humans, mutations in *PHOX2B* account for the full spectrum of symptoms that define Congenital Central Hypoventilation Syndrome (CCHS), in fact a complex dysautonomia. In turn, the involvement of *PHOX2B* in CCHS promises to yield insights in yet poorly understood aspects of the neuronal control of rhythmic breathing, that are defective in this disease. Further elucidation might come from conditional mutations of Phox2b in specific medullary interneurons.

References

1.   Adachi M, Lewis EJ (2002) The paired-like homeodomain protein, Arix, mediates protein kinase A-stimulated dopamine beta-hydroxylase gene transcription through its phosphorylation status. J Biol Chem 277: 22915-22924
2.   Adachi M, Browne D, Lewis EJ (2000) Paired-like homeodomain proteins Phox2a/Arix and Phox2b/NBPhox have similar genetic organization and independently regulate dopamine beta-hydroxylase gene transcription. DNA Cell Biol 19: 539-554

3.   Amiel J, Laudier B, Attié-Bitach T, Trang H, de Pontual L, Gener B, Trochet D, Simonneau M, Vekemans M, Munnich A, Gaultier C, Lyonnet S (2003) Polyalanine expansion and frameshift mutations of the paired-like homeobox gene PHOX2B in congenital central hypoventilation syndrome (Ondine's curse). Nat Genet 33: 459-461

4.   Aroca P, Lorente-Canovas B, Mateos FR, Puelles L (2006) Locus coeruleus neurons originate in alar rhombomere 1 and migrate into the basal plate: studies in chick and mouse embryos. J Comp Neurol 496: 802-818

5.   Bachetti T, Matera I, Borghini S, Di Duca M, Ravazzolo R, Ceccherini I (2005) Distinct pathogenetic mechanisms for PHOX2B associated polyalanine expansions and frameshift mutations in congenital central hypoventilation syndrome. Hum Mol Genet 14: 1815-1824

6.   Benjanirut C, Paris M, Wang WH, Hong SJ, Kim KS, Hullinger RL, Andrisani OM (2006) The cAMP pathway in combination with BMP2 regulates Phox2a transcription via cAMP response element binding sites. J Biol Chem 281: 2969-2981

7.   Bermingham NA, Hassan BA, Wang VY, Fernandez M, Banfi S, Bellen HJ, Fritzsch B, Zoghbi HY (2001) Proprioceptor pathway development is dependent on Math1. Neuron 30: 411-422

8.   Blanchi B, Kelly LM, Viemari JC, Lafon I, Burnet H, Bevengut M, Tillmanns S, Daniel L, Graf T, Hilaire G, Sieweke MH (2003) MafB deficiency causes defective respiratory rhythmogenesis and fatal central apnea at birth. Nat Neurosci 6: 1091-1100

9.   Blessing WW (1997a) Anatomy of the lower brainstem. In: The lower brainstem and bodily homeostasis, 1 Edition (Blessing WW, ed), pp 29-99. New York, Oxford University Press

10. Blessing WW (1997b) Breathing. In: The lower brainstem and bodily homeostasis, pp 101-164. New York, Oxford University Press

11. Bosley TM, Oystreck DT, Robertson RL, al Awad A, Abu-Amero K, Engle EC (2006) Neurological features of congenital fibrosis of the extraocular muscles type 2 with mutations in PHOX2A. Brain 129: 2363-2374

12. Brosenitsch TA, Katz DM (2002) Expression of Phox2 transcription factors and induction of the dopaminergic phenotype in primary sensory neurons. Mol Cell Neurosci 20: 447-457

13. Cargnin F, Flora A, Di Lascio S, Battaglioli E, Longhi R, Clementi F, Fornasari D (2005) PHOX2B regulates its own expression by a transcriptional autoregulatory mechanism. J Biol Chem 280: 37439-37448

14. Chen S, Ji M, Paris M, Hullinger RL, Andrisani OM (2005) The cAMP pathway regulates both transcription and activity of the paired homeobox transcription factor Phox2a required for development of neural crest-derived and central nervous system-derived catecholaminergic neurons. J Biol Chem 280: 41025-41036

15. Chen ZF, Rebelo S, White F, Malmberg AB, Baba H, Lima D, Woolf CJ, Basbaum AI, Anderson DJ (2001) The paired homeodomain protein DRG11 is required for the projection of cutaneous sensory afferent fibers to the dorsal spinal cord. Neuron 31: 59-73

16. Coppola E, Pattyn A, Guthrie SC, Goridis C, Studer M (2005) Reciprocal gene replacements reveal unique functions for Phox2 genes during neural differentiation. EMBO J 24: 4392-4403

17.   Cross SH, Morgan JE, Pattyn A, West K, McKie L, Hart A, Thaung C, Brunet JF, Jackson IJ (2004) Haploinsufficiency for Phox2b in mice causes dilated pupils and atrophy of the ciliary ganglion: mechanistic insights into human congenital central hypoventilation syndrome. Hum Mol Genet 13: 1433-1439

18.   Dauger S, Pattyn A, Lofaso F, Gaultier C, Goridis C, Gallego J, Brunet JF (2003) Phox2b controls the development of peripheral chemoreceptors and afferent visceral pathways. Development 130: 6635-6642

19.   Dubreuil V, Hirsch MR, Pattyn A, Brunet JF, Goridis C (2000) The Phox2b transcription factor coordinately regulates neuronal cell cycle exit and identity. Development 127: 5191-5201

20.   Dubreuil V, Hirsch MR, Jouve C, Brunet JF, Goridis C (2002) The role of Phox2b in synchronizing pan-neuronal and type-specific aspects of neurogenesis. Development 129: 5241-5253

21.   Dufour HD, Chettouh Z, Deyts C, de Rosa R, Goridis C, Joly JS, Brunet JF (2006) Pre-craniate origin of cranial motoneurons. Proc Natl Acad Sci USA 103: 8727-8732

22.   Durand E, Dauger S, Pattyn A, Gaultier C, Goridis C, Gallego J (2005) Sleep-disordered breathing in newborn mice heterozygous for the transcription factor Phox2b. Am J Respir Crit Care Med 172: 238-243

23.   Ericson J, Rashbass P, Schedl A, Brenner-Morton S, Kawakami A, van Heyningen V, Jessell TM, Briscoe J (1997) Pax6 controls progenitor cell identity and neuronal fate in response to graded Shh signaling. Cell 90: 169-180

24.   Flora A, Lucchetti H, Benfante R, Goridis C, Clementi F, Fornasari D (2001) Sp proteins and Phox2b regulate the expression of the human Phox2a gene. J Neurosci 21: 7037-7745

25.   Gaultier C, Trang H, Dauger S, Gallego J (2005) Pediatric disorders with autonomic dysfunction: what role for PHOX2B? Pediatr Res 58: 1-6

26.   Goldberg DS, Ludwig IH (1996) Congenital central hypoventilation syndrome: ocular findings in 37 children. J Pediatr Ophthalmol Strabismus 33: 175-180

27.   Goridis C, Rohrer H (2002) Specification of catecholaminergic and serotonergic neurons. Nat Rev Neurosci 3: 531-541

28.   Gozal D (1998) Congenital central hypoventilation syndrome: an update. Pediatr Pulmonol 26: 273-282

29.   Guillemot F, Lo LC, Johnson JE, Auerbach A, Anderson DJ, Joyner AL (1993) Mammalian achaete-scute homolog 1 is required for the early development of olfactory and autonomic neurons. Cell 75: 463-476

30.   Guo S, Brush J, Teraoka H, Goddard A, Wilson SW, Mullins MC, Rosenthal A (1999) Development of noradrenergic neurons in the Zebrafish hindbrain requires BMP, FGF8, and the homeodomain protein soulless/Phox2a. Neuron 24: 555-566

31.   Hilaire G, Viemari JC, Coulon P, Simonneau M, Bevengut M (2004) Modulation of the respiratory rhythm generator by the pontine noradrenergic A5 and A6 groups in rodents. Respir Physiol Neurobiol 143: 187-197

32.   Hirsch MR, Tiveron MC, Guillemot F, Brunet JF, Goridis C (1998) Control of noradrenergic differentiation and Phox2a expression by MASH1 in the central and peripheral nervous system. Development 125: 599-608

33.   Hirsch MR, Glover JC, Dufour HD, Brunet JF, Goridis C (2006) Forced expression of Phox2 homeodomain transcription factors induces a branchio-visceromotor axonal phenotype. Dev Biol 303: 687-702

34. Holzschuh J, Barrallo-Gimeno A, Ettl AK, Durr K, Knapik EW, Driever W (2003) Noradrenergic neurons in the zebrafish hindbrain are induced by retinoic acid and require tfap2a for expression of the neurotransmitter phenotype. Development 130: 5741-5754

35. Hong SJ, Kim CH, Kim KS (2001) Structural and functional characterization of the 5' upstream promoter of the human Phox2a gene: possible direct transactivation by transcription factor Phox2b. J Neurochem 79: 1225-1236

36. Hong SJ, Chae H, Kim KS (2004) Molecular cloning and characterization of the promoter region of the human Phox2b gene. Brain Res Mol Brain Res 125: 29-39

37. Howard M, Stanke M, Schneider C, Wu X, Rohrer H (2000) The transcription factor dHAND is a downstream effector of BMPs in sympathetic neuron specification. Development 127: 4073-4081

38. Howard MJ (2005) Mechanisms and perspectives on differentiation of autonomic neurons. Dev Biol 277: 271-286

39. Hsieh MM, Lupas G, Rychlik J, Dziennis S, Habecker BA, Lewis EJ (2005) ERK1/2 is a negative regulator of homeodomain protein Arix/Phox2a. J Neurochem 94: 1719-1727

40. Huber K, Karch N, Ernsberger U, Goridis C, Unsicker K (2005) The role of Phox2B in chromaffin cell development. Dev Biol 279: 501-508

41. Janczewski WA, Feldman JL (2006) Distinct rhythm generators for inspiration and expiration in the juvenile rat. J Physiol 570: 407-420

42. Kim H, Seo H, Yang C, Brunet JF, Kim K (1998) Noradrenergic-specific transcription of the dopamine b-hydroxylase gene requires synergy of multiple cis-acting elements including at least two Phox2a-binding sites. J Neurosci 18: 8247-8260

43. Lo L, Tiveron MC, Anderson DJ (1998) MASH1 activates expression of the paired homeodomain transcription factor Phox2a, and couples pan-neuronal and subtype-specific components of autonomic neuronal identity. Development 125: 609-620

44. Lo L, Morin X, Brunet JF, Anderson DJ (1999) Specification of neurotransmitter identity by Phox2 proteins in neural crest stem cells. Neuron 22: 693-705

45. Lucas ME, Muller F, Rudiger R, Henion PD, Rohrer H (2006) The bHLH transcription factor hand2 is essential for noradrenergic differentiation of sympathetic neurons. Development 133: 4015-4024

46. Moriguchi T, Takako N, Hamada M, Maeda A, Fujioka Y, Kuroha T, Huber RE, Hasegawa SL, Rao A, Yamamoto M, Takahashi S, Lim KC, Engel JD (2006) Gata3 participates in a complex transcriptional feedback network to regulate sympathoadrenal differentiation. Development 133: 3871-3881

47. Morikawa Y, Dai YS, Hao J, Bonin C, Hwang S, Cserjesi P (2005) The basic helix-loop-helix factor Hand 2 regulates autonomic nervous system development. Dev Dyn 234: 613-621

48. Morin X, Cremer H, Hirsch M-R, Kapur RP, Goridis C, Brunet JF (1997) Defects in sensory and autonomic ganglia and absence of locus coeruleus in mice deficient for the homeobox gene *Phox2a*. Neuron 18: 411-423

49. Mosse YP, Laudenslager M, Khazi D, Carlisle AJ, Winter CL, Rappaport E, Maris JM (2004) Germline PHOX2B mutation in hereditary neuroblastoma. Am J Hum Genet 75: 727-730

50. Mulkey DK, Stornetta RL, Weston MC, Simmons JR, Parker A, Bayliss DA, Guyenet PG (2004) Respiratory control by ventral surface chemoreceptor neurons in rats. Nat Neurosci 7: 1360-1369

51.   Nakano M, Yamada K, Fain J, Sener EC, Selleck CJ, Awad AH, Zwaan J, Mullaney PB, Bosley TM, Engle EC (2001) Homozygous mutations in ARIX (PHOX2A) result in congenital fibrosis of the extraocular muscles type 2. Nat Genet 29: 315-320

52.   Noden DM, Francis-West P (2006) The differentiation and morphogenesis of craniofacial muscles. Dev Dyn 235: 1194-1218

53.   Onimaru H, Homma I (2003) A novel functional neuron group for respiratory rhythm generation in the ventral medulla. J Neurosci 23: 1478-1486

54.   Paris M, Wang WH, Shin MH, Franklin DS, Andrisani OM (2006) The homeodomain transcription factor Phox2a, via cAMP-mediated activation, induces p27Kip1 transcription, coordinating neural progenitor cell cycle exit and differentiation. Mol Cell Biol 18: 18

55.   Pattyn A, Goridis C, Brunet JF (2000a) Specification of the central noradrenergic phenotype by the homeobox gene *Phox2b*. Mol Cell Neurosci 15: 235-243

56.   Pattyn A, Guillemot F, Brunet JF (2006) Delays in neuronal differentiation in Mash1/Ascl1 mutants. Dev Biol 295: 67-75

57.   Pattyn A, Hirsch M-R, Goridis C, Brunet JF (2000b) Control of hindbrain motor neuron differentiation by the homeobox gene *Phox2b*. Development 127: 1349-1358

58.   Pattyn A, Morin X, Cremer H, Goridis C, Brunet JF (1997) Expression and interactions of the two closely related homeobox genes Phox2a and Phox2b during neurogenesis. Development 124: 4065-4075

59.   Pattyn A, Morin X, Cremer H, Goridis C, Brunet JF (1999) The homeobox gene Phox2b is essential for the development of autonomic neural crest derivatives. Nature 399: 366-370

60.   Pattyn A, Simplicio N, van Doorninck JH, Goridis C, Guillemot F, Brunet J-F (2004) Mash1/Ascl1 is required for the development of central serotonergic neurons. Nat Neurosci 7: 589-595

61.   Pattyn A, Vallstedt A, Dias JM, Samad OA, Krumlauf R, Rijli FM, Brunet JF, Ericson J (2003) Coordinated temporal and spatial control of motor neuron and serotonergic neuron generation from a common pool of CNS progenitors. Genes Dev 17: 729-737

62.   Perri P, Bachetti T, Longo L, Matera I, Seri M, Tonini GP, Ceccherini I (2005) PHOX2B mutations and genetic predisposition to neuroblastoma. Oncogene 24: 3050-3053

63.   Pujol N, Torregrossa P, Ewbank JJ, Brunet JF (2000) The homeodomain protein CePHOX2/CEH-17 controls antero-posterior axonal growth in *C. elegans*. Development 127: 3361-3371

64.   Qian Y, Shirasawa S, Chen CL, Cheng L, Ma Q (2002) Proper development of relay somatic sensory neurons and D2/D4 interneurons requires homeobox genes Rnx/Tlx-3 and Tlx-1. Genes Dev 16: 1220-1233

65.   Rekling JC, Feldman JL (1998) PreBötzinger complex and pacemaker neurons: hypothesized site and kernel for respiratory rhythm generation. Annu Rev Physiol 60: 385-405

66.   Richerson GB (2004) Serotonergic neurons as carbon dioxide sensors that maintain pH homeostasis. Nat Rev Neurosci 5: 449-461

67.   Romer AS (1972) The vertebrate as a dual animal-somatic and visceral. Evol Biol 6: 121-156

68.   Rychlik JL, Gerbasi V, Lewis EJ (2003) The interaction between dHAND and Arix at the dopamine beta-hydroxylase promoter region is independent of direct dHAND binding to DNA. J Biol Chem 278: 49652-49660

69. Samad OA, Geisen MJ, Caronia G, Varlet I, Zappavigna V, Ericson J, Goridis C, Rijli FM (2004) Integration of anteroposterior and dorsoventral regulation of Phox2b transcription in cranial motoneuron progenitors by homeodomain proteins. Development 131: 4071-4083

70. Schreihofer AM, Guyenet PG (1997) Identification of C1 presympathetic neurons in rat rostral ventrolateral medulla by juxtacellular labeling *in vivo*. J Comp Neurol 387: 524-536

71. Stanke M, Stubbusch J, Rohrer H (2004) Interaction of Mash1 and Phox2b in sympathetic neuron development. Mol Cell Neurosci 25: 374-382

72. Stanke M, Junghans D, Geissen M, Goridis C, Ernsberger U, Rohrer H (1999) The Phox2 homeodomain proteins are sufficient to promote the development of sympathetic neurons. Development 126: 4087-4494

73. Stornetta RL, Moreira TS, Takakura AC, Kang BJ, Chang DA, West GH, Brunet JF, Mulkey DK, Bayliss DA, Guyenet PG (2007) Expression of Phox2b by brainstem neurons involved in chemosensory intergration in the adult rat. J Neurosci 20: 10305-10314

74. Swanson DJ, Adachi M, Lewis EJ (2000) The homeodomain protein Arix promotes protein kinase A-dependent activation of the dopamine beta-hydroxylase promoter through multiple elements and interaction with the coactivator cAMP-response element-binding protein-binding protein. J Biol Chem 275: 2911-2223

75. Talikka M, Stefani G, Brivanlou AH, Zimmerman K (2004) Characterization of Xenopus Phox2a and Phox2b defines expression domains within the embryonic nervous system and early heart field. Gene Expr Patterns 4: 601-607

76. Thaler JP, Koo SJ, Kania A, Lettieri K, Andrews S, Cox C, Jessell TM, Pfaff SL (2004) A postmitotic role for Isl-class LIM homeodomain proteins in the assignment of visceral spinal motor neuron identity. Neuron 41: 337-350

77. Tiveron MC, Hirsch MR, Brunet JF (1996) The expression pattern of the transcription factor Phox2 delineates synaptic pathways of the autonomic nervous system. J Neurosci 16: 7649-7660

78. Tiveron M-C, Pattyn A, Hirsch MR, Brunet JF (2003) Role of Phox2b and Mash1 in the generation of the vestibular efferent nucleus. Dev Biol 260: 46-57

79. Trochet D, Hong SJ, Lim JK, Brunet JF, Munnich A, Kim KS, Lyonnet S, Goridis C, Amiel J (2005a) Molecular consequences of PHOX2B missense, frameshift and alanine expansion mutations leading to autonomic dysfunction. Hum Mol Genet 14: 3697-3708

80. Trochet D, Bourdeaut F, Janoueix-Lerosey I, Deville A, de Pontual L, Schleiermacher G, Coze C, Philip N, Frebourg T, Munnich A, Lyonnet S, Delattre O, Amiel J (2004) Germline mutations of the paired-like homeobox 2B (PHOX2B) gene in neuroblastoma. Am J Hum Genet 74: 761-764

81. Trochet D, O'Brien LM, Gozal D, Trang H, Nordenskjold A, Laudier B, Svensson PJ, Uhrig S, Cole T, Niemann S, Munnich A, Gaultier C, Lyonnet S, Amiel J (2005b) PHOX2B genotype allows for prediction of tumor risk in congenital central hypoventilation syndrome. Am J Hum Genet 76: 421-426

82. Tsarovina K, Pattyn A, Stubbusch J, Muller F, van der Wees J, Schneider C, Brunet JF, Rohrer H (2004) Essential role of Gata transcription factors in sympathetic neuron development. Development 131: 4775-4786

83. Valarché I, Tissier-Seta JP, Hirsch MR, Martinez S, Goridis C, Brunet JF (1993) The mouse homeodomain protein Phox2 regulates *Ncam* promoter activity in concert with Cux/CDP and is a putative determinant of neurotransmitter phenotype. Development 119: 881-896

84.   Van Limpt V, Schramm A, van Lakeman A, Sluis P, Chan A, van Noesel M, Baas F, Caron H, Eggert A, Versteeg R (2004) The Phox2B homeobox gene is mutated in sporadic neuroblastomas. Oncogene 23: 9280-9288

85.   Viemari JC, Bevengut M, Burnet H, Coulon P, Pequignot JM, Tiveron MC, Hilaire G (2004) Phox2a gene, A6 neurons, and noradrenaline are essential for development of normal respiratory rhythm in mice. J Neurosci 24: 928-937

86.   Weese-Mayer DE, Berry-Kravis EM, Marazita ML (2005) In pursuit (and discovery) of a genetic basis for congenital central hypoventilation syndrome. Respir Physiol Neurobiol 149: 73-82

87.   Xu H, Firulli AB, Zhang X, Howard MJ (2003) HAND2 synergistically enhances transcription of dopamine-beta-hydroxylase in the presence of Phox2a. Dev Biol 262: 183-193

88.   Yazdani A, Chung DC, Abbaszadegan MR, Al-Khayer K, Chan WM, Yazdani M, Ghodsi K, Engle EC, Traboulsi EI (2003) A novel PHOX2A/ARIX mutation in an Iranian family with congenital fibrosis of extra-ocular muscles type 2 (CFEOM2). Am J Ophthalmol 136: 861-865

89.   Yang C, Kim HS, Seo H, Kim CH, Brunet JF, Kim K-S (1998) Paired-like homeodomain proteins, Phox2a and Phox2b, are responsible for noradrenergic cell-Specific transcription of the *dopamine-ß-hydroxylase* gene. J Neurochem 71: 1813-1826

90.   Zellmer E, Zhang Z, Greco D, Rhodes J, Cassel S, Lewis EJ (1995) A homeodomain protein selectively expressed in noradrenergic tissue regulates transcription of neurotransmitter biosynthetic genes. J Neurosci 15: 8109-8120

4. Congenital central hypoventilation syndrome: from patients to gene discovery

Ha TRANG

Service de Physiologie, Hôpital Robert Debré, 48 boulevard Sérurier 75019 Paris, France

4.1 Introduction

Central Congenital Hypoventilation Syndrome (CCHS) is a disorder characterized by congenital failure of the autonomic control of breathing which results in severe alveolar hypoventilation during sleep [24]. A broad range of autonomic nervous system (ANS) dysfunction is associated. The incidence of CCHS has been estimated to be at 1/200,000 live births [31]. Treatment is currently supportive by a lifelong dependence on ventilatory support.

Mellins et al. were the first to report a newborn with CCHS in 1970 [24]. In 1978, Haddad et al. brought to the medical community two crucial data: the association of CCHS and Hirschprung's disease (HSCR), an aganglionosis of the bowel, and familial cases of CCHS which support a genetic basis for the syndrome [16]. In the late 1990's, genes known to be involved in ANS development were probed in patients with CCHS by several independent research teams [2; 27; 37]. In 2003, Amiel et al. identified heterozygous mutations of the paired-like homeobox gene 2B (*PHOX2B*) as the main CCHS-causing mutation [1].

The purpose of this chapter is to provide a comprehensive history of CCHS from its first case report to the discovery of the *PHOX2B* mutation. Studies in transgenic mice (see Chapters 3 and 14) have guided the search of genetic mutations in humans and functional and anatomical studies in patients with CCHS allow insight into its physiopathology (see Chapter 5). The discovery of the CCHS-causing mutation further contributes to the recognition of new phenotypes and raises a great hope for a specific treatment in the future.

4.2 Clinical presentation of CCHS

The hallmark characteristic of CCHS is persistent alveolar hypoventilation during sleep due to absent or markedly reduced central hypercapnic ventilatory responses [12; 38]. However, the overall spectrum of CCHS phenotype involves a more global ANS dysfunction, with a large inter-individual variability among patients [35].

4.2.1 Respiratory phenotype

Clinical presentation

The clinical presentation is dominated by respiratory symptoms usually present at birth or during early infancy. Neonates with CCHS exhibit recurrent apneas, periods of cyanosis during sleep, and despite severe hypoxia and bradycardia, fail to increase their breathing. Fig.1 illustrates polysomnographic data in a female 8-day-old neonate. Polysomnography showed severe central alveolar hypoventilation during sleep due to shallow breathing and low respiratory rate which caused reduced minute ventilation. Alveolar hypoventilation was more severe during quiet sleep than during active sleep. Ventilatory responses to hypercapnia, to hypoxia were absent or markedly reduced in all states of alertness.

Abnormal central control of breathing persists during the whole life. With advancing postnatal age, central hypoventilation becomes more marked during non-rapid eye movement sleep. During wakefulness, most patients can breathe spontaneously with acceptable alveolar ventilation and gas exchange. However, those severely affected may hypoventilate both while asleep and while awake. Ventilatory support is required during sleep for most of the patients with CCHS, and also during the day for 5 to 10% of them [31; 35].

Mecanisms underlying respiratory drive in CCHS

Normal central control of breathing relies on the integrity of peripheral chemoreceptors, of central chemoreceptors, of integration processes mediating signals from chemoreceptors to the brainstem, and of the brainstem respiratory generator itself. A number of functional ventilatory studies, recently coupled to new imaging techniques, have aimed to investigate the range of deficient and intact components of control of breathing in CCHS.

Ventilatory responses to sustained hypercapnia, to sustained hypoxia, are absent or markedly reduced whatever the state of alertness in patients with CCHS (Fig. 1). Alveolar hypoventilation is most severe during non rapid eye movement sleep, a sleep state during which control of breathing depends quasi-exclusively upon central CO_2 level. Most of the patients are able to breathe adequately during wakefulness [31; 35], despite loss of perception of breathlessness [28].

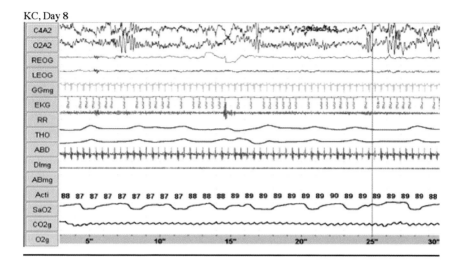

KC, Day 8

	WAKEFULNESS	ACTIVE SLEEP	QUIET SLEEP
Spontaneously breathing room air			
Tidal volume (ml/kg)	8.0	8.5	4.6 (↘)
Respiratory rate(per min)	45	27	20 (↘)
Minute ventilation (ml/min/kg)	359	227	90 (↘)
Hypercapnic challenge			
Change in minute ventilation by change in end-tidal P_{CO2} (ml/min/kg/mmHg)	6.7 (↘)	Not done	1.6 (↘)
Hypoxic challenge			
Change in minute ventilation by change in oxygen saturation (ml/min/kg/%)	-1.8 (↘)	-8.7 (↘)	Not done

Fig. 1. Upper panel. Infant with CCHS at 8 days of age. Polysomnographic 30 sec-window showing shallow breathing, low respiratory rate (~ 16 breaths/min) and desaturation while spontaneously breathing room air during sleep. Abbreviations: C4A2, O2A2: electroencephalogram; REOG, LEOG: right and left electro-oculograms; GGmg: genioglossus electromyogram; EKG: electrocardiogram; RR: cardiac period; THO, ABD: respiratory movements of the thorax and the abdomen; DImg, ABmg: electromyograms of diaphragm and abdominal muscles; SaO$_2$: oxygen saturation using pulse oximetry; CO2g: end-tidal PCO$_2$; O2g: plethysmographic signal of oxygen saturation. Lower panel. Hypercapnic and hypoxic challenges using the rebreathing technique (inspired fraction of CO$_2$ of 5%, and inspired fraction of oxygen of 15%, respectively) during wakefulness, active sleep and quiet sleep. As the infant spontaneously breathes room air, tidal volume, respiratory rate and minute ventilation dramatically decrease in active sleep, compared to wakefulness, and in quiet sleep compared to active sleep. Hypercapnic and hypoxic ventilatory responses are abnormally low in all states of alertness.

A number of evidences supports for a relatively normal central respiratory generator in CCHS. Patients with CCHS can breathe adequately if stimulated by mechanisms other than chemoreception, i.e. by common activities during wakefulness, by exercise at moderate intensity, or when stimulated by passive leg movements [14; 15; 26].

Suprapontine stimulation increases significantly respiratory drive during wakefulness in CCHS. Its impact may depend on the level of mental activity. Moderate mental activity has been shown to increase ventilation in a small group of patients with CCHS, whereas challenge with intense mental activity may result in decrease of breathing in some of them [28].

Passive motion of the lower limbs preferentially involves a mechanoreceptor feedback without activation of feed-forward centers. Gozal et al. have shown that passive leg movements increased alveolar ventilation during wakefulness and during non-rapid eye movement sleep in patients with CCHS [14; 15]. Thus, the dysfunctional brain structures that mediate the abnormal ventilatory response to hypercapnia or hypoxia do not appear to disrupt the on-response of reflex ventilatory changes elicited by passive motion.

Peripheral chemoreceptor drive has been shown to be active in a small group of patients with CCHS, presumably in those who can have adequate spontaneous ventilation during daytime [13]. However, structural abnormalities (reduced size and number of glomus cells) have been clearly identified in the carotid bodies of two children with CCHS [7].

In CCHS, the defect is located at the level of central chemoreception. Interestingly, despite the absence of ventilatory responses to hypercapnia, patients with CCHS arise from sleep during a hypercapnic challenge [22]. Two hypotheses may be suggested: (i) either ventilatory and arousal responses involve separate neural circuits, (ii) and/or major deficits in the integration processes of signals. Use of functional magnetic resonance imaging demonstrated that responses to hypercapnia, to hypoxia or to loaded breathing involved widespread brain areas not classically thought to mediate respiratory drive [17; 20; 21; 39]. Cerebellar and basal ganglia contributions in responding to hypercapnia are shown to be altered in patients with CCHS, whereas thalamic and midbrain regions fail to mediate the perception of breathlessness [17]. Moreover, imaging techniques using diffusion tensor imaging show increased mean diffusivity in regions in the brainstem, but in also the cerebellum, forebrain and temporal and frontal cortices [19]. Thus, these data suggest that there are both functional and structural impairments of these cerebral areas in CCHS (see Chapter 5).

In summary, the intrinsic deficit of central control of breathing is present in patients with CCHS whatever the state of alertness. However, it may be overcome under conditions in which other operative mechanisms are active, such as wakefulness and motion, at least in part of the CCHS population. One limitation of the studies cited above is the small number of patients included due to the rarity of the disease. Because of the phenotypic variability in CCHS, this may explain in part some of the conflicting data and therefore, it is likely that compensatory respiratory drive mechanisms may vary across patients.

4.2.2 Others phenotypes

Cardiac phenotype

Patients with CCHS sometimes experience dizziness or syncopes which may be ascribed to dysfunction of ANS controlling cardiovascular function. Heart rate variability is decreased [40]. In a survey, 19% of patients with CCHS have cardiac arythmias and 4% require a cardiac pacemaker [35].

Moreover, the circadian pattern of blood pressure is abnormal with loss of the physiological decrease in the values during transition from wakefulness to sleep [30]. The spontaneous baroreflex sensitivity is shown to be altered, with a predominant vagal dysfunction and a relatively preserved sympathetic function [25; 32]. The hypothesis is that impaired control of the baroreflex likely derives in affected brain regions from the ventral frontal, insular and cingulate cortices [20].

Enteric nervous dysfunction

HSCR, an aganglionosis of a variable colonic length due to failure of migration of neural crest-derived cells, is present in 15-20% of patients with CCHS. Interestingly, when HSCR is associated with CCHS, a long colonic segment is involved in 80% of the cases, and male-female ratio is 1, compared to 20% and 4, respectively for isolated HSCR [6; 31].

Other ANS dysfunction

Various manifestations of ANS dysfunction have been reported, such as ocular dysmotility [35], esophageal dysmotility [11], abnormal thermoregulation. About 3 to 5% of patients with CCHS develop neural crest tumors such as ganglioneuromas, ganglioneuroblastomas, or neuroblastomas [5; 31; 34; 35].

4.2.3 From genes back to patients: new phenotypes

The discovery of the CCHS-causing mutation allows recognition of new phenotypes. Central hypoventilation may occur during early childhood or adulthood (namely late onset central hypoventilation syndromes) [10; 23; 33]. CCHS can be associated with brainstem anomalies identified using magnetic resonance image, as shown in two patients, one with a Chiari I malformation and the other with hypoplasia of the pons [4].

4.3 Genetic mutations in CCHS

Before the discovery of the CCHS-causing gene

A genetic origin of CCHS has long been suspected based upon rare familial cases [16; 18: 29]. In the pursuit of the genes responsible for CCHS, several independent teams have used the candidate gene approach. Early studies focused upon genes related to HSCR, as this disorder occurs in about 15 to 20% of the patients with CCHS. Later, screening was expanded to genes thought to be involved in neural crest cell migration and in ANS development.

The first patient with CCHS (and HSCR) to be reported with a genetic mutation had a mutation in the rearranged during transfection gene (*RET*) [2]. Also, mutations in glial cell-line-derived neurotrophic factor (*GDNF*) [2; 27], in human achaete-scute homologue-1 (*HASH1*) [8; 27], in brain-derived neurotrophic factor (BDNF) [37], in *PHOX2A* [27], and in GDNF family receptor alpha 1 (*GFRA1*) [27] have been found in a few patients with CCHS.

Table 1. Distribution of mutations of the *PHOXB* gene in the CCHS population.

* Patients originated from France, Great-Britain, Germany, Sweden and the United States [33] and predominantly from the United States [5]. Data are number of patients (and percentage).

Reference	[33]	[5]
Number of patients screened *	188	184
Heterozygous mutations of *PHOX2B*	174	184
Alanine expansion mutations	161 (92%)	170 (92%)
Non-alanine expansion mutations	13 (8%)	14 (8%)
Frameshift mutations	10	9
Missense mutations	3	1
Nonsense mutations	/	4

Identification of PHOX-2B as the CCHS-causing gene

In 2003, for the first time, heterozygous mutations of the *PHOX2B* gene were found in most of a series of 29 patients with CCHS [1]. *PHOX2B* gene is a paired-like homeobox, located on chromosome 4p12 and encoding a highly conserved transcription factor that usually contains two polyalanine repeat sequences of 9 and 20 residues (see Chapters 3 and 6).

The *PHOX2B* gene is identified as the CCHS-causing gene [1; 23; 27; 36]. Data from the two largest series including 372 patients with CCHS worldwide report a *PHOX2B* mutation-detection rate of 92 to 100% [5; 34]. As shown in the Table, the most frequent mutations identified (92%) were polyalanine expansions (of +5 to +13 alanines within exon 3) with a predominance of small expansions (of +5 to +7 alanines). The other 8% were a frameshift, a missense, or a nonsense mutation (Table 1) (see Chapter 6).

Interestingly, an expansion of +5 alanines was found in some patients with late-onset central hypoventilation syndrome, suggesting that this syndrome may share similar genetic basis with CCHS [23; 33; 34].

Mode of inheritance of PHOX-2B mutations and CCHS

CCHS is a genetic disorder with an autosomal dominant mode of inheritance. The risk of CCHS would be 50% for offspring born to a parent with CCHS. Not all mutations of *PHOX2B* in patients with CCHS occured *de novo*: somatic mosaicisms were detected in 4.5% of unaffected parents of patients with CCHS [34]. Thus, the risk of CCHS would be up to 50% for offspring born to parents with mosaicisms and a child with CCHS.

Penetrance of the mutation is incomplete, as phenotype show elevated inter-variability among patients, and parents of patients with CCHS who carry the *PHOX2B* mutation have been shown to be unaffected [23]. The discovery of the CCHS-causing gene shows a high clinical relevance for genetic counselling and prenatal diagnosis.

4.4 Phenotype-genotype correlations

There are a number of evidences supporting a correlation between polyalanine expansion length and the severity of *in vitro* functions (see Chapter 6) and clinical phenotype. The longest alanine expansions are associated with the most severe respiratory deficiencies in patients with CCHS [23; 34; 36]. The smallest expansion of +5 alanines is the only mutation found in patients with late onset-central hypoventilation syndrome. However, it can also be found in patients with neonatal onset of central hypoventilation. Thus, a similar genetic mutation can be associated with variable onset of central hypoventilation, suggesting a role for epigenetic influences.

A high predisposition to neural crest tumors has been long recognized in patients with CCHS [23; 27]. The presence of non-alanine expansion mutations (i.e. frameshift, nonsense or missense mutations) is significantly associated with the occurrence of neural crest tumors [5; 34] (see Chapter 6). Hence, molecular testing may help to identify a subset of patients with CCHS at high risk for developing malignant tumors.

In contrast, no significant association is found between the type of *PHOX2B* mutations and the presence of HSCR in addition to CCHS. Nevertheless, the weak *RET* haplotype predisposing to HSCR is more frequent in patients with CCHS and HSCR than in those with isolated CCHS [9]. These data suggest that *RET* gene acts as a modifier gene for the HSCR phenotype to occur in patients with CCHS.

Synergistic effects of others modifier genes as the mechanisms underlying the variable phenotype of CCHS are likely. So far, only 89 patients with CCHS have been screened for mutations other than *PHOX2B* [1; 27; 36]. Ten per cent of them carry a mutation of others genes involved in the ANS development, such as *GDNF, GFRA1, HASH-1, PHOX2A*, or a mutation of the *BDNF* gene (2; 27; 37).

The phenotype reveals to be unremarkable for these patients with two genetic mutations, but their limited number should prevent from definite conclusions. There is a need for further genetic studies looking for mutations or polymorphisms of others genes associated with *PHOX2B* mutations.

4.5 Conclusion

Experimental studies on the pathogenetic mechanisms of CCHS-associated *PHOX2B* (see Chapter 6) allow insight into the physiopathology and perspectives of a specific treatment in the future.

However, functional brain deficits have been shown to extend far beyond the areas of *PHOX2B* expression during development. One hypothesis would be that impaired *PHOX2B* gene function targeting the autonomic ganglia exerts a direct effect on control of the cerebral vasculature, thereby altering the development of structures in brain areas that control respiratory, cardiovascular and other vital functions (see Chapter 5). Thus, to be efficient in patients with CCHS, the specific treatment should target altogether brain areas which express *PHOX2B* and those thought to be secondarily affected.

References

1.   Amiel J, Laudier B, Attié-Bitach T, Trang H, de Pontual L, Gener B, Trochet D, Etchevers H, Ray P, Simonneau M, Vekemans M, Munnich A, Gaultier C, Lyonnet S (2003) Polyalanine expansion and frame shift mutations of the paired-like homeobox gene *PHOX2B* in congenital central hypoventilation syndrome (Ondine's curse). Nat Genet 33: 440-442

2.   Amiel J, Salomon R, Attie T, Pelet A, Trang H, Mokhtari M, Gaultier C, Munnich A, Lyonnet S (1998) Mutations of the RET–GDNF signaling pathway in Ondine's curse. Am J Hum Genet 62: 715-717

3.   Antic NA, Malow BA, Lange N, McEvoy RD, Olson AL, Turkington P, Windisch W, Samuels M, Stevens CA, Berry-Kravis EM, Weese-Mayer DE (2006) PHOX2B mutation-confirmed congenital central hypoventilation syndrome: presentation in adulthood. Am J Respir Crit Care Med 174: 923-927

4. Bachetti T, Robbiano A, Parodi S, Matera I, Merello E, Capra V, Baglietto MP, Rossi A, Ceccherini I, Ottonello G (2006) Brainstem anomalies in two patients affected by congenital central hypoventilation syndrome. Am J Respir Crit Care Med 174: 706-709

5. Berry-Kravis EM, Zhou L, Rand CM, Weese-Mayer DE (2006) Congenital central hypoventilation syndrome: PHOX-2B mutations and phenotype. Am J Respir Crit Care Med 174: 1139-1144

6. Croaker GD, Shi E, Simpson E, Cartmill T, Cass DT (1998) Congenital central hypoventilation syndrome and Hirschsprung's disease. Arch Dis Child 78: 316-322

7. Cutz E, Ma TK, Perrin DG, Moore AM, Becker LE (1997) Peripheral chemoreceptors in congenital central hypoventilation syndrome. Am J Respir Crit Care Med 155: 358-363

8. De Pontual L, Nepote V, Attie-Bitach T, Al Halabiah H, Trang H, Elghouzzi V, Levacher B, Benihoud K, Auge J, Faure C, Laudier B, Vekemans M, Munnich A, Perricaudet M, Guillemot F, Gaultier C, Lyonnet S (2003) Noradrenergic neuronal development is impaired by mutation of proneural HASH-1 gene in congenital central hypoventilation syndrome (Ondine's curse). Hum Mol Genet 12: 3173-3180

9. De Pontual L, Pelet A, Trochet D, Jaubert F, Espinosa-Parrilla Y, Munnich A, Brunet JF, Goridis C, Feingold J, Lyonnet S, Amiel J (2006) Mutations of the RET gene in isolated and syndromic Hirschsprung's disease in human disclose major and modifier alleles at a single locus. J Med Genet 43: 419-423

10. Doherty LS, Kiely JL, Deegan PC, Nolan G, McCabe S, Green AJ, Ennis S, Mc Nicholas WT (2007) Late-onset central hypoventilation syndrome: a family genetic study. Eur Respir J 29: 312-316

11. Faure C, Viarme F, Cargill G, Navarro J, Gaultier C, Trang H (2002) Abnormal esophageal motility in children with congenital central hypoventilation syndrome. Gastroenterology 122: 1258-1263

12. Gozal D (1998) Congenital central hypoventilation syndrome: an update. Pediatr Pulmonol 26: 273-82

13. Gozal D, Marcus CL, Shoseyov D, Keens TG (1993) Peripheral chemoreceptor function in children with the congenital central hypoventilation syndrome. J Appl Physiol 74: 379-387Gozal D, Marcus CL, Ward SL, Keens TG (1996) Ventilatory responses to passive leg motion in children with congenital central hypoventilation syndrome. Am J Respir Crit Care Med 153: 761-768

15. Gozal D, Simakajornboon N (2000) Passive motion of the extremities modifies alveolar ventilation during sleep in patients with congenital central hypoventilation syndrome. Am J Respir Crit Care Med 162: 1747-1751

16. Haddad GG, Mazza NM, Defendini R, Blanc WA, Driscoll JM, Epstein MA, Epstein RA, Mellins RB (1978) Congenital failure of automatic control of ventilation, gastrointestinal motility and heart rate. Medicine 57: 517-526

17. Harper RM, Macey PM, Woo MA, Macey KE, Keens TG, Gozal D, Alger JR (2005) Hypercapnic exposure in congenital central hypoventilation syndrome reveals CNS respiratory control mechanisms. J Neurophysiol 93: 1647-1658

18. Khalifa MM, Flavin MA, Wherrett BA (1998) Congenital central hypoventilation syndrome in monozygotic twins. J Pediatr 113: 853-855

19. Kumar R, Macey PM, Woo MA, Alger JR, Harper RM (2006) Elevated mean diffusivity in widespread brain regions in congenital central hypoventilation syndrome. J Magn Reson Imaging 24: 1252-1258

20.   Macey PM, Valderama C, Kim AH, Woo MA, Gozal D, Keens TG, Harper RK, Harper RM (2004) Temporal trends of cardiac and respiratory responses to ventilatory challenges in congenital central hypoventilation syndrome. Pediatr Res 55: 953-959

21.   Macey PM, Woo MA, Macey KE, Keens TG, Saeed MM, Alger JR, Harper RM (2005) Hypoxia reveals posterior thalamic, cerebellar, midbrain, and limbic deficits in congenital central hypoventilation syndrome. J Appl Physiol 98: 958-969

22.   Marcus CL, Bautista DB, Amihyia A, Ward SL, Keens TG (1991) Hypercapneic arousal responses in children with congenital central hypoventilation syndrome. Pediatrics 88: 993-998

23.   Matera I, Bachetti T, Puppo F, Di Duca M, Morandi F, Casiraghi GM, Cilio MR, Hennekam R, Hofstra R, Schober JG, Ravazzolo R, Ottonello G, Ceccherini I (2004) PHOX2B mutations and polyalanine expansions correlate with severity of respiratory phenotype and associated symptoms in both congenital and late onset central hypoventilation syndrome. J Med Genet 41: 373-80

24.   Mellins RB, Balfour HH, Turino GM, Winters RW (1970) Failure of automatic control of ventilation (Ondine's curse). Report of an infant born with this syndrome and review of the literature. Medicine 49: 487-504

25.   O'Brien LM, Holbrook CR, Vanderlaan M, Amiel J, Gozal D (2005) Autonomic function in children with congenital central hypoventilation syndrome and their families. Chest 128: 2478-2484

26.   Paton JY, Swaminathan S, Sargent CW, Hawksworth A, Keens TG (1993) Ventilatory response to exercise in children with congenital central hypoventilation syndrome. Am Rev Respir Dis 147: 1185-1191

27.   Sasaki A, Kanai M, Kijima K, Akaba K, Hashimoto M, Hasegawa H, Otaki S, Koizumi T, Kusuda S, Ogawa Y, Tuchiya K, Yamamoto W, Nakamura T, Hayasaka K (2003) Molecular analysis of congenital central hypoventilation syndrome. Hum Genet 114: 22-26

28.   Spengler CM, Gozal D, Shea SA (2001) Chemoreceptive mechanisms elucidated by studies of congenital central hypoventilation syndrome. Respir Physiol 129: 247-255

29.   Sritippayawan S, Hamutcu R, Kun SS, Ner Z, Ponce M, Keens TG (2002) Mother-daughter transmission of congenital central hypoventilation syndrome. Am J Respir Crit Care Med 166: 367-369

30.   Trang H, Boureghda S, Denjoy I, Alia M, Kabaker M. (2003) 24-hour BP in children with congenital central hypoventilation syndrome. Chest 124: 1393-1399

31.   Trang H, Dehan M, Beaufils F, Zaccaria I, Amiel J, Gaultier C, the French CCHS Working Group (2005) The French Congenital Central Hypoventilation Syndrome Registry: general data, phenotype, and genotype. Chest 127: 72-79

32.   Trang H, Girard A, Laude D, Elghozi JL (2005) Short-term blood pressure and heart rate variability in congenital central hypoventilation syndrome (Ondine's curse). Clin Sci (Lond) 108: 225-230

33.   Trang H, Laudier B, Trochet D, Munnich A, Lyonnet S, Gaultier C, Amiel J (2004) PHOX2B gene mutation in a patient with late-onset central hypoventilation. Pediatr Pulmonol 38: 349-351

34.   Trochet D, O'Brien LM, Gozal D, Trang H, Nordenskjold A, Laudier B, Svensson PJ, Uhrig S, Cole T, Niemann S, Munnich A, Gaultier C, Lyonnet S, Amiel J (2005) PHOX2B genotype allows for prediction of tumor risk in congenital central hypoventilation syndrome. Am J Hum Genet 76: 421-426

35.   Vanderlaan M, Holbrook CR, Wang M, Tuell A, Gozal D (2004) Epidemiologic survey of 196 patients with congenital central hypoventilation syndrome. Pediatr Pulmonol 37: 217-229

36.   Weese-Mayer DE, Berry-Kravis EM, Zhou L, Maher BS, Silvestri JM, Curran ME, Marazita ML (2003) Idiopathic congenital central hypoventilation syndrome: analysis of genes pertinent to early autonomic nervous system embryologic development and identification of mutations in PHOX-2B. Am J Med Genet 123: 3173-3180

37.   Weese-Mayer DE, Bolk S, Silvestri JM, Chakravarti A (2002) Idiopathic congenital central hypoventilation syndrome: evaluation of brain-derived neurotrophic factor genomic DNA sequence variation. Am J Med Genet 107: 306-310

38.   Weese-Mayer DE, Shannon DC, Keens TG, Silvestri JM, American Thoracic Society (1999) Idiopathic congenital central hypoventilation syndrome. Diagnosis and management. Am J Respir Crit Care Med 160: 368-373

39.   Woo MA, Macey PM, Macey KE, Keens TG, Woo MS, Harper RK, Harper RM (2005) fMRI responses to hyperoxia in congenital central hypoventilation syndrome. Pediatr Res 57: 510-508

40.   Woo MS, Woo MA, Gozal D, Jansen MT, Keens TG, Harper RM (1992) Heart rate variability in congenital central hypoventilation syndrome. Pediatr Res 31: 291-296

5. Structural and functional brain abnormalities in Congenital Central Hypoventilation Syndrome

Ronald M. HARPER, Mary A. WOO, Paul M. MACEY and Rajesh KUMAR

David Geffen School of Medicine and School of Nursing, UCLA, Los Angeles, CA 90095-1763, U.S.A.

5.1 Introduction

The name "congenital central hypoventilation syndrome" (CCHS) implies central nervous system dysfunction, with the primary characteristics of the syndrome, a reduced drive to breathe during sleep and reduced CO_2 and O_2 sensitivity [19; 39], linked to an as-yet-undescribed failure of particular brain structures. The purported genetic process underlying the disorder, mutation of the *PHOX2B* gene [4; 6; 44; 52; 53], is associated with injury to specific regions of the medulla and peripheral autonomic ganglia [12]. However, a complete description of the symptomology of CCHS points to a broad range of neural influences being deficient, implying that more than just brain stem functions are affected.

CCHS children show a range of physiologic and especially autonomic aberrations in addition to the loss of ventilatory drive during sleep and reduced CO_2 or O_2 ventilatory sensitivity (see Chapter 4). Although symptoms vary among patients, deficiencies in autonomic nervous system control are especially prominent, with both sympathetic and parasympathetic components of that control affected. The sympathetic aspects include profuse sweating [50], disturbed cardiovascular control, as manifested by syncope and inappropriate heart rate changes to blood pressure challenges [27], and an absence of blood pressure lowering at night [49]. Parasympathetic deficiencies include unequal pupillary constriction, defects in glandular secretion [50], and impaired vagal influences on heart rate variability [56]. Other altered physiologic and emotional measures are pronounced, including poor thermoregulation [50], a loss of affect (including emotions associated with the urge to breathe accompanying high CO_2 or hypoxia) [42; 45], difficulties in initiating urination, and alterations in fluid regulation (M. Vanderlaan, personal communication).

Many of the physiological and emotional functions that are deficient are regulated in rostral brain areas, and especially in limbic structures. Fluid regulation depends on structures in the anterior hypothalamus near the lamina terminalis, and thermoregulation is principally mediated by anterior hypothalamic structures. Voluntary initiation of urination depends on the anterior cingulate cortex [7; 8], and anecdotal evidence indicates that a number of CCHS children show a deficit in such control. Since the perception of suffocation is a powerful drive for inspiratory effort, and low O_2 or high CO_2 leads to that perception, loss of this urge to breathe in CCHS represents yet another significant removal of influence on respiratory effort. Negative perceptions, such as dyspnea from loaded breathing or from hypoxia or hypercapnia, are not regulated by medullary structures, except perhaps for final common path output of autonomic action to strong emotions. Instead, these perceptions are mediated by limbic areas, including the amygdala, insular cortex, and cingulate cortex [2; 5; 15; 40].

In addition to perception of affect associated with sensory processing of ventilatory stimuli, expression of emotion on motor systems is an issue in CCHS. Such expression can operate independently of traditional motor pathways [22-24], as evidenced by an inability to trigger appropriate facial muscle action to emotional stimuli (with retention of voluntary action [51]), a common occurrence in CCHS. Since affect exerts marked effects on breathing musculature, defects in forebrain emotional structures can have a profound effect on ventilation.

An obvious question stemming from the affective and autonomic findings is how such influences associated with rostral brain functions are related to the expression of the *PHOX2B* gene. *PHOX2B* principally affects autonomic ganglia and autonomic nervous system structures in medullary sites, rather than forebrain structures [12]. A recent study [47] also suggests that *PHOX2B* is expressed in chemosensory integration neurons of the retrotrapezoid nucleus. Mutant *Phox2b* preparations also show impaired ventilation during sleep [14] and to hyperoxic exposure [43]. The mutant preparations demonstrate that *Phox2b* exerts a significant role on ventilation, although the investigators caution that several characteristics of breathing differ from the human CCHS condition (see Chapter 14). A principal issue is that some of the breathing characteristics are especially deficient in rapid eye movement (REM) sleep rather than quiet sleep, as is the case in CCHS. Thus, a relationship between *Phox2b* and breathing is established, but these mutant demonstrations do not explain the forebrain-mediated physiological symptoms, the emotional characteristics of the condition, or the state-related variations in breathing regulation in CCHS. The objective of this chapter is to outline effects of the syndrome on multiple brain structures which may contribute to the characteristics of the syndrome.

5.2 Structural injury and functional deficits in CCHS

Given the physiological and affective characteristics of CCHS, underlying structural neural deficits must be present. However, the American Thoracic

Society definition of CCHS [3] indicates that no structural deficits appear on routine clinical evaluation of magnetic resonance or computed tomography brain images which could underlie the major characteristics of the syndrome; consequently, any structural deficit must not be gross, i.e., no tumor or significant lesion is apparent. The injury or maldevelopment may be revealed by specialized imaging techniques combined with statistical assessment of brain tissue which demonstrates group differences between patients and control populations. If brain structure is altered through injury or developmental, the manner in which neural structures respond to a task may also differ. Functional magnetic resonance imaging (fMRI) procedures can reveal responses of brain areas to autonomic or ventilatory stimuli which differentiate CCHS patients from the normal population, and assist demonstration of deficits in the manner in which neural structures mediate such functions.

Use of specialized imaging techniques has revealed multiple areas of neural structural damage or maldevelopment in CCHS. As would be expected from the physiological and emotional characteristics of the syndrome, many of the deficient regions lie in forebrain areas, although the brainstem is also affected. The forebrain sites include limbic and hypothalamic areas, and isolated regions serving specific physiological functions that are deficient in CCHS. In addition, significant damage appears in white matter, often in areas which form principal communication links with structures showing impaired functioning in the condition. When the entire brain is evaluated using fMRI for responses to autonomic or ventilatory manipulations, inadequate patterns to hypercapnia, hypoxia, hyperoxia, expiratory loading, and cold pressor challenges emerge in multiple brain regions that are also structurally affected. The combined findings highlight deficits in specific sites with roles in mediating the physiological and emotional responses to the challenges, and provide insights into the processes underlying the syndrome's characteristics.

5.3 Imaging findings

5.3.1 Structural injury

Description of fiber and neuronal damage can be obtained by T2 relaxometry, which measures the relative proportion of free to bound water content [1], an index that is impacted by cell loss, demyelination, or loss of cellular membranes, thus providing an indicator of tissue injury [1; 13; 25]. Such assessment reveals damage in CCHS to the basal forebrain in a restricted area extending from the lamina terminalis caudally through the anterior hypothalamus, and the anterior and medial thalamus. Additional injury appears in the anterior cingulate and in adjacent white matter, ventral frontal cortex, putamen and globus pallidus, hippocampus, and cerebellar cortex and deep nuclei (Fig. 1) [30].

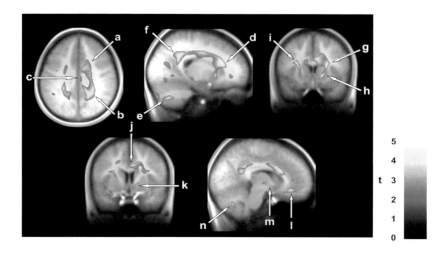

Fig. 1. Overlays of deficient areas (colored), indicated by T2 relaxometry measures, in 12 CCHS patients (mean age = 15 yrs) compared to 28 control cases. Backgrounds are averages of the 40 subjects' structural scans; axial and coronal images are in neurological convention, i.e., left side of image represents left side of brain. Color-coding indicates significance level (t-statistic; see colorbar). Deficits extend in anterior cingulate cortex (a, d, j) and white matter (a, b, d), including the mid cingulate (c) to posterior cingulate (b, f), cerebellar cortex (e) and deep nuclei (n), internal capsule (g, h, i) extending to the nucleus accumbens/septal area (k), the basal forebrain and hypothalamus (m), and ventral frontal cortex (l). Derived from [30], with permission.

These findings of structural injury provide a basis for many of the physiological signs in CCHS. Cerebellar structures, especially the deep fastigial nuclei, can limit the range of blood pressure elevation and lowering, providing a compensatory mechanism to overcome, for example, an extreme loss in blood pressure [32]; such compensation can include extensor motor action useful for restoring blood pressure in hypotension [21]. The cerebellar injury may underlie the syncope often found in CCHS. The cerebellum, however, is not a structure targeted by *PHOX2B*; thus, the injury is likely triggered by some other process.

Another magnetic resonance imaging procedure, diffusion tensor imaging, measures water diffusivity in three dimensions, and the average diffusivity of water, mean diffusivity (MD), can also indicate possible tissue damage. Although MD reflects similar pathologies as T2 relaxometry in this condition, the measure appears to be more sensitive to injury in CCHS. Mean diffusivity shows additional injury in the ventral diencephalon, dorsal midbrain and ventral pons, more-extensive regions of the anterior cingulate and frontal cortex, septum and caudate, ventral temporal cortex, and dorsal and caudal thalamus (Fig. 2) [31].

The T2 relaxometry and MD structural findings indicate extensive injury in CCHS, despite little obvious injury on routine clinical scans (T1- and T2-weighted imaging). Brain stem damage, which appears in ventral regions with diffusion tensor imaging-based MD, was expected, given the genetic and

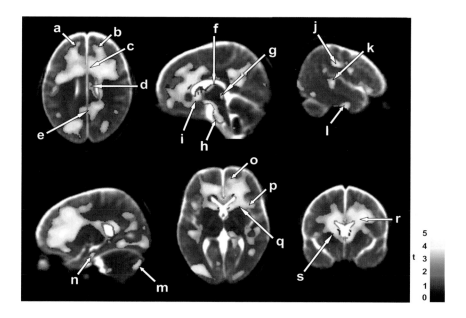

Fig. 2. Deficient areas (in color) indicated by higher MD in 15 CCHS patients (mean age = 15 yrs) compared to 30 age- and gender-matched control subjects. Backgrounds are averages of the 45 subjects' MD images; other conventions as in Fig. 1. Multiple clusters overlay those found with T2 relaxometry measures, but are more extensive; the deficits near the anterior cingulate (c) extend to the frontal cortex (a, b), mid (d) and posterior cingulate (e). Regions in the anterior-superior thalamus (f), dorsal midbrain (g), and lateral cortical regions (j, k ,l) also show injury. Clusters extending from the septal region (i) to the ventral surfaces of the diencephalon, midbrain and pons (h) are apparent. The anterior portion of the hippocampus/amygdala (n), the cerebellum (m), and the basal ganglia (q, s, r) to insula (p) also differ. Derived from [31], with permission.

CO_2-sensitivity links with this region, but the extensive injury in basal forebrain, caudate, anterior cingulate, and septal areas was not anticipated.

5.3.2 Functional deficits

Although structural injury is apparent from both T2 relaxometry and diffusion tensor imaging MD procedures, the findings do not necessarily imply underlying dysfunction in affected sites. Functional magnetic resonance imaging allows non-invasive examination of responses over the entire brain to autonomic and ventilatory challenges. Of particular interest for CCHS is the determination of central responses to CO_2; central chemosensitivity is especially affected in the condition, with peripheral chemoreception possibly intact [18].

Normally, hypercapnia elicits fMRI responses in multiple brain regions of control subjects, with the cerebellum responding early and transiently to increased CO_2, together with posterior thalamic and other additional rostral brain regions. The dorsal medulla in the region of the nucleus of the solitary tract, as expected, also responds to hypercapnia. CCHS patients differ in brain responses to elevated CO_2, especially in the deep cerebellar nuclei and dorsolateral cerebellar cortex (Fig. 3). In addition, a cluster extending from the posterior thalamus through the medial midbrain to the dorsolateral pons, as well as an area in the right caudate nucleus, continuing ventrolaterally through the putamen and ventral insula to the mid-hippocampus differ in CCHS cases. Other sites of aberrant function include the midline dorsal medulla, bilateral amygdala, right dorsal-posterior temporal

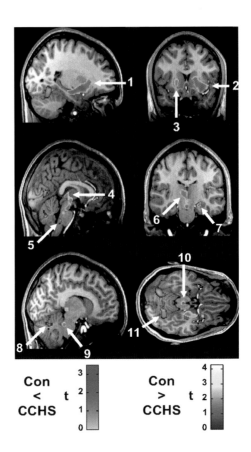

Fig. 3. fMRI signal differences (colored clusters) between 14 CCHS patients and 14 control children (mean age = 11 yrs) during a 5% CO_2/95% O_2 challenge. Signal differences are overlaid on a single control subject's anatomical image; other conventions as in Fig. 1. Responses are higher in CCHS in a cluster extending from the right caudate nucleus through the insula to the mid-hippocampus (1, 2, 6); a smaller region in the contralateral basal ganglia is also affected (3). An additional cluster (4) extends from the posterior thalamus through the ventral midbrain to the mid and dorsal pons/anterior cerebellar peduncle region (6, 9, 10), with signals larger in CCHS over controls. Control signals are larger than those in CCHS in the cerebellar cortex (8, 11) and the dorsal medulla (5). Derived from [20], with permission.

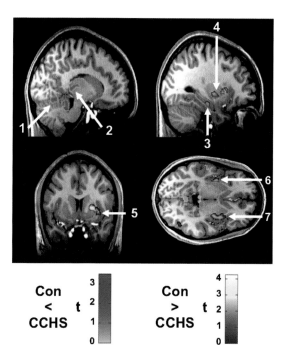

Fig. 4. fMRI signal differences between 14 CCHS patients and 14 age- and gender-matched control children (mean age = 11 yrs) acquired during a 15% O_2/85% N_2 challenge. Conventions as in Figure 3. Signals increased in CCHS over controls in the cerebellar cortex (1) and posterior thalamus (2), and were lower in CCHS than controls in the hippocampus (3), insular cortex (4, 6, 7), and putamen (5). Derived from [36], with permission.

cortex, and left anterior insula [20]. These sites consist of both medullary regions expected to be affected by *PHOX2B* and forebrain areas, and overlap many of the areas appearing with structural injury to T2 relaxometry and MD assessment (e.g., cerebellar, insular, amygdala and thalamic structures).

Inappropriate ventilatory responses to hypoxia are also characteristic of CCHS, and the neural processes underlying this defect are unclear as well. A 15% O_2/85% N_2 mixture elicited significant differences in response (magnitude and timing) in CCHS children over controls in the cerebellar cortex and deep nuclei, posterior thalamus, insula, amygdala, ventral anterior thalamus, right hippocampus, dorsal and ventral midbrain, caudate, claustrum, and putamen (Fig. 4) [36]. The nature of the response patterns in CCHS differ from controls in multiple ways, including late responses, absent early transient responses, and signal changes in the opposite direction from controls. No signal changes emerge in dorsal medullary sites, despite the classical association of that region's association with chemoreceptor reflexes and the demonstrated targeting by *PHOX2B*. Brain areas which respond inappropriately to hypoxia are those previously identified in posterior thalamic, midbrain, and cerebellar sites for normal mediation of low O_2 found in animal fetal and adult preparations [29; 46], and overlap multiple areas of structural injury found in CCHS.

Patients afflicted with CCHS also do not respond normally to high O_2. Brain responses to hyperoxia differ between control subjects and CCHS patients, as do respiratory rates, which initially fall in CCHS, but increase in controls. Multiple cerebellar, midbrain, and pontine sites respond similarly to hyperoxia in controls and CCHS cases, but differ early in the right amygdala, concurrently with respiratory rate, and late in the right insula, paralleling aberrant heart rate changes in CCHS. The medial and anterior cingulate, hippocampus, basal ganglia, as well as pontine and midbrain structures and regions within the superior temporal and inferior frontal cortical gyri differ in responses [35; 55]. Again, although medullary and pontine areas targeted by *PHOX2B* are affected, rostral brain areas that modify sympathetic outflow and alter respiratory timing, such as the amygdala and right insula, respond inappropriately in CCHS cases.

5.4 Cardiovascular control deficits

Although abnormal ventilatory characteristics are primary features of CCHS, autonomic nervous system-related aberrations are also prominent. Brain responses to autonomic challenges are deficient in multiple cardiovascular regulatory sites of CCHS patients. A cold pressor challenge to the forehead, which elicits a range of autonomic effects, results in diminished responses in primary sensory thalamic and cortical areas. Altered responses also appear in cerebellar cortex and deep nuclei, basal ganglia, and mid-to-posterior cingulate, insular, frontal and temporal cortices [34].

These affected thalamic and limbic structures play significant roles in mediating autonomic responses. Reduced primary thalamic sensory responses likely indicate reduced influence of afferent cold signals. The insular cortex acts as a gain control for the baroreflex [58], and the ventral frontal cortex and cingulate cortex help to mediate blood pressure responses [11; 28]. The limbic structures are closely interconnected [9; 26; 37], and have pronounced projections with the rostral ventral respiratory group [17]. Deficits in limbic areas could thus interact with respiratory regions, providing a means by which state-related action on forebrain regions (most prominently modified in quiet sleep, a state of most-marked breathing pathology in CCHS) can act on respiratory drive.

Again, no brain responses to the cold pressor challenge appear in either control or CCHS group in the dorsal medulla, but medial and ventral medullary areas show enhanced signals in CCHS patients. Expiratory loading, although a breathing task, also elicits substantial blood pressure increases, and shows diminished signals in the cingulate and right parietal cortex, cerebellar cortex and fastigial nucleus, and basal ganglia, whereas anterior cerebellar cortical sites and deep nuclei, dorsal midbrain, and dorsal pons show increased signals in CCHS [33].

Timing of responses is also affected; the dorsal and ventral medulla show delayed responses in CCHS patients.

5.5 Potential mechanisms in injury

Both the structural and functional imaging findings, as well as the expression of physiological and affective characteristics in the syndrome, show that neural deficits in CCHS are not confined to structures in the brain stem, likely targets of *PHOX2B*. The processes leading to the forebrain deficits are unclear, but several mechanisms could be operating. One possibility is that *PHOX2B* may affect structures further along the neuroaxis than previously demonstrated or that expression of the mutation can vary substantially and exert different degrees of pathology. In addition, CCHS children are exposed to multiple episodes of hypoxia, some of which may stem from the earliest manifestations of the syndrome. CCHS is seldom recognized until episodes of failed breathing are detected, and since REM sleep is somewhat protective for the syndrome (i.e., breathing drive is relatively higher during that state), and infants spend a disproportionate amount of time in REM sleep, the failure of sleep-related breathing may not be determined until quiet sleep emerges, allowing repeated hypoxic exposure.

Hypoxic processes may be instrumental in creating the injury found in the cerebellar cortex and deep nuclei. The unique anatomical arrangement of olivary climbing fibers with cerebellar Purkinje neurons provides ideal conditions for excitotoxic injury elicited by hypoxia [54]. The rapidity with which intermittent hypoxia can elicit damage in cerebellar Purkinje cells and deep nuclei in animal models (as little as 5 hrs, [38]) suggests that neural injury can emerge rapidly from failed breathing in early life. Even after recognition of CCHS in affected infants, episodes of hypoxic exposure continue, either from accidental hypoventilation through ventilatory equipment failure, ventilatory reduction by core temperature rises from infections, or through inadequate attention to breathing during waking sedentary periods. The neural structures most frequently injured during acute hypoxia or carbon monoxide exposure have been well-described, and typically include cerebellar, basal ganglia, thalamic and white matter structures [10; 16].

However, many of the structures showing injury in CCHS are not classically associated with hypoxic exposure. Among these sites, an area of deficit extending from the lamina terminalis through the anterior hypothalamus is prominent. This region plays an especially important role for CCHS symptomology, since thermoregulation and fluid regulation are primary concerns in the syndrome. Other areas are also not usually associated with hypoxic exposure, such as the anterior cingulate cortex and the septal region, as well as the amygdala. These limbic sites mediate different aspects of affect and play roles in blood pressure regulation.

Since some affected sites are not among structures typically found injured to hypoxic exposure, and have not been associated with *PHOX2B* expression, other processes must be operating. A speculative mechanism is that targets susceptible to *PHOX2B*, the autonomic ganglia of the medulla, modify perfusion of selected forebrain structures in CCHS. These ganglia are responsible for regulating vascular flow, and if not adequately developed, may not allow vascularization sufficient to provide requisite perfusion of rostral brain or other areas. Expression of many of the characteristics of CCHS requires a period of time to emerge, and such late expression may result from accumulated under development or injury due to sustained inadequate perfusion.

This speculation, however, does not account for the remarkable specificity in structures injured in CCHS, with limbic regions being more-heavily targeted. In particular, the column of injury from the lamina terminalis, extending through the hypothalamus, is not an area usually found to be damaged to hypoxic or generalized ischemic exposure. The proposed inadequate perfusion mechanism suggests deficiencies in specific vascular areas supplying unique regions. These aspects clearly must be explored further.

5.6 Conclusion

CCHS patients show both structural and functional deficits when compared to age- and gender-matched control children, with the deficits appearing using a range of magnetic resonance imaging procedures. The structural deficits are evident in multiple brain sites from the brain stem and cerebellum to the forebrain, and are especially prominent in limbic areas of the rostral brain which play significant roles in affect and autonomic control. The rostral brain structural deficiencies can be related to significant characteristics of the syndrome, including abnormal thermoregulation, fluid regulation, initiation of urination, sympathetic nervous system control, and emotional responses to hypercapnia and hypoxia. Cerebellar deficits likely contribute to a portion of the sympathetic control and coordination of respiratory musculature pathology in the condition, while insular, ventral frontal, and cingulate cortices also contribute, possibly interacting with deep cerebellar projections to the thalamus and then to limbic areas [26; 41]. Primary characteristics of the syndrome, including the loss of ventilatory drive during sleep and diminished ventilatory responses to CO_2 and O_2, could develop from several sites of structural deficit, including O_2 integrative sites in the posterior thalamus [29], CO_2-sensitive sites in the cerebellum [57], hypothalamic and basal forebrain areas [48], and pontine areas which also show functional impairment to hypercapnic or hypoxic challenges [20; 36]. Many of the affected brain structures lie outside targets normally associated with *PHOX2B* expression, suggesting that the deficits may develop from hypoxia-related injury from a primary breathing deficiency initiated by mutant *PHOX2B* action, or from perfusion deficits associated with altered *PHOX2B* expression on medullary autonomic ganglia, or by as-yet-unknown processes.

Acknowledgements

This research was supported by NIH HD-22695. We thank Dr. Jeffry Alger, Dr. Shantanu Sinha, Ms. Amy Kim, Dr. Katherine Macey, Ms. Rebecca Harper and Dr. Mohammad Saeed.

References

1. Abernethy LJ, Klafkowski G, Foulder-Hughes L, Cooke RW (2003) Magnetic resonance imaging and T2 relaxometry of cerebral white matter and hippocampus in children born preterm. Pediatr Res 54: 868-874
2. Amaral DG (2003) The amygdala, social behavior, and danger detection. Ann N Y Acad Sci 1000: 337-347
3. American Thoracic Society (1999) Idiopathic congenital central hypoventilation syndrome: diagnosis and management. Am J Crit Care Med 160: 368-373
4. Amiel J, Laudier B, Attie-Bitach T, Trang H, de Pontual L, Gener B, Trochet D, Etchevers H, Ray P, Simonneau M, Vekemans M, Munnich A, Gaultier C, Lyonnet S (2003) Polyalanine expansion and frameshift mutations of the paired-like homeobox gene PHOX2B in congenital central hypoventilation syndrome. Nat Genet 33: 459-461
5. Banzett RB, Mulnier HE, Murphy K, Rosen SD, Wise RJ, Adams L (2000) Breathlessness in humans activates insular cortex. Neuroreport 11: 2117-2120
6. Berry-Kravis EM, Zhou L, Rand CM, Weese-Mayer DE (2006) Congenital central hypoventilation syndrome: PHOX2B mutations and phenotype. Am J Respir Crit Care Med 174: 1139-1144
7. Blok BF, Sturms LM, Holstege G (1998) Brain activation during micturition in women. Brain 121: 2033-2042
8. Blok BF, Willemsen AT, Holstege G (1997) A PET study on brain control of micturition in humans. Brain 120 : 111-121
9. Carmichael ST, Price JL (1995) Limbic connections of the orbital and medial prefrontal cortex in macaque monkeys. J Comp Neurol 363: 615-641
10. Chang KH, Han MH, Kim HS, Wie BA, Han MC (1992) Delayed encephalopathy after acute carbon monoxide intoxication: MR imaging features and distribution of cerebral white matter lesions. Radiology 184: 117-122
11. Critchley HD, Mathias CJ, Josephs O, O'Doherty J, Zanini S, Dewar BK, Cipolotti L, Shallice T, Dolan RJ (2003) Human cingulate cortex and autonomic control: converging neuroimaging and clinical evidence. Brain 126: 2139-2152
12. Dauger S, Pattyn A, Lofaso F, Gaultier C, Goridis C, Gallego J, Brunet JF (2003) Phox2b controls the development of peripheral chemoreceptors and afferent visceral pathways. Development 130: 6635-6642
13. Di Costanzo A, Di Salle F, Santoro L, Bonavita V, Tedeschi G (2001) T2 relaxometry of brain in myotonic dystrophy. Neuroradiology 43: 198-204
14. Durand E, Dauger S, Pattyn A, Gaultier C, Goridis C, Gallego J (2005) Sleep-disordered breathing in newborn mice heterozygous for the transcription factor Phox2b. Am J Respir Crit Care Med 172: 238-243
15. Evans KC, Banzett RB, Adams L, McKay L, Frackowiak RS, Corfield DR (2002) BOLD fMRI identifies limbic, paralimbic, and cerebellar activation during air hunger. J Neurophysiol 88: 1500-1511

16.   Faro MD, Windle WF (1969) Transneuronal degeneration in brains of monkeys asphyxiated at birth. Exp Neurol 24: 38-53

17.   Gaytan SP, Pasaro R (1998) Connections of the rostral ventral respiratory neuronal cell group: an anterograde and retrograde tracing study in the rat. Brain Res Bull 47: 625-642

18.   Gozal D, Marcus CL, Shoseyov D, Keens TG (1993) Peripheral chemoreceptor function in children with the congenital central hypoventilation syndrome. J Appl Physiol 74: 379-387

19.   Haddad GG, Mazza NM, Defendini R, Blanc WA, Driscoll JM, Epstein MA, Epstein RA, Mellins RB (1978) Congenital failure of automatic control of ventilation, gastrointestinal motility and heart rate. Medicine 57: 517-526

20.   Harper RM, Macey PM, Woo MA, Macey KE, Keens TG, Gozal D, Alger JR (2005) Hypercapnic exposure in congenital central hypoventilation syndrome reveals CNS respiratory control mechanisms. J Neurophysiol 93: 1647-1658

21.   Harper RM, Richard CA, Rector DM (1999) Physiological and ventral medullary surface activity during hypovolemia. Neuroscience 94: 579-586

22.   Holstege G (2002) Emotional innervation of facial musculature. Mov Disord 17 Suppl 2: S12-16

23.   Holstege G, Meiners L, Tan K (1985) Projections of the bed nucleus of the stria terminalis to the mesencephalon, pons, and medulla oblongata in the cat. Exp Brain Res 58: 379-391

24.   Hopkins DA, Holstege G (1978) Amygdaloid projections to the mesencephalon, pons and medulla oblongata in the cat. Exp Brain Res 32: 529-547

25.   Jackson GD, Connelly A, Duncan JS, Grunewald RA, Gadian DG (1993) Detection of hippocampal pathology in intractable partial epilepsy: increased sensitivity with quantitative magnetic resonance T2 relaxometry. Neurology 43: 1793-1799

26.   Kaitz SS, Robertson RT (1981) Thalamic connections with limbic cortex. II. Corticothalamic projections. J Comp Neurol 195: 527-545

27.   Kim AH, Macey PM, Woo MA, Yu PL, Keens TG, Gozal D, Harper RM (2002) Cardiac responses to pressor challenges in congenital central hypoventilation syndrome. Somnologie 6: 109-115

28.   King AB, Menon RS, Hachinski V, Cechetto DF (1999) Human forebrain activation by visceral stimuli. J Comp Neurol 413: 572-582

29.   Koos BJ, Chau A, Matsuura M, Punla O, Kruger L (1998) Thalamic locus mediates hypoxic inhibition of breathing in fetal sheep. J Neurophysiol 79: 2383-2393

30.   Kumar R, Macey PM, Woo MA, Alger JR, Keens TG, Harper RM (2005) Neuroanatomic deficits in congenital central hypoventilation syndrome. J Comp Neurol 487: 361-371

31.   Kumar R, Macey PM, Woo MA, Alger JR, Harper RM (2006) Elevated mean diffusivity in widespread brain regions in congenital central hypoventilation syndrome. J Magn Reson Imaging 24: 1252-1258

32.   Lutherer LO, Lutherer BC, Dormer KJ, Janssen HF, Barnes CD (1983) Bilateral lesions of the fastigial nucleus prevent the recovery of blood pressure following hypotension induced by hemorrhage or administration of endotoxin. Brain Res 269: 251-257

33.   Macey KE, Macey PM, Woo MA, Harper RK, Alger JR, Keens TG, Harper RM (2004) fMRI signal changes in response to forced expiratory loading in congenital central hypoventilation syndrome. J Appl Physiol 97: 1897-1907

34. Macey PM, Macey KE, Woo MA, Keens TG, Harper RM (2005) Aberrant neural responses to cold pressor challenges in congenital central hypoventilation syndrome. Pediatr Res 57: 500-509

35. Macey PM, Valderama C, Kim AH, Woo MA, Gozal D, Keens TG, Harper RK, Harper RM (2004) Temporal trends of cardiac and respiratory responses to ventilatory challenges in congenital central hypoventilation syndrome. Pediatr Res 55: 953-959

36. Macey PM, Woo MA, Macey KE, Keens TG, Saeed MM, Alger JR, Harper RM (2005) Hypoxia reveals posterior thalamic, cerebellar, midbrain, and limbic deficits in congenital central hypoventilation syndrome. J Appl Physiol 98: 958-969

37. Martin LJ, Powers RE, Dellovade TL, Price DL (1991) The bed nucleus-amygdala continuum in human and monkey. J Comp Neurol 309: 445-485

38. Pae EK, Chien P, Harper RM (2005) Intermittent hypoxia damages cerebellar cortex and deep nuclei. Neurosci Lett 375: 123-128

39. Paton JY, Swaminathan S, Sargent CW, Keens TG (1989) Hypoxic and hypercapnic ventilatory responses in awake children with congenital central hypoventilation syndrome. Am Rev Respir Dis 140: 368-372

40. Peiffer C, Poline JB, Thivard L, Aubier M, Samson Y (2001) Neural substrates for the perception of acutely induced dyspnea. Am J Respir Crit Care Med 163: 951-957

41. Person RJ, Andrezik JA, Dormer KJ, Foreman RD (1986) Fastigial nucleus projections in the midbrain and thalamus in dogs. Neuroscience 18: 105-120

42. Pine DS, Weese-Mayer DE, Silvestri JM, Davies M, Whitaker AH, Klein DF (1994) Anxiety and congenital central hypoventilation syndrome. Am J Psychiatry 151: 864-870

43. Ramanantsoa N, Vaubourg V, Dauger S, Matrot B, Vardon G, Chettouh Z, Gaultier C, Goridis C, Gallego J (2006) Ventilatory response to hyperoxia in newborn mice heterozygous for the transcription factor Phox2b. Am J Physiol Regul Integr Comp Physiol 290: R1691-1696

44. Sasaki A, Kanai M, Kijima K, Akaba K, Hashimoto M, Hasegawa H, Otaki S, Koizumi T, Kusuda S, Ogawa Y, Tuchiya K, Yamamoto W, Nakamura T, Hayasaka K (2003) Molecular analysis of congenital central hypoventilation syndrome. Hum Genet 114: 22-26

45. Shea SA, Andres LP, Shannon DC, Guz A, Banzett RB (1993) Respiratory sensations in subjects who lack a ventilatory response to CO_2. Respir Physiol 93: 203-219

46. Sica AL, Greenberg HE, Ruggiero DA, Scharf SM (2000) Chronic-intermittent hypoxia: a model of sympathetic activation in the rat. Respir Physiol 121: 173-184

47. Stornetta RL, Moreira TS, Takakura AC, Kang BJ, Chang DA, West GH, Brunet JF, Mulkey DK, Bayliss DA, Guyenet PG (2006) Expression of Phox2b by brainstem neurons involved in chemosensory integration in the adult rat. J Neurosci 26: 10305-10314

48. Teppema LJ, Dahan A (2005) Central chemoreceptors. In: Ward, DS (ed) Pharmacology and pathophysiology of the control of breathing. Taylor & Francis, Boca Raton, FL, pp 21-69

49. Trang H, Boureghda S, Denjoy I, Alia M, Kabaker M (2003) 24-hour BP in children with congenital central hypoventilation syndrome. Chest 124: 1393-1399

50. Vanderlaan M, Holbrook CR, Wang M, Tuell A, Gozal D (2004) Epidemiologic survey of 196 patients with congenital central hypoventilation syndrome. Pediatr Pulmonol 37: 217-229

51.   Waxman SG (1996) Clinical observations on the emotional motor system. Prog Brain Res 107: 595-604

52.   Weese-Mayer DE, Berry-Kravis EM, Marazita ML (2005) In pursuit (and discovery) of a genetic basis for congenital central hypoventilation syndrome. Respir Physiol Neurobiol 149: 73-82

53.   Weese-Mayer DE, Berry-Kravis EM, Zhou L, Maher BS, Silvestri JM, Curran ME, Marazita ML (2003) Idiopathic congenital central hypoventilation syndrome: analysis of genes pertinent to early autonomic nervous system embryologic development and identification of mutations in PHOX2b. Am J Med Genet 123: 267-278

54.   Welsh JP, Yuen G, Placantonakis DG, Vu TQ, Haiss F, O'Hearn E, Molliver ME, Aicher SA (2002) Why do Purkinje cells die so easily after global brain ischemia? Aldolase C, EAAT4, and the cerebellar contribution to posthypoxic myoclonus. Adv Neurol 89: 331-359

55.   Woo MA, Macey PM, Macey KE, Keens TG, Woo MS, Harper RK, Harper RM (2005) fMRI responses to hyperoxia in congenital central hypoventilation syndrome. Pediatr Res 57: 510-518

56.   Woo MS, Woo MA, Gozal D, Jansen MT, Keens TG, Harper RM (1992) Heart rate variability in congenital central hypoventilation syndrome. Pediatr Res 31: 291-296

57.   Xu F, Frazier DT (1997) Respiratory-related neurons of the fastigial nucleus in response to chemical and mechanical challenges. J Appl Physiol 82: 1177-1184

58.   Zhang ZH, Dougherty PM, Oppenheimer SM (1999) Monkey insular cortex neurons respond to baroreceptive and somatosensory convergent inputs. Neuroscience 94: 351-360

6. *In vitro* studies of *PHOX2B* gene mutations in congenital central hypoventilation syndrome

Tiziana BACHETTI and Isabella CECCHERINI

Laboratorio di Genetica Molecolare, Istituto Giannina Gaslini – 16148 Genova, Italy

6.1 Introduction

Heterozygous frameshift mutations, polyalanine expansions and, at a lesser extent, missense and non sense nucleotide substitutions of the *PHOX2B* gene have been identified in the vast majority of CCHS individuals [4; 19; 23; 32] (Table 1).

The PHOX2B protein is known to act as a tissue-specific transcription factor, mainly expressed during the development of the autonomous nervous system [11] (see Chapter 3). PHOX2B is therefore responsible for the expression regulation of target genes involved in this developmental pathway. In particular, direct PHOX2B binding has been demonstrated to the regulatory regions of the dopamine-β-hydroxilase (*DβH*), *PHOX2A* and more recently *TLX-2*, genes encoding an enzyme acting in the cathecholamine synthesis, a transcription factor necessary for specification of noradrenergic neuronal phenotype and a transcription factor expressed in the developing enteric nervous system, respectively [2; 8; 14]. Based on these acquisitions, a functional approach has become feasible, and thus undertaken, to explore the molecular mechanisms underlying CCHS pathogenesis.

To this end, the two most abundant classes of *PHOX2B* defects, polyalanine expansions and frameshift mutations, have been tested so far in two different laboratories for their possible effects on the correct function of the protein, namely transactivation activity, DNA binding efficiency and subcellular localization [5; 27]. In addition, the cellular response to *PHOX2B* polyalanine expansions has been investigated to assess the possible existence of mechanisms involved in limiting their pathogenetic effects [6].

Table 1. *PHOX2B* mutations detected, and functionally tested, in CCHS patients

[1]numbers in square brackets correspond to references listed at the end of the chapter

POLYALANINE EXPANSIONS				
Extra alanine residues	Extra nucleotides	Total aminoacids	Mutation detection first report	Functional test first report
5	15	319	[1]	
6	18	320		
7	21	321		
8	24	322		-
9	27	323		
11	33	325		
12	36	326		
13	39	327		
FRAMESHIFT MUTATIONS				
Mutation	Reading frame			
c.618insC	2	358		-
c.721-758del38nt	2	345		
c.862-866insG	2	358		
c.614-618delC	3	307		
c.606insA	2	358		-
c.693-700del8nt	2	355		-
c.689-696dup8	3	310		-
c.721-756del35	2	346		-
c.930insG	2	358		
c.936insT	2	358		
c.931del5	2	356		
c.722-738del17	2	352		-
c.945A>C	1	355		-
MISSENSE MUTATIONS				
Nucleotide substitution	Amino-acid change			
c.421C>G	R141G	314		
c.299G>T	R100L	314		
c.422G>A	Q141R	314		-
c.428A>G	Q143R	314		-
NONSENSE MUTATIONS				
c.463A>T	K155X	154		-

6.2 PHOX2B polyalanine expansions

6.2.1 Impaired transactivation activity of PHOX2B proteins carrying polyalanine expansions

At today, in frame duplications occurred within the nucleotide stretch coding for the 20-alanine repeat of the *PHOX2B* gene, leading to expansions from +5 to +13Ala residues, have been identified in the vast majority of the CCHS patients carrying *PHOX2B* mutations (from 81.5% to 98.4% depending on the series) [4; 19; 23; 28; 32].

To investigate how *PHOX2B* polyalanine expansions can induce CCHS pathogenesis, expression constructs containing each of the mutations were generated and their ability to regulate the transcription of three known target genes compared to a wild type *PHOX2B* construct. In particular, reporter gene constructs in which the regulatory regions under analysis were cloned upstream of the *Luciferase* gene have been used to quantify the effect of each of the above wild type and mutant plasmids on driving expression of the reporter gene.

In particular, when different polyalanine *PHOX2B* expression constructs were co-transfected with the *DβH* and *PHOX2A* regulatory regions, a strict inverse correlation between the Luciferase activity and the length of the polyalanine tract was observed, showing that the transcriptional regulation of these two genes is directly dependent on the correct structure of the PHOX2B domain including the 20-alanines tract [6]. On the other hand, the weak effect displayed by mutant PHOX2B versions on the *TLX-2* promoter suggests that the polyalanine domain is not probably crucial for *TLX-2* transcriptional activation [8].

Polyalanine regions are present in several proteins and in-frame duplications in the corresponding genes are emerging causes of human genetic diseases [9; 18]. Expansions affecting the polyalanine tracts of different proteins have been shown to exert their pathogenetic effect by inducing intracellular aggregation of mutant products, thus hampering their correct sub-cellular localization [3; 10; 12; 21]. This has prompted to investigate whether mutant PHOX2B versions carrying additional alanine residues had also similar effects, able to explain their impaired transcriptional activity.

6.2.2 Polyalanine expansions cause PHOX2B cellular mislocalization

The possibility that polyalanine expansions could affect the PHOX2B correct subcellular localization was firstly hypothesized after detecting, in nuclear extracts of cells transfected with mutant *PHOX2B* expression constructs carrying additional 5, 9 and 13 alanine residues, decreasing amounts of the protein, inversely correlating with the length of the polyalanine tract [6].

Fluorescence microscope analysis of COS-7 cells expressing PHOX2B proteins fused to a green fluorescent molecule showed large amounts of mutant proteins retained in the cytoplasm with formation of aggregates, as exemplified in

Fig. 1A. In particular, a correlation between increasing length of the polyalanine repeat and percentage of cells characterized by a complete or partial cytoplasmic localization was demonstrated (Fig. 1B, left), thus confirming that impaired sub-cellular PHOX2B localization could be ascribed to the aggregation-prone effect of the polyalanine expansion [6]. Similar experiments performed in HeLa cells induced formation of PHOX2B polyalanine aggregates, though in different amounts compared to what observed in COS-7 cells [27], suggesting that mislocalization of the mutant protein is a common pathogenetic mechanism leading to impaired transcriptional activity of mutant PHOX2B carrying alanine expanded tracts.

A

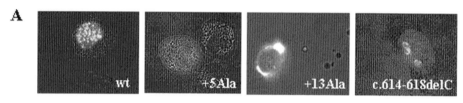

B

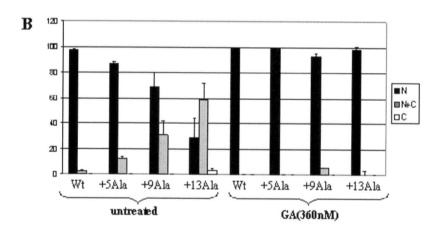

Fig. 1 Cellular localization of mutant PHOX2B analysed by fluorescence microscopy.
A: from left to right, the nuclear localisation of the wild type protein, the nuclear+cytoplasmic localization of the mutant protein carrying the smallest polyalanine expansion (+5Ala), the exclusively cytoplasmic localization of the mutant protein carrying the largest polyalanine expansion (+13Ala) and the nucleolar localization of the c-614-618delC mutant are shown. B: the diagram shows the cellular localization of transfected PHOX2Bwt-GFP, PHOX2B+5Ala-GFP, PHOX2B +9Ala-GFP and PHOX2B +13Ala-GFP without any treatment (left) and after addition of geldanamycin (GA 360nM). In particular, percentage of cells with N (nuclear), N+C (nuclear + cytoplasmic) and C (cytoplasmic) localization is shown for each protein. Values are the means of at least three independent experiments.

6.2.3 Impaired DNA binding due to PHOX2B polyalanine expansions

Although the PHOX2B polyalanine stretch lies outside the homeodomain, the region necessary to bind the PBD2 element of the *DβH* promoter [2], electrophoretic mobility shift assays (EMSA) performed by using radiolabelled probes including the PBD2 sequence and *in vitro* translated PHOX2B proteins have shown decreased DNA binding of mutant proteins with respect to wild type. In particular, binding activity is reduced when testing the +9 alanine expansion and barely detectable with the longest expansions (+12, +13), while it seems not to be affected in the case of the smallest expansions (+5, +7) [27]. The hypothesis that reduced DNA binding might be due to aggregation of PHOX2B mutant proteins, and therefore to unavailability of the transcription factor, has been confirmed *in vitro*, by the same authors, by showing that PHOX2B proteins with expanded alanine tracts spontaneously form oligomers whose size correlates with the length of the polyalanine repeat.

6.2.4 Negative dominant effect of PHOX2B carrying polyalanine expansions

PHOX2B mutations detected in CCHS patients are heterozygous, thus suggesting a dominant inheritance of the disease. As interaction between wild-type and mutant proteins, leading to dominant negative effects, has already been reported for polyalanine expanded proteins in human disease [3; 10; 12; 21], a series of experiments, aiming to verify whether such a pathogenetic mechanism could play a role also in CCHS associated *PHOX2B* mutations, was carried out. In particular, fluorescence microscope analysis has shown that the +13Ala PHOX2B protein does interact with the wild-type transcription factor in the nucleus, with limited but significant formation of nuclear aggregates [6]. In addition, a significant impairment of the wild type induced transactivation was demonstrated in the presence of a 10-fold excess of mutant protein, after assaying the Luciferase activity in cells expressing different ratios of the wild type and +13Ala PHOX2B proteins [27]. Therefore, besides causing functional haploinsufficiency, PHOX2B poly Ala expanded tracts can exert a partial dominant negative effect, with a minor fraction of the wild type PHOX2B interacting with the misfolded +13Ala mutant.

2.5 Cellular response to aggregates containing PHOX2B with polyalanine expansions

It has recently been observed that aggregate-prone proteins with polyalanine expansions are degraded by autophagy and by the proteasome machinery [1; 13; 22], two mechanisms responsible for the maintenance of protein balance in Eukaryotic cells [17; 24]. It has also been shown that components of the heat shock response pathway and members of the proteasome machinery co-localize with aggregates of misfolded proteins characterised by the expansion of polyalanine and polyglutamine residues [1; 21; 25].

Fluorescence miscroscope analysis has shown that the heat shock protein HSP70, already observed to suppress protein aggregation due to polyglutamine and polyalanine expansions [3; 16; 25] and to reduce aggregation of PABPN1 polyalanine expanded mutants when overexpressed and recruited in the nucleus [31], co-localizes with PHOX2B polyalanine aggregates [27]. After observing that specific inhibition of autophagy and ubiquitin-proteasome systems in cells expressing the PHOX2B +13Ala led to an increase in cellular aggregates, suggesting the involvement of both mechanisms in the clearance of aberrant PHOX2B protein [5], several analyses have been performed to investigate the role of chaperones in their folding. In particular, treatment of cells expressing PHOX2B mutants carrying polyalanine expansions with geldanamycin (GA), a naturally occurring antibiotic that interacts with HSP90 and leads to up-regulation of HSP40 and HSP70 [21], has allowed to investigate whether activation of the heat shock response could interfere with polyalanine PHOX2B aggregation. Fluorescence microscope analysis showed that cells expressing mutant PHOX2B carrying polyalanine expansions, and treated with GA at the time of transfection, displayed a dose-dependent progressive increase in nuclear localisation of the mutant protein, together with a parallel GA dose-dependent decrease in aggregates formation (see the right side of Fig. 1B for the effect of a very high dose of GA), thus suggesting that overexpression of chaperones could prevent formation of PHOX2B aggregates. In addition, geldanamycin has turned out to be efficient also in inducing clearance of PHOX2B pre-formed polyalanine aggregates and ultimately also in rescuing the PHOX2B ability to transactivate the DβH promoter [5].

Such a GA-induced correct folding of mutant PHOX2B proteins seems to be not only due to drastic increase of HSP70 expression, but also mediated at some extent by the proteasome activity. Indeed, as suggested by impairment of correct nuclear localisation following treatments of cells with geldanamycin + proteasome inhibitor, analysis of mutant protein levels performed by western blot has shown that geldanamycin induces a decrease in the amount of PHOX2B(+13Ala), likely due to proteasome-mediated degradation [5].

The important role of proteasome in cleaning cells from PHOX2B polyalanine aggregates has been confirmed in cells expressing the mutant protein carrying the largest polyalanine expansion by demonstrating association between the amounts of PHOX2B aggregates and apoptosis, while no significant apoptosis was detected in cells transfected with plasmids expressing the shortest PHOX2B mutants. Analysis of apoptosis has also revealed a role of geldanamycin in preventing cell death; despite its toxic effects, geldanamycin treatment of cells expressing PHOX2B (+13Ala) induced a selective decrease of the percentage of cells in advanced stages of apoptosis, suggesting that this drug protects damaged cells against the final fragmentation step, probably by delaying their progression toward the death process. Nevertheless, the low level of toxicity detected in association with PHOX2B polyalanine aggregates suggests that CCHS pathogenesis is likely due to neuronal dysfunction rather than to a massive cell loss [5].

6.3 *PHOX2B* frameshift mutations

Although frameshift mutations characterize a minor fraction of CCHS patients bearing a defective *PHOX2B* allele, they are not less relevant than polyalanine expansions.

According to the detection of *PHOX2B* missense and frameshift mutations in sporadic and familial cases of neuroblastoma [20; 26; 29], a recent study has confirmed that *PHOX2B* frameshift mutations predispose CCHS patients to develop tumors of the sympathethic nervous system [28].

Differently from polyalanine expansions, this group of mutations produces proteins characterised by one of two different aberrant C-terminal regions, depending on how the frame shift is achieved. The aberrant frame-2 is read in the presence of either insertion of one nucleotide, or one nucleotide plus a multiple of three nucleotides, or deletions of two nucleotides, or two nucleotides plus a multiple of three nucleotides. Conversely, the aberrant frame-3 is read in the presence of either deletion of one nucleotide, or one nucleotide plus a multiple of three nucleotides, or insertions of two nucleotides, or two nucleotides plus a multiple of three nucleotides. Functional studies have focused so far on mutations c.614-618delC, c.721-758del38nt, c.862-866insG, c.931del5, c.936insT, as reported in Table 1. In particular, among these mutations, the c.614-618delC is the only one producing a truncated protein characterized by the frame-3 aminoacid sequence from the one nucleotide-deletion site, while the others lead to the synthesis of an additional and larger frame-2 aminoacidic sequence starting at the PHOX2B mutant site (frame 2).

6.3.1 Transactivation activity of PHOX2B proteins carrying frameshift mutations

Similarly to experiments performed for the study of polyalanine expansions, expression constructs containing each of the above mutations were generated and analysed for their ability to regulate the transcription of one or more target genes, to investigate the effect of *PHOX2B* frameshift mutations in CCHS pathogenesis.

All mutant proteins have turned out to show a severely compromised activity on the *DβH* promoter [6; 27], with a progressive decrease in the amount of Luciferase activity induced by the c.614-618delC, c.721-58del38nt, c.862-866insG and c.930insG mutations, an observation suggestive of a correlation between transactivation activity and length of the disrupted C-terminal sequence resulting from the frame-2 shift.

These four expression constructs have also been tested for their ability in transactivating the *PHOX2A* and *TLX-2* 5'flanking regions. In particular, cotransfections of each construct with the *TLX-2* promoter in SK-N-BE cells have shown an effect similar to that observed on the *DβH* regulatory sequence; further analysis have confirmed this result showing in addition that, while overexpression of wild-type PHOX2B could give a marked increase in endogenous *TLX-2* mRNA level, under the same conditions each of the four *PHOX2B* frameshift mutants failed to induce a significant change in endogenous TLX-2 expression [8].

Finally, differently from what observed in the transactivation experiments using the *DβH* and *TLX-2* promoters, all *PHOX2B* frameshift mutations have surprisingly shown a 10%-30% increased activation of the *PHOX2A* regulatory region with respect to the wild type protein, which has resulted statistically significant for those mutant proteins characterised by the frame-2 shift.

6.3.2 Analysis of mutant PHOX2B-DNA binding

PHOX2B proteins, each carrying one of the three frameshift mutations c.721-758del38nt, c.931del5 and c.936insT, have been produced *in vitro* and analysed for their ability to bind DNA. Electromobility shift assays (EMSA) have been set up also in this case to evaluate the affinity of PHOX2B abnormal proteins for the PBD2 sequence; in particular, the c.721-758del38 mutant showed no binding to this specific region of the *DβH* promoter, whereas the 931del5 and 936insT mutants seemed to maintain the ability to bind this domain. However, the formation of a diffuse protein-DNA complex when testing these two latter mutations has suggested that they may be characterised by an anomalous conformation of the mutant protein [27].

6.3.3 Subcellular localization of PHOX2B proteins carrying frameshift mutations

Differently from what observed for mutants characterised by polyalanine expansions, PHOX2B proteins carrying a shift in the reading frame could correctly be localized in the nucleus. In particular, subcellular localization of the c.930insG mutant, fused to a green fluorescent protein (GFP), has been analysed in COS-7 cells by fluorescence microscopy and no difference could be detected with respect to the wild type protein [6]. This result has been further confirmed by analysis of subcellular localization of the c.721-758del38 and c.931del5 mutants, both characterised by translational frame 2, in HeLa cells by using a PHOX2B specific antibody [27].

In addition, the c.614-618delC mutant, characterised by an anomalous truncated C-terminal region due to frame-shift type 3, has also been localized in the nucleus, but almost exclusively into the nucleolus, as confirmed by silver staining of nucleoli which exactly matched the GFP fluorescence of the c.614-618delC mutant (Fig. 1A) [6]. This has been explained by the presence of an arginine-rich stretch inside the new C-terminal sequence typically generated by a PHOX2B reading frame type 3, in agreement with reported evidence that basic amino acid regions are required for nucleolar localization [30].

6.4 Conclusion

After the identification of *PHOX2B* as the Congenital Central Hypoventilation Syndrome gene, *in vitro* functional analysis of mutations has led to a marked progress in disclosing molecular mechanisms underlying the pathogenesis of this disorder.

On the basis of results thus obtained, distinct CCHS pathogenetic mechanisms must be postulated for *PHOX2B* polyalanine expansions and frameshift mutations, as summarized in Table 2. In particular, functional studies of PHOX2B polyalanine triplet expansions are in accordance with what already obtained for similar genetic defects found in different nuclear proteins [3; 12; 21].

The genotype-phenotype correlation between extension of the PHOX2B polyalanine tract and clinical severity of CCHS, already postulated on a clinical base [19; 32], has been confirmed by observing that a progressive decrease in *DβH* and *PHOX2A* promoter transactivation is associated with increasing lengths of the expanded polyalanine tract [6; 27]. Such an effect has been ascribed, with different degrees, to both anomalous retention of the mutant protein in the cytosolic compartment, partial interaction with the wild type protein in the nucleus and intrinsic defects of the mutant proteins in DNA binding. However, the direct involvement of *PHOX2B* polyalanine expansions on gene transcriptional activity cannot be generalized to all the *PHOX2B* target genes as such defects have turned out to affect only very weakly the transcriptional regulation of the *TLX-2* gene [8].

In the light of the notion that CCHS pathogenesis is likely due to neuronal dysfunction rather than to cell loss, and bearing in mind that generation of a proper mouse model will provide a straightforward mean to clarify the cellular responses to PHOX2B polyalanine aggregates, the present identification of the effect of the ansamycin geldanamycin in rescuing *in vitro* the correct subcellular localization and the full functionality of the mutant proteins carrying a polyalanine expanded tract, and the involvement in such a process of two well known cellular mechanisms of elimination of misfolded products, might become crucial when aiming at the prevention of the cellular toxicity associated with PHOX2B polyalanine aggregates.

Though less frequent than polyalanine expansions, PHOX2B frameshift mutations have been observed to produce more severe disruption of PHOX2B function [7]. The identification, in a French study, of an association between presence of these mutations and prediction of tumor risk in CCHS [28], in addition to an American study reporting a remarkable proportion of CCHS patients found to carry either missense, nonsense or frameshift mutations, requiring continuous ventilatory dependence and affected with HSCR and/or neural crest tumors [7], has suggested the existence of a specific molecular mechanism(s) accounting for such a PHOX2B mediated compound CCHS pathogenesis. While proteins expressed by constructs bearing frameshift mutations show correct nuclear localization, their aberrant C-terminal regions are indeed likely to play an active role by either missing the ability to establish correct protein-protein interactions with their molecular partners or gaining the possibility to interact with wrong molecules, an hypothesis very attractive especially in the light of the strict association of frameshift and missense PHOX2B mutations with risk of neuroblastoma development.

Table 2: Effect of *PHOX2B* polyalanine expansions and frameshift mutations on various cellular targets

		Transactivation of target genes	DNA binding	Sub-cellular localisation	Cellular response
Polyalanine expansions	*DβH*	Correlation between impaired activation and length of the polyalanine tract	+5Ala and +7Ala: normal	Percentage of cells with a sub-cellular non-exclusively nuclear correlates with length of the polyalanine tract	Toxicity due to expression of the +13Ala mutation is counteracted by activation of the heat shock response mediated by the ubiquitin-proteasome system
	PHOX2A		+9Ala: decreased		
	TLX-2	Weak impairment of transactivation with no correlation with the length of the polyalanine tract	+12Ala and +13Ala: abolished		
Frame-shift mutations	*DβH*	Frame 2: correlation between impaired activation and length of the disrupted C-terminal region.		Frame 2:c.930insG: nuclear	
		Frame 3: weak effect on transactivation	c.931del5: decreased	c.931del5: nuclear	
	PHOX2A	Marked wactivation	c.936insT: decreased	c.721-758del38nt: nuclear	Not investigated yet
	TLX-2	Frame 2: correlation between impaired activation and length of the disrupted C-terminal region.	c.721758del38nt: abolished		
		Frame 3: weak effect on transactivation		Frame 3: c.614-618delC: nucleolar	

In conclusion, studies undertaken so far on the pathogenetic mechanism of CCHS associated PHOX2B defects have already pointed out a marked difference between polyalanine expansions and frameshift mutations in terms of target promoters transactivation, aggregates formation and sub-cellular localization. Future experimental *in vitro* investigation will further provide clues on pathogenesis and possible therapeutic hints.

References

1. Abu-Baker A, Messaed C, Laganiere J, Gaspar C, Brais B, Rouleau GA (2003) Involvement of the ubiquitin-proteasome pathway and molecular chaperones in oculopharyngeal muscular dystrophy. Hum Mol Genet 12: 2609-2623
2. Adachi M, Browne D, Lewis EJ (2000) Paired-like homeodomain proteins Phox2a/Arix and Phox2b/NBPhox have similar genetic organization and independently regulate dopamine beta-hydroxylase gene transcription. DNA Cell Biol 19: 539-554
3. Albrecht AN, Kornak U, Boddrich A, Suring K, Robinson PN, Stiege AC, Lurz R, Stricker S, Wanker EE, Mundlos S (2004) A molecular pathogenesis for transcription factor associated polyalanine tract expansions. Hum Mol Genet 13: 2351-2359
4. Amiel J, Laudier B, Attiè-Bitach T, Trang H, De Pontual L, Gener B, Trochet D, Etchevers H, Ray P, Simonneau M, Vekemans M, Munnich A, Gaultier C, Lyonnet S (2003) Polyalanine expansion and frameshift mutations of the paired like homeobox gene *PHOX2B* in congenital central hypoventilation syndrome. Nat Genet 33: 459-461
5. Bachetti T, Bocca P, Borghini S, Matera I, Prigione I, Ravazzolo R, Ceccherini I (2006) Geldanamycin promotes nuclear localisation and clearance of PHOX2B misfolded proteins containing polyalanine expansions. Int J Biochem Cell Biol
6. Bachetti T, Matera I, Borghini S, Di Duca M, Ravazzolo R, Ceccherini I (2005) Distinct pathogenesis mechanisms for PHOX2B associated polyalanine expansions and frameshift mutations in Congenital Central Hypoventilation Syndrome. Hum Mol Genet 14: 1815-1824
7. Berry-Kravis EM, Zhoul L, Rand CM, Weese-Mayer D (2006) Congenital Central Hypoventilation Syndrome: PHOX2B mutations and phenotype. Am J Respir Crit Care Med 174: 1139-1144
8. Borghini S, Bachetti T, Fava M, Di Duca M, Cargnin F, Fornasari D, Ravazzolo R, Ceccherini I (2006) The TLX-2 homeobox gene is a transcriptional target of PHOX2B in neural-crest-derived cells. Biochem J 395: 355-361
9. Brown LY, Brown SA (2004) Alanine tracts: the expanding story of human illness and trinucleotide repeats. Trends Genet 20: 51-58
10. Brown L, Paraso M, Arkell R, Brown S (2005) *In vitro* analysis of partial loss-of-function ZIC-2 mutations in Holoprosencephaly: alanine tract expansion modulates DNA binding and transactivation. Hum Mol Genet 14: 411-420
11. Brunet JF, Pattyn A (2002) Phox2 genes – from patterning to connectivity. Curr Opin Genet Dev 12: 435-440
12. Caburet S, Demarez A, Moumme L, Fellows M, De Baere E, Veitia RA (2004) A recurrent polyalanine expansion in the transcription factor FOXL2 induces extensive nuclear and cytoplasmic protein aggregation. J Med Genet 41: 931-946
13. Davies JE, Sarkar S, Rubinsztein DC (2006) Trehalose reduces aggregate formation and delays pathology in a transgenic mouse model of oculopharyngeal muscular dystrophy. Hum Mol Genet 15: 23-31
14. Flora A, Lucchetti H, Benfante R, Goridis C, Clementi F, Fornasari D (2001) Sp proteins and Phox2b regulate the expression of the human Phox2a gene. J Neurosci 21: 7037-7045

15. Holzinger A, Mittal RA, Kachel W, Priessmann H, Hammel M, Ihrler S, Till H, Münch H-G (2005) A novel 17 bp deletion in the *PHOX2B* gene causes congenital central hypoventilation syndrome with total aganglionosis of the small and large intestine. Am J Med Genet 139: 50-51

16. Jana NR, Tanaka M, Wang G-H, Nukina N (2000) Polyglutamine length-dependent interaction of Hsp40 and Hsp70 family chaperones with truncated N-terminal huntingtin: their role in suppression of aggregation and cellular toxicity. Hum Mol Genet 9: 2009-2018

17. Layfield R, Cavey JR, Lowe J (2003) Role of ubiquitin-mediated proteolysis in the pathogenesis of neurodegenerative disorders. Ageing Res Rev 2: 343-356

18. Lavoie H, Debeane F, Trinh O-D, Turcotte J-F, Corbeil-Girard L-P, Dicaire M-J, Saint-Denis A, Pagé M, Rouleau GA, Brais B (2003) Polymorphism, shared functions and convergent evolution of genes with sequences coding for polyalanine domains. Hum Mol Genet 12: 2967-2979

19. Matera I, Bachetti T, Puppo F, Di Duca M, Morandi F, Casiraghi GM, Cilio MR, Hennekam R, Hofstra R, Schober JG, Ravazzolo R, Ottonello G, Ceccherini I (2004) PHOX2B mutations and polyalanine expansions correlate with the severity of the respiratory phenotype and associated symptoms in both congenital and late onset central hypoventilation syndrome. J Med Genet 41: 373-380

20. Mosse YP, Laudenslager M, Khazi D, Carlisle AJ, Winter CL, Rappaport E, Maris JM (2004) Germline PHOX2B mutations in hereditary neuroblastoma. Am J Hum Genet 75: 727-730

21. Nasrallah IM, Minarcik JC, Golden JA (2004) A polyalanine tract expansion in Arx forms intranuclear inclusions and results in increased cell death. J Cell Biol 167: 411-416

22. Ravikumar B, Duden R, Rubinsztein DC (2002) Aggregated-prone proteins with polyglutamine and polyalanine expansions are degraded by autophagy. Hum Mol Genet 11: 1107-1117

23. Sasaki A, Kanai M, Kijima K, Akaba K, Hashimoto M, Otaki S, Koizumi T, Kusuda S, Ogawa Y, Tuchiya K, Yamamoto W, Nakamura T, Hayasaka K (2003) Molecular analysis of congenital central hypoventilation syndrome. Hum Genet 114: 22-26

24. Shintani T, Klionsky D (2004) Autophagy in health and disease: a double-edged sword. Science 306: 990-995

25. Sittler A, Lurz R, Lueder G, Priller J, Hayer-Hartl MK, Hartl FU, Lehrach H, Wanker E (2001) Geldanamycin activates a heat shock response and inhibits huntingtin aggregation in a cell culture model of Huntington's disease. Hum Mol Genet 12: 1307-1315

26. Trochet D, Bourdeaut F, Janoueix-Lerosey I, Deville A, de Pontual L, Schleier-macher G, Coze C, Philip N, Frebourg T, Munnich A, Lyonnet S, Delattre O, Amiel J (2004) Germline mutations of the paired-like homeobox 2B (PHOX2B) gene in neuroblastoma. Am J Hum Genet 74: 761-764

27. Trochet D, Hong SJ, Lim JK, Brunet JF, Munnich A, Kim KS, Goridis C, Amiel J (2005a) Molecular consequences of PHOX2B missense frameshift and alanine expansion mutations leading to autonomic dysfunction. Hum Mol Genet 14: 3697-3708

28. Trochet D, O'Brien LM, Gozal D, Trang H, Nordenskjold A, Laudier B, Svensson PJ, Uhrig S, Cole T, Niemann S, Munnich A, Gaultier C, Lyonnet S, Amiel J (2005b) PHOX2B genotype allows for prediction of tumoral risk in congenital central hypoventilation syndrome. Am J Hum Genet 76: 421-426

29. Van Limpt V, Schramm A, Lakeman A, Van Sluis P, Chan A, Van Noesel M, Baas F, Caron H, Eggert A, Versteeg R (2004) The Phox2b homeobox gene is mutated in sporadic neuroblastomas. Oncogene 23: 9280-9288
30. Visintin R, Amon A (2000) The nucleolus: the magician's hat for cell cycle tricks. Curr Opin Cell Biol 12: 372-377
31. Wang Q, Mosser DD, Bag J (2005) Induction of HSP70 expression and recruitment of HSC70 and HSP70 in the nucleus reduce aggregation of a polyalanine expansion mutant of PABPN1 in HeLa cells. Hum Mol Genet 14: 3673-3684
32. Weese-Mayer DE, Berry-Kravis EM, Zhou L, Maher BS, Silvestri J-M, Curran ME, Marazita ML (2003) Idiopathic congenital central hypoventilation syndrome: analysis of genes pertinent to early autonomic nervous system embryologic development and identification of mutations in *PHOX2B*. Am J Med Genet 123: 267-278

7. Sudden infant death syndrome: study of genes pertinent to cardiorespiratory and autonomic regulation

Debra E. WEESE-MAYER[1], Michael J. ACKERMAN[2], Mary L. MARAZITA[3] and Elizabeth M. BERRY-KRAVIS[4]

[1]Pediatric Respiratory Medicine, Rush Children's Hospital, Rush University Medical Center, Chicago Illinois, 1653 West Congress Parkway, Chicago, IL 60612, [2] Department of Pediatrics, Medicine, Molecular Pharmacology & Experimental Therapeutics, Mayo Clinic College of Medicine, Rochester Minnesota, U.S.A. 200 First St. S.W., Rochester, Minnesota 55905 U.S.A [3] Center for Craniofacial and Dental Genetics, Division of Oral Biology, Oral and Maxillofacial Surgery, Human Genetics, Graduate School of Public Health, University of Pittsburgh/School of Dental Medicine. 100 Technology Drive, Suite 500, Cellomics Building, Pittsburgh, Pennsylvania 55905 U.S.A [4] Departments of Neurology, Pediatrics, and Biochemistry, Rush University Medical Center. 1653 W. Congress Parkway, Chicago Illinois 60612 U.S.A.

7.1 Introduction

The "Back to Sleep" campaign, introduced in the U.S. in 1992, identified modifiable environmental risk factors for SIDS and led to a decrease in SIDS incidence from 1.2/1000 live births [46] to 0.529/1000 live births in 2003 [36]. Despite this decline, the final 2003 National Vital Statistics indicate a 2.7 fold increase in SIDS rate among African American infants relative to Caucasian infants (1.152/1,000 livebirths vs. 0.424/1,000 livebirths) [36]. Because of this ethnic disparity and the continued occurrence of SIDS deaths despite improved compliance with modifiable risk factors, investigators considered the possibility of genetic and gene-by-environment interaction to explain the remaining 2,162 SIDS cases in the U.S., alone [36]. Thus far, all genetic studies have been based upon clinical, neuropathological, and epidemiological observations in SIDS victims, with subsequent identification and study of candidate genes. The focus of this chapter will be exclusively on those genes pertinent to cardiorespiratory or autonomic regulation.

7.2 Cardiac channelopathy genes in SIDS

Cardiac channelopathies represent heritable arrhythmia syndromes due to defective cardiac channels, and include congenital long QT syndrome (LQTS), Brugada syndrome (BrS), and catecholaminergic polymorphic ventricular tachycardia (CPVT).

7.2.1 The "Schwartz-QT hypothesis"

The "Schwartz-QT" hypothesis proposed that abnormal cardiac repolarization and QT prolongation might result in SIDS [61; 93], and was supported by results from an 18-year prospective study of 34,000 infants with day 3/day 4 of life ECGs [97]: ECGs in 50% of the 24 SIDS cases had a corrected QT interval (QTc) >440 ms vs. 2.5% of the entire cohort (odds ratio = 41.3, 95% CI 17.3-98.4). These findings suggest QTc prolongation in week 1 of life is either a marker for generalized autonomic instability/vulnerability or an indicator of an infant with a potentially lethal, LQTS-predisposing genetic substrate.

7.2.2 Molecular evidence linking the Schwartz-QT hypothesis and LQTS

Subsequently, 3 independent, molecular genetics case reports described sporadic *de novo* germline mutations in the 3 most common LQTS-susceptibility genes, suggesting that some SIDS cases may have terminal rhythm ventricular fibrillation and/or a LQTS-causing cardiac channel mutations [16; 95; 96].

Congenital LQTS is characterized by QT interval prolongation and "torsades de pointes". Roughly half of genotype-positive subjects have QT prolongation and symptoms including syncope, seizures and sudden death, while the others have non-penetrant/concealed LQTS with normal/borderline QT intervals at rest. To date, 10 LQTS-susceptibility genes have been discovered (Table 1). Over 75% of LQTS stems from mutations in either the *KCNQ1*-encoded I_{Ks} potassium channel (LQT1, 30-35%), the *KCNH2*-encoded I_{Kr} potassium channel (LQT2, 25-30%), or the *SCN5A*-encoded I_{Na} sodium channel (LQT3, 5-10%). Numerous genotype-phenotype relationships have emerged including the observation of relatively genotype-specific arrhythmogenic triggers like swimming/exertion/LQT1, auditory/LQT2, and sleep/LQT3 [5; 68; 96; 114]. 5-10% of cases involve *de novo* mutations like those in the anecdotal SIDS molecular case reports, with the rest inherited.

Ackerman et al. provided the first genetic epidemiology studies investigating the hypothesis of LQTS-associated cardiac channel mutations in SIDS by post-mortem mutational analysis of the 5 major LQTS disease genes (*KCNQ1*, *KCNH2*, *SCN5A*, *KCNE1*, and *KCNE2*) in 93 SIDS cases and 400 controls [2; 101]. The investigators targeted the *SCN5A*-encoded sodium channel because of known association between LQT3 and sleep events with a high lethality/event rate, and identified 2/58 Caucasian SIDS cases (3.4%) (but no African American

SIDS cases) with rare, novel missense mutations that conferred a marked gain-of-function with accentuation and persistence of late sodium current [2]. Subsequently, Tester and Ackerman [101] identified 2 more cases of probable LQTS-mediated SIDS in their study of the 4 potassium channel genes. Later, this observation was validated in a SIDS cohort of 201 Norwegian infants [8], and reported probable LQTS-causing mutations in ~9% of cases (half in sodium channel, *SCN5A*, gene) [105].

Table 1. Summary of LQTS-susceptibility genes

AD – autosomal dominant; AR – autosomal recessive; ATS – Andersen Tawil syndrome; JLNS – Jervell and Lange-Nielsen syndrome; LQTS – long QT syndrome; TS – Timothy syndrome

LQTS subtype	Locus	Gene	Mode of inheritance	Current	Frequency (%)
LQT1 (JLNS1)	11p15.5	*KCNQ1*	AD (AR in JLNS)	$I_{Ks\,(\alpha)}$	30 - 35
LQT2	7q35-36	*KCNH2*	AD	$I_{Kr(\alpha)}$	25 - 30
LQT3	3p21-p24	*SCN5A*	AD	I_{Na}	5 - 10
LQT4	4q25-q27	*ANKB*	AD	Na/Ca	< 1
LQT5 (JLNS2)	21q22.1	*KCNE1*	AD (AR in JLNS)	$I_{Ks(\beta)}$	~ 1
LQT6	21q22.1	*KCNE2*	AD	$I_{Kr(\beta)}$	< 1
ATS1 (LQT7)	17q23	*KCNJ2*	AD	$I_{K1(\alpha)}$	50 of ATS; < 1 of LQTS
TS1 (LQT8)	12p13.3	*CACNA1C*	Sporadic	$I_{Ca.L(\alpha)}$	50 of TS; < 1 of LQTS
CAV3-LQTS (LQT9)	3p25	*CAV3*	Sporadic	Caveolin-3 (I_{Na})	< 1
SCN4B-LQTS (LQT10)	11q23.3	*SCN4B*	AD	$I_{Na(\beta4)}$	< 1

Most recently, putative SIDS-causing mutations were identified in 3/34 African American infants involving one of the newest LQTS-susceptibility genes (LQT9), *CAV3*-encoded caveolin-3 [19; 104]. Ackerman et al. also explored the *RyR2*-encoded cardiac ryanodine receptor/calcium release channel as a candidate gene for SIDS, with mutations in *RyR2* underlying the pathogenic basis for type 1 catecholaminergic polymorphic ventricular tachycardia (CPVT1). Though CPVT has exertional/sympathetic triggered cardiac events in a structurally normal heart, the resting ECG is always normal. Tester et al. discovered that 2 of the 93 infants harbored gain-of-function mutations in *RyR2* [102]. Even after excluding the most common channel polymorphisms: K897T-KCNH2, H558R-SCN5A, and G38S-KCNE1, nearly 1/3 of infants possessed at least one genetic variant noted previously in ethnic-matched reference alleles in one of the 5 cardiac channel genes [3; 4; 101]. Whether or not these channel polymorphisms reduce repolarization reserve and/or facilitate adrenergically-mediated cardiac arrhythmias requires further investigation. For example, 5 of the SIDS victims were positive for R1047L-KCNH2, a polymorphism previously identified as an independent risk factor for drug (dofetilide)-induced torsades [99; 101]. In addition, the African American-specific sodium channel common polymorphism, S1103Y-SCN5A, was over-represented among a cohort of 133 African-American infants and a mexiletine-sensitive increased late sodium current was elicited by cellular acidosis [83].

7.3 Serotonergic system genes in SIDS

7.3.1 Rationale for study of serotonin (5HT) genes

Panigrahy et al. [76] reported a decrease in serotonergic receptor binding in the arcuate nucleus, n. raphé obscurus, and other medullary regions that contain serotonergic cell bodies (n. paragigantocellularis lateralis, n. gigantocellularis, and intermediate reticular zone) in primarily Hispanic and Caucasian SIDS cases vs. controls. Subsequently, Ozawa and Okado [75] reported a decrease in serotonergic receptor binding in the dorsal nucleus of the vagus, solitary nucleus and ventro-lateral medulla in SIDS cases in Japan. Kinney et al. [45] later confirmed their prior observations of altered serotonin (5HT) receptor binding in medullary regions, but in Native American Indians. Most recently, Paterson et al. [77] described an increase in number and density of serotonin neurons, and a lower density of serotonin 1A receptor binding sites in medullary regions of homeostatic control in primarily Hispanic and Caucasian SIDS cases vs. controls. These neuropathological reports motivated studies focused on genes involved in the serotonergic system.

7.3.2 Serotonin transporter (5HTT) gene

Although the 5HT system is extensive, its function is regulated globally by the action of a single protein, the serotonin transporter (5HTT). The *5HTT* gene, located at 17q11.1-q12 [87], controls the duration and strength of the interactions between 5HT and its receptors by regulating membrane re-uptake of 5HT from the extracellular space [23; 33; 51; 58].

Two polymorphisms in the 5'regulatory region of the *5HTT* gene differentially modulate gene expression. One of these involves an insertion-deletion in a repeat sequence in the promoter region of the *5HTT* gene and the other is a variable number tandem repeat (VNTR) in intron 2 of the *5HTT* gene. The two most common alleles of the promoter polymorphism account for over 95% of alleles in most populations [33; 51]. The short allele (S) corresponds to 14 copies of the 20-23 base pair repeat unit and the long allele (L) corresponds to 16 copies. The long allele is a more effective promoter within cell transfection models [33]. Subjects with the L/L genotype have an increased availability of raphé serotonin transporters on *in vivo* neuroimaging studies [34], as well as increased midbrain 5HTT binding and 5HTT mRNA levels in human postmortem brain [52] when compared with individuals carrying at least one S allele. The *5HTT* promoter allele distribution varies widely by ethnicity [18; 21; 28; 29; 33; 39; 43; 49; 70; 74]. Likewise, SIDS incidence varies widely by ethnicity, with an incidence of 1.13 per 1,000 live births for African Americans and 0.49 per 1,000 live births for Caucasian Americans in 2000 [65], an incidence of 1.15 per 1,000 live births for African Americans and 0.42 per 1,000 live births for Caucasians in 2003, and an incidence of 0.248 per 1,000 live births in Japan in 2001 [42].

A polymorphic variable number tandem repeat (VNTR) containing 9, 10, or 12 copies of a 16-17 bp repeat sequence in intron 2 of *5HTT* [73] has also been shown to differentially regulate gene expression. Fiskerstrand et al. [23] reported increased expression in promoter-driven reporter gene constructs containing 12 repeats, in comparison to those containing 10 repeats. Specifically, the 12 repeat construct was a stronger enhancer in differentiating embryonic stem cells; and it was suggested that the intron 2 VNTR may affect distribution and rate of transcriptional control [23]. MacKenzie and Quinn [58] introduced the VNTR enhancer region coupled to a reporter gene into transgenic mice. They noted increased expression levels in the vicinity of the developing rostral hindbrain during embryonic development in mice expressing the 12 repeat construct as compared to those expressing the 10 repeat construct. Lovejoy et al. [55] identified specific sequence variants within individual repeats in the VNTR that were responsible for high variability in transcriptional efficiency in transfected murine embryonic stem cells and repeat number-dependent variation in VNTR promoter activity in human JAR cells. This suggested that both repeat copy number and the primary sequence of the repeat units within the intron 2 VNTR may play a role in tissue-specific *5HTT* expression, leading to variation in *5HTT* expression in the nervous system and association with disease susceptibility.

A 3'Untranslated Region (UTR) single nucleotide polymorphism in *5HTT* was identified by Battersby et al. [10] and located within a putative polyadenylation signal for one of the commonly used polyadenylation sites for the *5HTT* mRNA. Although allelic variation at the site does not substantially influence polyadenylation site usage [10], the 3'RACE assay used to assess polyadenylation was not quantitative and it remains possible that minor abnormalities in polyadenylation *in vivo* might affect stability of 5HTT mRNA and/or transport into the cytoplasm. A subsequent study [63] did not identify any effect of this 3'UTR polymorphism on the platelet serotonin transporter expression assayed by [3H] paroxetine binding.

7.3.3 5HTT Promoter region polymorphism in SIDS

Narita et al. [71] examined the role of a functional polymorphism in the promoter region of the *5HTT* gene in SIDS risk among 27 Japanese SIDS cases and 115 age-matched controls. Narita et al. [71] genotyped the promoter insertion/deletion polymorphism in cases and controls and demonstrated significant differences in genotype distribution and allele frequency, with an excess of the L/L genotype and L allele in the SIDS group relative to controls (7.4% vs. 1.7% for L/L; 22.2% vs. 13.5% for L allele). Additionally, they found 3 extra long (XL, 18 repeats) alleles in SIDS cases (5.6%) *vs.* 1 in controls (0.4%). This study provided the first highly significant evidence for the role of a specific gene in SIDS risk.

Weese-Mayer et al. [107] replicated the finding of an increase in frequency of the L allele of the *5HTT* promoter insertion/deletion polymorphism in SIDS cases in an independent sample of 87 SIDS cases (43 African American, 44 Caucasian) and 87 gender/ethnicity-matched controls from the U.S. They found significant differences in both genotype distribution and allele frequency in the combined (African American and Caucasian) dataset and for allele frequency in the Caucasian dataset. Specifically, there was an excess of the L/L genotype and the L allele in the SIDS group relative to controls (54.0% vs. 39.1% for L/L; 73.0% vs. 58.6% for L allele). Further, significantly fewer SIDS cases vs. controls with no L allele (S/S genotype) were reported in the entire cohort (8.0% vs. 21.8%) and within the Caucasian subgroup (13.6% vs. 34.1%). While the results were not statistically significant within the African American subgroup (small sample size) there was a trend toward increased frequency of the long allele in the African American SIDS cases.

In addition to the case-control results, Weese-Mayer et al. examined allele and genotype frequency differences by ethnicity in an additional set of 334 control subjects [107]. The frequency of the long allele was increased in African Americans (73.9%) vs. Caucasians (53%). Weese-Mayer et al. concluded that the promoter polymorphism in *5HTT* may play an important role in SIDS risk and may explain, in part, the ethnic differences in SIDS risk [107]. Specifically, SIDS rates are high among African Americans and low among Japanese, and the *5HTT* L allele frequency is high among African Americans (see above) and low among Japanese controls (13.5 % in Narita et al. study, [71]), thereby potentially explaining the heightened SIDS incidence among African Americans in contrast to the low incidence among the Japanese. Despite this ethnic variation in L allele frequency, SIDS cases were more likely than controls to have a L allele in the Japanese, Caucasian, and African American study samples.

7.3.4 5HTT intron 2 VNTR and SIDS

Weese-Mayer et al. [113] subsequently studied the *5HTT* intron 2 VNTR genotype in a cohort of 90 pairs of SIDS cases and gender/ethnicity matched controls (46 Caucasian, 44 African American). Genotype distribution, allele frequency for the 12-repeat allele, and frequency of the 12/12 genotype differed significantly between African American SIDS cases and controls, but were not in the overall dataset or the Caucasian subgroup. The association of the 12-allele with SIDS in the African American group was driven predominantly by a significant increase in 12-alleles in African American male SIDS cases which was not observed in females. Similar to the promoter variant, allele frequencies and genotype distribution also varied across ethnic groups, with a higher frequency of the 12-allele in the African American population, in both cases and controls.

Examination of both promoter polymorphism genotype and intron 2 VNTR genotype in the cohort revealed a significant association between SIDS and the combined "L/L or L/S and 12/12" genotype in the total dataset and the African

American subgroup, but not the Caucasian subgroup. Further, haplotype analysis demonstrated a significant difference in the overall haplotype frequencies between SIDS cases and controls in the entire cohort as well as each ethnic subgroup. Finally, the "L-12" haplotype ("long" allele present at the promoter and "12" allele present at intron 2 on the same chromosome) was significantly more frequent in SIDS cases than controls, and also in African American SIDS cases vs. controls, but not Caucasian SIDS cases vs. controls. These studies established an association between SIDS and the 12 repeat allele of the intron 2 VNTR and the L-12 haplotype in the African American subgroup.

7.3.5 5HTT 3'Untranslated region and SIDS

Despite association of two functional polymorphisms in *5HTT* with SIDS, a polymorphism in a putative polyadenylation site in the 3' UTR of *5HTT* [10] was not found to be associated with SIDS [59] in 92 pairs of gender/ethnicity-matched SIDS cases and controls. Specifically, genotype distribution did not differ between the SIDS and control groups in the overall dataset, or the Caucasian or African American subgroups. Analyses performed for haplotypes spanning the *5HTT* gene (promoter, intron 2 and 3' UTR variants) revealed no significant differences in haplotype frequency distribution either for the total dataset or for the Caucasian or African American subgroups.

7.4 Autonomic Nervous System (ANS) genes in SIDS

7.4.1 Rationale for studying ANS genes

Serotonin influences a broad range of physiological systems including the regulation of breathing, the cardiovascular system, temperature, and the sleep-wake cycle [40], all involved in regulation of the ANS. Observations consistent with ANS dysfunction have been reported in SIDS including profuse sweating [41], elevated body temperature [24; 84], tachycardia then bradycardia preceding the terminal event [64], reduced heart rate variability [48; 92; 100], drenching sweats and facial pallor [100], and decreased responses to obstructive sleep events [26]. Accordingly, genes pertinent to the early embryology of the ANS were considered to potentially confer SIDS risk. This approach has been successful in clarifying the genetic basis of Congenital Central Hypoventilation Syndrome (CCHS), known to have associated ANS dysregulation [60; 110; 111] and thought to be related to SIDS [112]. The gene responsible for CCHS has recently been identified as paired-like homeobox protein (PHOX)2B and virtually all individuals with typical CCHS are heterozygous for mutations in *PHOX2B* or have a non-polyalanine mutation in the *PHOX2B* gene [6; 11; 62; 91; 103; 109].

7.4.2 ANS genes and SIDS

Weese-Mayer et al. [108] examined several genes thought to play a role in ANS development, including bone morphogenic protein-2 (*BMP2*), mammalian achaete-scute homolog-1 (*MASH1*), *PHOX2A*, rearranged during transfection factor (*RET*), endothelin converting enzyme-1 (*ECE1*), endothelin-1 (*EDN1*), T-cell leukemia homeobox protein (*TLX3*), and engrailed-1 (*EN1*). Two distinct groups were investigated in this study: 92 SIDS cases and 92 matched control subjects. DNA from all 92 SIDS cases and from 26 of the 92 matched controls was sequenced for exon and splice site mutations in *BMP2*, *MASH1*, *PHOX2A*, *RET*, *ECE1*, *EDN1*, *TLX3*, and *EN1*. Any base change expected to affect a splice site or result in modification of the protein sequence which was identified in SIDS subjects or controls was further screened in all 92 controls.

Data in Table 2 indicate protein-changing variants in 92 SIDS cases and 92 matched controls for the candidate genes. Sequence data from *PHOX2A*, *RET*, *ECE1*, *TLX3*, and *EN1* revealed 11 rare protein-changing polymorphisms in 14 SIDS cases (15.2% of SIDS cases) and subsequent genotyping for these polymorphisms in controls identified 1 polymorphism in 2 controls (2.2% of controls). Each mutation occurred in 1 SIDS case with the exception of the *TLX3* base change that occurred in 4 SIDS cases and 2 controls. African American infants accounted for 10 of the SIDS cases and the 2 controls with protein-changing mutations. No protein-changing alterations were identified for *MASH1* or *EDN1*.

Four common protein-changing polymorphisms were identified in *BMP2*, *RET*, *ECE1*, and *EDN1*. The allele frequency did not differ between SIDS cases and controls. Among SIDS and control groups, the allele frequencies for the *BMP2* common polymorphism were significantly different between Caucasian and African American infants. Among controls the allele frequencies for the *BMP2* and *ECE1* polymorphisms were significantly different between Caucasian and African American infants.

7.4.3 Rationale for study of the PHOX2B Gene

PHOX2B encodes a highly conserved homeobox domain transcription factor with 2 stable polyalanine repeats of 9 and 20 residues and is a key gene in ANS development with a role in early embryologic development as a transcriptional activator in promotion of pan-neuronal differentiation including upregulation of proneural genes, *MASH1* expression and motoneural differentiation [54]. *PHOX2B* has a separate role by a different pathway wherein it represses expression of inhibitors of neurogenesis [53]. Further, *PHOX2B* is required to express tyrosine hydroxylase, dopamine beta hydroxylase [35], and *RET*, and to maintain *MASH1*, thereby regulating noradrenergic neuronal specification in vertebrates [79]. Finally, *PHOX2B* knock-out mice ($^{-/-}$) do not survive as ANS circuits do not form or degenerate [79]. Further, recent studies indicate that *PHOX2B* plays a regulatory role in the selection between motor neuron or serotonergic neuronal fate in the development of the central nervous system [80; 81].

Table 2. Protein-changing variants in genes pertinent to the early embryologic origin of the ANS for 92 SIDS and 92 control subjects

Cauc=Caucasian; Afr. Amer.=** African American. Reproduced with permission [108]

Gene	Genotype	Amino acid effect	Rare polymorphisms # of cases with variant						Common polymorphisms allele frequency of variant					
			SIDS			Controls			SIDS			Controls		
			Cauc.*	Afr. Amer.**	Total	Cauc.	Afr. Amer.	Total	Cauc.	Afr. Amer.	Total	Cauc.	Afr. Amer.	Total
BMP2	T570A	S190R							0.32	0.07	0.19	0.38	0.06	0.22
PHOX2A	C287A	T96K	1	0	1	0	0	0						
RET	G35A	R12H	1	0	1	0	0	0						
	C166A	L56M	1	0	1	0	0	0						
	C1157T	A386V	0	1	1	0	0	0						
	G1253A	R418Q	1	0	1	0	0	0						
	G2071A	G691S							0.07	0.05	0.06	0.11	0.12	0.12
	A2147C	K716T	0	1	1	0	0	0						
ECE1	C1022T	T341I							0.04	0.01	0.02	0.09	0.01	0.05
	A1060G	T354A	0	1	1	0	0	0						
EDN1	G594T	K198N							0.27	0.28	0.26	0.22	0.16	0.19
TLX3	C196T	P66S	0	4	4	0	2	2						
	G152A	R51H	0	1	1	0	0	0						
EN1	C719T	T240I	0	1	1	0	0	0						
	C986A	T329K	0	1	1	0	0	0						

Loss of function experiments in mice have shown that for the transition from motor neuron production to 5HT neuron production to commence, downregulation of *PHOX2B* is required [81]. Recognizing the identified importance of the 5HT system in SIDS as described above, these loss of function experiments identify a role for *PHOX2B* in the development of the 5HT system and potentially a relationship between 5HT system development and *PHOX2B* in SIDS risk.

7.4.4 PHOX2B gene and SIDS

Based on the established relationship between SIDS, *5HTT*, and ANS dysregulation coupled with the recognized role of *PHOX2B* in ANS and 5HT system development, Weese-Mayer et al. [108] studied a cohort of 91 SIDS cases and 91 matched controls for the polyalanine expansion mutation characteristic of CCHS. None of the study subjects demonstrated the *PHOX2B* polyalanine mutation characteristic of CCHS.

Subsequently, Rand et al. [88] sequenced the coding regions and intron-exon boundaries of *PHOX2B* in the same SIDS cohort along with 91 gender/ethnicity matched control subjects, and identified a single common polymorphism (IVS2+101A>G; g.1364A>G) in intron 2 of the *PHOX2B* gene located 100 base pairs downstream of the exon 2 splice site. The frequency of subjects carrying the variant G allele (genotype GG or GA) of this polymorphism was significantly higher in the SIDS group than in the matched control group, and also higher in Caucasian SIDS cases than in matched control subjects, but did not reach significance in the African American SIDS vs control comparison. Likely the result is non-significant in the African American group because of the high baseline frequency of this polymorphism in the African American population, which significantly exceeds the frequency of the variant in the Caucasian population as seen in the control groups. The allele frequency of the variant G allele for the intron 2 polymorphism was not significantly increased in the SIDS group relative to controls, although there was a strong trend toward higher G allele frequency in the entire SIDS cohort and Caucasian SIDS group compared to controls that was not seen in the African American SIDS cases and controls. The difference in allele frequency did not reach significance because the homozygous GG genotype was more frequent in the control group, suggesting that the effect of this polymorphism on SIDS risk is relevant to presence or absence of the G allele and thus is similar in the homozygous and heterozygous state.

Eight polymorphisms located in the third exon of the *PHOX2B* gene (Table 3) occurred significantly more frequently among SIDS cases (34 occurrences observed in 27/91 cases) than controls (19 occurrences observed in 16/91 controls). Likewise, the number of occurrences among SIDS cases in the Caucasian and African American subgroups was nearly double the number among their respective controls. Among SIDS cases containing a polymorphism in exon 3, 6/27 (22%) contained 2 or more polymorphisms compared to 2/16 (12%) controls (Table 3). Each of the 8 samples with 2 or more polymorphisms in exon 3 were African American. Two of the 8 polymorphisms identified were protein-altering missense mutations (F153L and S176T), occurring in 9 SIDS cases and 4 controls (10% and 4%, respectively; Table 3).

Gene-gene interaction between the *PHOX2B* exon 3 polymorphisms and the *5HTT* promoter L/L genotype or the L allele, or the *5HTT* intron 2 polymorphisms, was not found for SIDS risk, in comparisons of SIDS cases with controls or when the cohort was divided into ethnicity-specific subgroups. Gene interaction analysis revealed that of the 27 SIDS cases containing a *PHOX2B* exon 3 polymorphism(s), 3 also contained a *RET* mutation compared to 1/61 SIDS cases that contained no exon 3 polymorphism and were also tested for *RET* mutations. Significantly more Caucasian SIDS cases (3/11) with a *PHOX2B* exon 3 polymorphism had an additional *RET* mutation compared to 0/34 Caucasian SIDS cases containing no exon 3 polymorphism.

Kijima et al. [44] also sequenced the *PHOX2B* gene in 23 Japanese SIDS cases and 50 controls and identified 1 polymorphism in exon 2 of *PHOX2B* and 2 intron 2 polymorphisms, none of which were identified in the Rand et al. [88] study. These polymorphisms were identified in 1, 1, and 9% of subjects, respectively, but the authors do not clarify if these were identified in SIDS cases or controls. Conversely, none of the *PHOX2B* exon 3 polymorphisms that Rand et al. [88] described in the Caucasian and African Americans were reported in the Japanese cases.

7.5 Nicotine metabolizing genes in SIDS

7.5.1 Rationale for study of nicotine metabolizing genes

Exposure to tobacco, both prenatal as well as postnatal, has been identified as a key risk factor in the etiology of SIDS [7; 12; 13; 57; 66]. A relationship between tobacco exposure and altered ANS function has long been recognized for adults with both chronic [47; 56; 72; 82] and acute [85] exposure and more recently for infants exposed to smoke prenatally [25]. Based on these relationships between SIDS, tobacco exposure, and ANS dysregulation, genes involved in nicotine metabolism were identified as possible candidate genes for further study of the genetic basis for SIDS.

The ability to convert toxic metabolites in cigarette smoke to less harmful compounds is key to minimizing the adverse health effects of exposure to tobacco. Polycyclic aromatic hydrocarbons (PAHs), some of the most important carcinogens in cigarette smoke, are metabolized through a two-stage process. In phase 1 inhaled PAHs are activated by converting the hydrophobic compounds into hydrophilic compounds which are reactive and have electrophilic intermediates capable of binding DNA. Cytochrome P-450 1A1 (*CYP1A1*) encodes aryl hydrocarbon hydroxylase, a major enzyme responsible for phase 1 metabolism of PAHs. During the second phase of the metabolism of PAHs detoxification occurs through enzymes

Table 3. *PHOX2B* Exon 3 polymorphisms in 91 SIDS cases and 91 matched control subjects

*These polymorphisms were identified 34 times in the SIDS group compared to 19 times in the control group (p = 0.01) and in 27/91 SIDS cases compared to 16/91 controls (p = 0.05) Reproduced with permission from [88]

Location	Genotype	Amino acid effect	SIDS			Control		
			Caucasian	African American	Total	Caucasian	African American	Total
Exon 3	c.459T>G	F153L	0	1	1	2	0	2
Exon 3	c.526T>A	S176T	6	2	8	2	0	2
Exon 3	c.552C>T	silent	1	2	3	0	2	2
Exon 3	c.642C>T	silent	0	2	2	0	0	0
Exon 3	c.726A>G	silent	0	2	2	0	1	1
Exon 3	c.750G>A	silent	1	3	4	0	2	2
Exon 3	c.762A>C	silent	1	7	8	1	5	6
Exon 3	c.870C>A	silent	2	4	6	1	3	4
Total occurrences of polymorphisms*			11	23	34	6	13	19

such as glutathione S-transferases (*GSTs*) or uridine diphosphate (*UDP*)-glucuronosyltransferase through transformation into compounds that can be excreted from the body. GSTT1 is encoded by the *GSTT1* gene, and is a major enzyme in phase 2 of cigarette smoke metabolism [9; 32; 69]. Polymorphisms in both the *CYP1A1* and *GSTT1* genes [9; 38], have been reported to impact the metabolic detoxification process for cigarette smoke. Thus, expression of polymorphisms in these genes have been associated with low birth weight [106], and may account for the varying susceptibility to other adverse health consequences of cigarette smoke exposure, including SIDS.

7.5.2 Nicotine metabolizing genes and SIDS

Rand et al. [89] reported on frequency of known *CYP1A1* and *GSTT1* polymorphisms in 106 SIDS cases and 106 control subjects matched for gender and ethnicity. The frequency of the *GSTT1* homozygous deletion genotype did not differ between SIDS cases (22/106; 20%) and matched controls (32/106; 30%) in either the complete sample or the Caucasian or African American subgroups. Likewise, no association with SIDS was observed for genotype distribution or allele frequencies at any of three *CYP1A1* polymorphisms. When multiple alleles were considered in combination, no association was found between cases containing one or more of the *CYP1A1* rare polymorphic alleles and the SIDS phenotype. Further, no association was observed between cases containing both the *GSTT1* deletion genotype and a *CYP1A1* polymorphism with the SIDS phenotype. Higher frequencies of variant allele combinations were observed in the African American subgroups compared to Caucasian subgroups likely due to the higher genetic heterogeneity in that population.

7.6 Clinical significance

7.6.1 Significance of cardiac channelopathies in SIDS

In summary, a primary cardiac channelopathy is estimated to cause 5–15% of SIDS. Besides rare, pathogenic LQTS/BrS/CPVT susceptibility mutations, these results raise the possibility that cardiac channel polymorphisms may contribute to the SIDS Triple-Risk Hypothesis [22]. Under this hypothesis, SIDS requires 1) an environmental trigger, 2) a critical developmental period, and 3) a vulnerable host. Even after excluding the most common channel polymorphisms (K897T-*KCNH2*, H558R-*SCN5A*, and G38S-*KCNE1*), nearly one-third of infants possessed at least one genetic variant noted previously in ethnic-matched reference alleles in one of the 5 cardiac channel genes [1; 2; 101]. Whether or not these cardiac channel polymorphisms reduce repolarization reserve and/or facilitate adrenergically-mediated cardiac arrhythmias requires further investigation.

Routine newborn LQTS genetic screening has not been implemented despite an incidence of 1:2,500. Further studies are needed to determine whether or not polymorphism-specific genotyping for S1103Y should be performed routinely in African American infants. On the other hand, strategies to detect the presence of LQTS preemptively must continue to be sought. To this end, Schwartz et al. continue to investigate the utility and cost-effectiveness of universal ECG screening of Italian infants at 3-4 weeks of age [8; 30; 86]. Importantly, the objective of such an ECG surveillance program is *not* the identification of infants at risk for SIDS (an unrealistic target), but is instead the *early* identification of infants affected by LQTS, a potentially lethal, highly treatable condition.

7.6.2 Significance of 5HTT studies in SIDS

In summary, the promoter variant long alleles and VNTR 12-repeat alleles are associated with SIDS. As these alleles are both the more effective promoters [23; 33] and both associated with increased expression of *5HTT* transporters in various brain regions [34; 52], synaptic serotonin levels would be expected to be lower in those infants with a long or 12-repeat allele, and perhaps lowest in those with both variants. Increased prevalence of the more effective promoter alleles in SIDS cases would suggest that lower synaptic serotonin levels are associated with SIDS risk, and decreased serotonergic receptor binding may thus occur through down-regulation of presynaptic autoreceptors. Alternatively, the long allele, 12-repeat allele, or the combination may relate to SIDS through a developmental effect on raphé neurons. The serotonin transporter is expressed early in ontogenesis in the mouse and the rat [115], and may influence serotonin synapse formation and serotonin-dependent patterning of neuronal networks [14]. Moisewitsch et al. [67] have shown that 5HT, acting through 5HT1A receptors, stimulates migration of cranial neural crest cells in a dose-dependent fashion in the mouse embryo and in cultured cranial explants. Thus, lower 5HT levels due to higher *5HTT* expression during development might lead to alterations in medullary serotonergic neuronal numbers and synaptic connections, with resultant lower serotonin binding in SIDS medulla. Recognizing that the *5HTT* short allele has been associated with anxiety, phobias, and an increased fear response [21; 31; 37; 43; 51; 74], one might also postulate that SIDS would be less likely in infants with the S/S genotype due to an exaggerated fear response and increased arousability. Differences in risk alleles and haplotypes between Caucasian and African American populations would likely result from racial differences in modifier genes or other interacting factors which influence 5HT levels. Alternatively, the risk conferred by the L-12 haplotype identified in this study may not have directly to do with regulation of *5HTT* expression by the combined effect of these loci, but could relate to another yet undefined functional polymorphism in linkage disequilibrium with the L-12 haplotype.

Despite localization to a polydenylation signal, the *3'UTR* SNP polymorphism has not been found to have any effect on expression of the 5HTT transporter protein. Likewise, no association was detected between this polymorphism and SIDS. These data suggest that the *3'UTR* may not really be a functional polymorphism, although it is still expected to be a useful tool for assessment of the contribution of genetic variation in the serotonin transporter gene to disease. Similar to the findings with the SIDS cohort presented here, the *5HTT* promoter polymorphism was recently shown to be associated with Attention Deficit Hyperactivity Disorder, while the 3' UTR was not associated [20]. Findings in both studies likely relate to the difference in functional effects of the promoter polymorphism and the 3'UTR SNP, such that SIDS risk is specifically related to functional variants that result in increased expression of 5HT transporter protein. Thus, future investigations on the influence of *5HTT* on SIDS risk should focus on polymorphisms which directly impact regulation of transporter protein expression or function.

7.6.3 Significance of ANS genes in SIDS

The presented studies represent the available information on the role of homeobox and signal transduction genes important in specifying cell fate in ANS differentiation in SIDS risk. The finding of specific protein-changing mutations in conserved residues [108] of *PHOX2A*, *RET*, *ECE1*, *TLX3*, and *EN1* among 15.2% of SIDS cases vs. 2.2% of controls suggests that specific polymorphisms in these genes may confer some SIDS risk. The observation that 71% of the SIDS cases with these mutations were African American may be consistent with the observed ethnic disparity in SIDS, although African populations tend to exhibit higher levels of genetic variation. The greatest number of rare mutations was identified in the *RET* gene. This is of particular interest because of the relationship of *RET* to Hirschsprung disease and to CCHS (both diseases of neural crest origin), and because of the *RET* knockout model with a depressed ventilatory response to inhaled carbon dioxide with decreased frequency and tidal volume [15]. The knock-out models for *ECE1* [90] and *TLX3* [98] also include impaired breathing and/or early death in the mouse phenotype, with suggestion of a central respiratory deficit. Based on these findings, further research is necessary to better understand the role of these and other genes in the SIDS phenotype and in explaining the ethnic disparity in SIDS.

The observation that none of the SIDS cases demonstrated the *PHOX2B* polyalanine expansion mutation previously identified in CCHS indicates less specific overlap between the two diseases than previously considered. However, as families of CCHS probands have a higher incidence of SIDS history in a family member [112], and as the anticipated incidence of this *PHOX2B* mutation is low in the general population, our sample size may not have been adequate to detect a case. Therefore, it may still be appropriate to evaluate infants with SIDS for the CCHS *PHOX2B* mutation in order to ascertain that CCHS was not the cause of death. However, specific (non-CCHS) polymorphisms in *PHOX2B* are more common in SIDS cases [88] and may confer SIDS risk independently or when present in combination with other mutations of ANS genes. Particularly, the *MASH1-PHOX-RET* pathway, in which *PHOX2B* is needed for the expression of *RET*, has been shown to be an integral part of the development of both the sympathetic and enteric nervous systems [78]. The interaction of *PHOX2B* and *RET* in this pathway is consistent with the finding of a possible interaction between poly-morphisms in *PHOX2B* in *RET* in mediating SIDS risk, suggesting that genetic changes at multiple points in the pathway could combine to amplify risk. Al-though *PHOX2B* plays a key role in the differentiation of central 5HT neurons [80; 81], the absence of significant interactions between *PHOX2B* and *5HTT* polymorphisms, suggests the two genes exert independent effects on SIDS risk, potentially by acting on different aspects of 5HT system function.

Although the *PHOX2B* intron 2 polymorphism (IVS2+101A>G; g.1364A>G) identified in this study is a silent transition in the non-coding region of *PHOX2B*, it was recently linked with Hirschsprung disease (HSCR) [27], an-other disease of ANS dysregulation. The Garcia-Barcelo report [27] of a decreased frequency of the intron 2 polymorphism among 91 ethnic Chinese HSCR cases (19%) compared to 71 unmatched ethnic Chinese controls (36%) strengthens the conclusion that this intron 2 polymorphism is related to ANS dysfunction. Al-though not directly involved in splicing [27], the intron 2 polymorphism could have other regulatory effects and may act in combination with mutations in genes involved in the *RET* and/or *EDNRB* signaling pathway to produce ANS dysfunc-tion.

The identification of polymorphisms in genes pertinent to the embryolo-gic origin of the ANS in SIDS cases lends support to the overriding hypothesis that infants who succumb to SIDS have an underlying genetic predisposition. The low rate of occurrence of mutations in ANS genes studied suggests that there are yet unidentified genes that are responsible for the SIDS phenotype, either directly or in conjunction with the polymorphisms identified in *PHOX2B, RET, 5HTT* and/or other genes involved in ANS or 5HT system development. Sequencing of additional genes involved in ANS or 5HT development in a larger group of SIDS cases will be expected to yield insight into the relationship between *PHOX2B*, ad-ditional candidate genes, and SIDS.

7.6.4 Significance of nicotine metabolizing genes in SIDS

An increased risk of SIDS has been observed in infants exposed to ciga-rette smoke. Although genetic traits that alter the metabolic efficiency of, and thereby increase susceptibility to damage from, toxins encountered through to-bacco exposure could serve as important risk factors for SIDS, no defined genetic link has been demonstrable in the one study of variations of the *GSTT1* and *CYP1A1* genes carried out thus far [89]. This study was limited by lack of infor-mation and objective testing for tobacco exposure in the SIDS cohort. The SIDS cohort is expected to represent a combination of cases with exposure and without exposure, so the presence of the no exposure cases will be expected to limit sig-nificance. Thus, it is possible that an association could be found in a larger group of cases or a group containing only those cases with cigarette smoke exposure. The recognized relationship between tobacco exposure and SIDS risk indicates that examination of these and additional genes involved in tobacco metabolism in a prospective SIDS cohort known to have a history/testing of confirmed smoke exposure may shed light on genetic factors that mediate the role that tobacco exposure plays in susceptibility to SIDS.

7.7 Conclusion and directions for future research on genetic factors in SIDS

As this review clearly indicates, a number of genetically controlled pathways appear to be involved in at least some cases of SIDS. Given the diversity of results to date, genetic studies support the clinical impression that SIDS is heterogeneous with more than one possible genetic etiology. Future studies should consider expanded phenotypic features that might help clarify the heterogeneity and improve the predictive value of the identified genetic factors. Such features should be evaluated to the extent possible in both SIDS victims and their family members.

Genetic studies of SIDS to date have been very limited in terms of the numbers of cases, numbers of candidate genes, and the types of studies with virtually all studies focusing on case-control studies of a few candidate genes. Further progress will require an attempt to develop a large database of individuals and affected families, with consistent phenotyping and large numbers of genes evaluated in all study subjects. It may well be that gene-by-gene and/or gene-by-environment interactions may play a role in SIDS, and having a large dataset with all individuals assessed for the same phenotypes, environmental factors, and genotypes will facilitate detection of any such effects. With 2,162 infants dying from SIDS in 2003 in the U.S. alone [36], and improved but still imperfect parent and caretaker compliance with known modifiable risk factors for SIDS, it behooves the scientific community to join in a collaborative multi-center study of candidate genes and/or genomics to expedite the discover of the genetic profile of the infant at risk for SIDS.

References

1. Ackerman MJ (2004) Cardiac channelopathies: it's in the genes. Nat Med 10: 463-464
2. Ackerman MJ, Siu BL, Sturner WQ, Tester DJ, Valdivia CR, Makielski JC, Towbin JA (2001) Postmortem molecular analysis of SCN5A defects in sudden infant death syndrome. JAMA 286: 2264-2269
3. Ackerman MJ, Splawski I, Makielski JC, Tester DJ, Will ML, Timothy KW, Keating MT, Jones G, Chadha M, Burrow CR, Stephens JC, Xu C, Judson R, Curran ME (2004) Spectrum and prevalence of cardiac sodium channel variants among black, white, Asian, and Hispanic individuals: implications for arrhythmogenic susceptibility and Brugada/long QT syndrome genetic testing. Heart Rhythm 1: 600-607

4.   Ackerman MJ, Tester DJ, Jones GS, Will ML, Burrow CR, Curran ME (2003) Ethnic differences in cardiac potassium channel variants: implications for genetic susceptibility to sudden cardiac death and genetic testing for congenital long QT syndrome. Mayo Clin Proc 78: 1479-1487

5.   Ackerman MJ, Tester DJ, Porter CJ (1999) Swimming, a gene-specific arrhythmogenic trigger for inherited long QT syndrome. Mayo Clin Proc74: 1088-1094

6.   Amiel J, Laudier B, Attie-Bitach T, Trang H, de Pontual L, Gener B, Trochet D, Etchevers H, Ray P, Simonneau M, Vekemans M, Munnich A, Gaultier C, Lyonnet S (2003) Polyalanine expansion and frameshift mutations of the paired-like homeobox gene *PHOX2B* in congenital central hypoventilation syndrome. Nat Genet 33: 459-461

7.   Anderson HR, Cook DG (1997) Passive smoking and sudden infant death syndrome: Review of the epidemiological evidence. Thorax 52: 1003-1009

8.   Arnestad M, Crotti L, Rognum TO, Insolia R, Pedrazzini M, Ferrandi C, Vege A, Wang DW, Rhodes TE, George AL, Schwartz PJ (2007) Prevalence of long-QT syndrome gene variants in sudden infant death syndrome. Circulation 115: 361-367

9.   Bartsch H, Nair U, Risch A, Rojas M, Wikman H, Alexandrov K (2000) Genetic polymorphism of CYP genes, alone or in combination, as a risk modifier of tobacco-related cancers. Cancer Epideml Biomar 9: 3-28

10.   Battersby S, Ogilvie AD, Blakwood DH, Shen S, Muqit MM, Muir WJ, Teague P, Goodwin GM, Harmar AJ (1999) Presence of multiple functional polyadenylation signals and a single nucleotide polymorphism in the 3' untranslated region of the human serotonin transporter gene. J Neurochem 72: 1384-1388

11.   Berry-Kravis EM, Zhou L, Rand CM, Weese-Mayer DE (2006) Congenital central hypoventilation syndrome: *PHOX2B* mutations and phenotype. Am J Respir Crit Care Med 174: 1139-1144

12.   Blair PS, Fleming PJ, Bensley D, Smith I, Bacon C, Taylor E, Berry J, Golding J, Tripp J (1996) Smoking and the sudden infant death syndrome: results from 1993-5 case-control study for confidential inquiry into stillbirths and deaths in infancy. Confidential enquiry into stillbirths and deaths regional coordinators and researchers. Brit Med J 313: 195-198

13.   Brooke H, Gibson A, Tappin D, Brown H (1997) Case-control study of sudden infant death syndrome in Scotland, 1992-5. Brit Med J 314: 1516-1520

14.   Bruning G, Liangos O, Baumgarten HG (1997) Prenatal development of the serotonin transporter in mouse brain. Cell Tissue Res 289: 211-221

15.   Burton MD, Kawashima A, Brayer JA, Kazemi H, Shannon DC, Schuchardt A, Costantini F, Pachnis V, Kinane TB (1997) RET proto-oncogene is important for the development of respiratory CO_2 sensitivity. J Autonom Nerv Syst 63: 137-143

16.   Christiansen M, Tonder N, Larsen LA, Andersen PS, Simonsen H, Oyen N, Kanters JK, Jacobsen JR, Fosdal I, Wettrell G, Kjeldsen (2005) Mutations in the HERG K^+-ion channel: a novel link between long QT syndrome and sudden infant death syndrome. Am J Cardiol 95: 433-434

17.   Collier DA, Arranz MJ, Sham P, Battersby S, Vallada H, Gill P, Aitchison KJ, Sodhi M, Li T, Roberts GW, Smith G, Morton J, Murray RM, Smith D, Kirov G (1996) The serotonin transporter is a potential susceptibility factor for bipolar affective disorder. Neuroreport 7: 1675-1679

18. Collier DA, Stober G, Li T, Heils A, Catalano M, Di Bella D, Arranz MJ, Murray RM, Vallada HP, Bengel D, Muller CR, Roberts GW, Smeraldi E, Kirov G, Sham P, Lesch KP (1996) A novel functional polymorphism within the promoter of the serotonin transporter gene: possible role in susceptibility to affective disorders. Mol Psychiatr 1: 453-460

19. Cronk LB, Ye B, Kaku T, Tester DJ, Vatta M, Makielski JC, Ackerman MJ (2007) Novel Mechanism for Sudden Infant Death Syndrome (SIDS): Persistent late sodium current secondary to mutations in Caveolin-3. Heart Rhythm 4: 161-166

20. Curran S, Purcell S, Craig I, Asherson P, Sham P (2005) The serotonin transporter gene as a QTL for ADHD. Am J Med Genet B Neuropsychiatr Genet 134: 42-47

21. Du L, Bakish D, Hrdina PD (2000) Gender differences in association between serotonin transporter gene polymorphism and personality traits. Psychiatr Gene 10: 159-164

22. Filiano JJ, Kinney HC (1994) A perspective on neuropathologic findings in victims of the sudden infant death syndrome: the triple-risk model. Biol Neonate 65: 194-197

23. Fiskerstrand CE, Lovejoy EA, Quinn JP (1999) An intronic polymorphic domain often associated with susceptibility to affective disorders has allele dependent differential enhancer activity in embryonic stem cells. FEBS Lett 458: 171-174

24. Fleming PJ, Gilbert R, Azaz Y, Berry PJ, Rudd PT, Stewart A, Hall E (1990) Interaction between bedding and sleeping position in the sudden infant death syndrome: a population based case-control study. Brit Med J 301: 85-89

25. Franco P, Chabanski S, Szliwowski H, Dramaix M, Kahn A (2000) Influence of maternal smoking on autonomic nervous system in healthy infants. Pediatr Res 47: 215-20.

26. Franco P, Szliwowski H, Dramaix M, Kahn A (1999) Decreased autonomic responses to obstructive sleep events in future victims of sudden infant death syndrome. Pediatr Res 46: 33-39

27. Garcia-Barcelo M, Sham MH, Lui VC, Chen BL, Ott J, Tam PK (2003) Association study of PHOX2B as a candidate gene for Hirschsprung's disease. Gut 52: 563-567

28. Gelernter J, Cubells JF, Kidd JR, Pakstis AJ, Kidd KK (1999) Population studies of polymorphisms of the serotonin transporter protein gene. Am J Med Genet 88: 61-66

29. Gelernter J, Kranzler H, Coccaro EF, Siever LJ, New AS (1998) Serotonin transporter protein gene polymorphism and personality measures in African American and European American subjects. Am J Psychiatr 155: 1332-1338

30. Goulene K S-BM, Crotti L, Priori SG, Salice P, Mannarino S, Rosati E, Schwartz PJ (2005) Neonatal electrocardiographic screening of genetic arrhythmogenic disorders and congenital cardiovascular diseases: prospective data from 31,000 infants. Eur Heart J 26: 214

31. Hariri AR, Mattay VS, Tessitore A, Kolachana B, Fera F, Goldman D, Egan MF, Weinberger DR (2002) Serotonin transporter genetic variation and the response of the human amygdala. Science. 297: 400-403

32. Hayashi S, Watanabe J, Kawajiri K (1992) High susceptibility to lung cancer analyzed in terms of combined genotypes of P450IA1 and Mu-class glutathione S-transferase genes. Jpn J Cancer Res 83: 866-870

33. Heils A, Teufel A, Petri S, Stober G, Riederer P, Bengel D, Lesch KP (1996) Allelic variation of human serotonin transporter gene expression. J Neurochem 66: 2621-2624

34. Heinz A, Jones DW, Mazzanti C, Goldman D, Ragan P, Hommer D, Linnoila M, Weinberger DR (2000) A relationship between serotonin transporter genotype and in vivo protein expression and alcohol neurotoxicity. Biol Psychiat 47: 643-649

35. Hirsch MR, Tiveron MC, Guillemot F, Brunet JF, Goridis C (1998) Control of noradrenergic differentiation and Phox2a expression by MASH1 in the central and peripheral nervous system. Development 125: 599-608

36. Hoyert DL, Heron MP, Murphy SL, Kung HC (2006) Deaths: final data for 2003. Natl Vital Stat Rep 54: 1-120

37. Hu S, Brody CL, Fisher C, Gunzerath L, Nelson ML, Sabol SZ, Sirota LA, Marcus SE, Greenberg BD, Murphy DL, Hamer DH (2000) Interaction between the serotonin transporter gene and neuroticism in cigarette smoking behavior. Mol Psychiatr 5: 181-188

38. Ishibe N, Wiencke JK, Zuo ZF, McMillan A, Spitz M, Kelsey KT (1997) Susceptibility to lung cancer in light smokers associated with CYP1A1 polymorphisms in Mexican and African-Americans. Cancer Epiderm Bimar 6: 1075-1080

39. Ishiguro H, Saito T, Akazawa S, Mitushio H, Tada K, Enomoto M, Mifune H, Toru M, Shibuya H, Arinami T (1999) Association between drinking-related antisocial behavior and a polymorphism in the serotonin transporter gene in a Japanese population. Alcohol Clin Exp Res 23: 1281-1284

40. Jacobs BL, Azmitia EC (1992) Structure and function of the brain serotonin system. Physiol Rev 72: 165-229

41. Kahn A, Groswasser J, Rebuffat E, Sottiaux M, Blum D, Foerster M, Franco P, Bochner A, Alexander M, Bachy A (1992) Sleep and cardiorespiratory characteristics of infant victims of sudden death: a prospective case-control study. Sleep 15: 287-292

42. Kai B (2003) Maternal and child health statistics of Japan

43. Katsuragi S, Kunugi H, Sano A, Tsutsumi T, Isogawa K, Nanko S, Akiyoshi J (1999) Association between serotonin transporter gene polymorphism and anxiety-related traits. Biol Psychiat 45: 368-370

44. Kijima K, Sasaki A, Niki T, Umetsu K, Osawa M, Matoba R, Hayasaka K (2004) Sudden infant death syndrome is not associated with the mutation of PHOX2B gene, a major causative gene of congenital central hypoventilation syndrome. Tohoku J Exp Med 203: 65-68

45. Kinney HC, Randall LL, Sleeper LA, Willinger M, Belliveau RA, Zec N, Rava LA, Dominici L, Iyasu S, Randall B, Habbe D, Wilson H, Mandel F, McClain, Welty TK (2003) Serotonergic brainstem abnormalities in Northern Plains Indians with the sudden infant death syndrome. J Neuropathol Exp Neurol 62: 1178-1191

46. Kochanek KD, Hudson BL (1995) Advance Report of Final Mortality Statistics, 1992 Hyattsville, MD: National Center for Health Statistics

47. Kotamaki M (1995) Smoking induced differences in autonomic responses in military pilot candidates. Clin Auton Res 5: 31-36

48. Ledwidge M, Fox G, Matthews T (1998) Neurocardiogenic syncope: a model for SIDS. Arch Dis Child 78: 481-483

49.   Lerman C, Shields PG, Audrain J, Main D, Cobb B, Boyd NR, Caporaso N (1998) The role of the serotonin transporter gene in cigarette smoking. Cancer Epidem Biomar 7: 253-255

50.   Lesch KP, Balling U, Gross J, Strauss K, Wolozin BL, Murphy DL, Riederer P (1994) Organization of the human serotonin transporter gene. J Neural Trans Gen Sec 95: 157-162

51.   Lesch KP, Bengel D, Heils A, Sabol SZ, Greenberg BD, Petri S, Benjamin J, Muller CR, Hamer DH, Murphy DL (1996) Association of anxiety-related traits with a polymorphism in the serotonin transporter gene regulatory region. Science 274: 1527-1531

52.   Little KY, McLaughlin DP, Zhang L, Livermore CS, Dalack GW, McFinton PR, Del Proposto ZS, Hill E, Cassin BJ, Watson SJ, Cook EH (1998) Cocaine, ethanol, and genotype effects on human midbrain serotonin transporter binding sites and mRNA levels. Am J Psychiat 155: 207-213

53.   Lo L, Morin X, Brunet JF, Anderson DJ (1999) Specification of neurotransmitter identity by Phox2 proteins in neural crest stem cells. Neuron 22: 693-705

54.   Lo L, Tiveron MC, Anderson DJ (1998) MASH1 activates expression of the paired homeodomain transcription factor Phox2a, and couples pan-neuronal and subtype-specific components of autonomic neuronal identity. Development 125: 609-620

55.   Lovejoy EA, Scott AC, Fiskerstrand CE, Bubb VJ, Quinn JP (2003) The serotonin transporter intronic VNTR enhancer correlated with a predisposition to affective disorders has distinct regulatory elements within the domain based on the primary DNA sequence of the repeat unit. Eur J Neurosci 17: 417-420

56.   Lucini D, Bertocchi F, Malliani A, Pagani M (1996) A controlled study of the autonomic changes produced by habitual cigarette smoking in healthy subjects. Cardiovasc Res 31: 633-639

57.   MacDorman MF, Cnattingius S, Hoffman HJ, Kramer MS, Haglund B (1997) Sudden infant death syndrome and smoking in the United States and Sweden. Am J Epidemiol 146: 249-257

58.   MacKenzie A, Quinn J (1999) A serotonin transporter gene intron 2 polymorphic region, correlated with affective disorders, has allele-dependent differential enhancer-like properties in the mouse embryo. Proc Natl Acad Sci USA 96: 15251-15255

59.   Maher BS, Marazita ML, Rand C, Zhou L, Berry-Kravis EM, Weese-Mayer DE (2006) 3' UTR polymorphism of the serotonin transporter gene and sudden infant death syndrome: haplotype analysis. Am J Med Genet 140: 1453-1457

60.   Marazita ML, Maher BS, Cooper ME, Silvestri JM, Huffman AD, Smok-Pearsall SM, Kowal MH, Weese-Mayer DE (2001) Genetic segregation analysis of autonomic nervous system dysfunction in families of probands with idiopathic congenital central hypoventilation syndrome. Am J Med Genet 100: 229-236

61.   Maron BJ, Clark CE, Goldstein RE, Epstein SE (1976) Potential role of QT interval prolongation in sudden infant death syndrome. Circulation 54: 423-430

62.   Matera, I, Bachetti T, Puppo F, Di Duca M, Morandi F, Casiraghi GM, Cilio MR, Hennekam R, Hofstra R, Schober JG, Ravazzolo R, Ottonello G, Ceccherini I (2004) *PHOX2B* mutations and polyalanine expansions correlate with the severity of the respiratory phenotype and associated symptoms in both congenital and late onset central hypoventilation syndrome. J Med Genet 41: 373-380

63.   Melke J, Westberg L, Landen M, Sundblad C, Eriksson O, Baghei F, Rosmond R, Eriksson E, Ekman A (2003) Serotonin transporter gene polymorphisms and platelet [3H] paroxetine binding in premenstrual dysphoria. Psychoneuroendocrino 28: 446-458

64.   Meny RG, Carroll JL, Carbone MT, Kelly DH (1994) Cardiorespiratory recordings from infants dying suddenly and unexpectedly at home. Pediatrics 93: 44-49

65.   Mitchell EA, Tuohy PG, Brunt JM, Thompson JM, Clements MS, Stewart AW, Ford RP, Taylor BJ (1997) Risk factors for sudden infant death syndrome following the prevention campaign in New Zealand: a prospective study. Pediatrics 100: 835-840

66.   Minino AM, Smith BL (2001) Deaths: preliminary data for 2000. National Vital Statistics Reports 49: 1-40

67.   Moiseiwitsch JR, Lauder JM (1995) Serotonin regulates mouse cranial neural crest migration. Proc Natl Acad Sci USA 92: 7182-7186

68.   Moss AJ, Robinson JL, Gessman L, Gillespie R, Zareba W, Schwartz PJ, Vincent GM, Benhorin J, Heilbron EL, Towbin JA, Priori SG, Napolitano C, Zhang L, Medina A, Andrews ML, Timothy K (1999) Comparison of clinical and genetic variables of cardiac events associated with loud noise versus swimming among subjects with the long QT syndrome. Am J Cardiol 84: 876-879

69.   Nakachi K, Imai K, Hayashi S, Kawajiri K (1993) Polymorphisms of the CYP1A1 and glutathione S-transferase genes associated with susceptibility to lung cancer in relation to cigarette dose in a Japanese population. Cancer Res 53: 2994-2999

70.   Nakamura T, Muramatsu T, Ono Y, Matsushita S, Higuchi S, Mizushima H, Yoshimura K, Kanba S, Asai M (1997) Serotonin transporter gene regulatory region polymorphism and anxiety-related traits in the Japanese. Am J Med Genet 74: 544-545

71.   Narita N, Narita M, Takashima S, Nakayama M, Nagai T, Okado N (2001) Serotonin transporter gene variation is a risk factor for sudden infant death syndrome in the Japanese population. Pediatrics 107: 690-692

72.   Niedermaier ON, Smith ML, Beightol LA, Zukowska-Grojec Z, Goldstein DS, Eckberg DL (1993) Influence of cigarette smoking on human autonomic function. Circulation 88: 562-571

73.   Ogilvie AD, Battersby S, Bubb VJ, Fink G, Harmar AJ, Goodwim GM, Smith CA (1996) Polymorphism in serotonin transporter gene associated with susceptibility to major depression. Lancet 347: 731-733

74.   Osher Y, Hamer D, Benjamin J (2000) Association and linkage of anxiety-related traits with a functional polymorphism of the serotonin transporter gene regulatory region in Israeli sibling pairs. Mol Psychiatr 5: 216-219

75.   Ozawa Y, Okado N (2002) Alteration of serotonergic receptors in the brainstems of human patients with respiratory disorders. Neuropediatrics 33: 142-149

76.   Panigrahy A, Filiano J, Sleeper LA, Mandell F, Valdes-Dapena M, Krous HF, Rava LA, Foley E, White WF, Kinney HC (2000) Decreased serotonergic receptor binding in rhombic lip-derived regions of the medulla oblongata in the sudden infant death syndrome. J Neuropathol Exp Neurol 59: 377-384

77.   Paterson DS, Trachtenbert FL, Thompson EG, Belliveau RA, Beggs AH, Darnall RA, Chadwick AE, Krous HF, Kinney HC (2006) Multiple serotonergic brainstem abnormalities in the Sudden Infant Death Syndrome. JAMA 296: 2124-2132

78. Pattyn A, Morin X, Cremer H, Goridis C, Brunet JF (1997) Expression and inter-
     actions of the two closely related homeobox genes Phox2a and Phox2b during
     neurogenesis. Development 124: 4065-4075
79. Pattyn A, Morin X, Cremer H, Goridis C, Brunet JF (1999) The homeobox gene
     Phox2b is essential for the development of autonomic neural crest derivatives.
     Nature 399: 366-370
80. Pattyn A, Simplicio N, van Doorninck JH, Goridis C, Guillemot F, Brunet JF
     (2004) Ascl1/Mash1 is required for the development of central serotonergic neu-
     rons. Nat Neurosci 7: 589-595
81. Pattyn A, Vallstedt A, Dias JM, Samad OA, Krumlauf R, Rijli FM, Brunet JF,
     Ericson J (2003) Coordinated temporal and spatial control of motor neuron and
     serotonergic neuron generation from a common pool of CNS progenitors. Gene
     Dev 17: 729-737
82. Piha SJ (1994) Cardiovascular autonomic reflexes in heavy smokers. J Auton
     Nerv Syst 48: 73-77
83. Plant LD, Bowers PN, Liu Q, Morgan T, Zhang T, State MW, Chen W, Kittles
     RA, Goldstein SA (2006) A common cardiac sodium channel variant associated
     with sudden infant death in African Americans, SCN5A S1103Y. J Clin Invest
     116: 430-435
84. Ponsonby AL, Dwyer T, Gibbons LE, Cochrane JA, Jones ME, McCall MJ
     (1992) Thermal environment and sudden infant death syndrome: case-control
     study. Brit Med J 304: 277-282
85. Pope CA, 3rd, Eatough DJ, Gold DR, Pang Y, Nielsen KR, Nath P, Verrier RL,
     Kanner RE (2001) Acute exposure to environmental tobacco smoke and heart rate
     variability. Environ Health Perspect 109: 711-716
86. Quaglini S, Rognoni C, Spazzolini C, Priori SG, Mannarino S, Schwartz PJ
     (2006) Cost-effectiveness of neonatal ECG screening for the long QT syndrome.
     Eur Heart J 27: 1824-1832
87. Ramamoorthy S, Bauman AL, Moore KR, Han H, Yang-Feng T, Chang AS, Ga-
     napathy V, Blakely RD (1993) Antidepressant- and cocaine-sensitive human se-
     rotonin transporter: molecular cloning, expression, and chromosomal localization.
     Proc Natl Acad Sci U A 90: 2542-2546
88. Rand CM, Weese-Mayer DE, Maher BS, Cooper ME, Marazita ML, Berry-
     Kravis EM (2006) Sudden infant death syndrome: Case-control frequency differ-
     ences in paired like homeobox (PHOX) 2B gene. Am J Med Genet 140: 1687-
     1691
89. Rand CM, Weese-Mayer DE, Maher BS, Zhou L, Marazita ML, Berry-Kravis
     EM (2006) Nicotine metabolizing genes GSTT1 and CYP1A1 in sudden infant
     death syndrome. Am J Med Genet 140: 1447-1452
90. Renolleau S, Dauger S, Vardon G, Levacher B, Simonneau M, Yanagisawa M,
     Gaultier C, Gallego J (2001) Impaired ventilatory responses to hypoxia in mice
     deficient in endothelin-converting-enzyme-1. Pediatr Res 49: 705-712
91. Sasaki A, Kanai M, Kijima K, Akaba K, Hashimoto M, Hasegawa H, Otaki S,
     Koizumi T, Kusuda S, Ogawa Y, Tuchiya K, Yamamoto W, Nakamura T, Ha-
     yasaka K (2003) Molecular analysis of congenital central hypoventilation syn-
     drome. Hum Genet 114: 22-26
92. Schechtman VL, Harper RM, Kluge KA, Wilson AJ, Hoffman HJ, Southall DP
     (1988) Cardiac and respiratory patterns in normal infants and victims of the sud-
     den infant death syndrome. Sleep 11: 413-424
93. Schwartz PJ (1976) Cardiac sympathetic innervation and the sudden infant death
     syndrome. A possible pathogenetic link. Am J Med 60: 167-172

94. Schwartz PJ, Priori SG, Bloise R, Napolitano C, Ronchetti E, Piccinini A, Goj C, Breithardt G, Schulze-Bahr E, Wedekind H, Nastoli J (2001) Molecular diagnosis in a child with sudden infant death syndrome. Lancet 358: 1342-1343

95. Schwartz PJ, Priori SG, Dumaine R, Napolitano C, Antzelevitch C, Stramba-Badiale M, Richard TA, Berti MR, Bloise R (2000) A molecular link between the sudden infant death syndrome and the long-QT syndrome. N Engl J Med 343: 262-267

96. Schwartz PJ, Priori SG, Spazzolini C, Moss AJ, Vincent GM, Napolitano C, Denjoy I, Guicheney P, Breithardt G, Keating MT, Towbin JA, Beggs AH, Brink P, Wilde AA, Toivonen L, Zareba W, Robinson JL, Timothy KW, Corfield V, Wattanasirichaigoon D, Corbett C, Haverkamp W, Schulze-Bahr E, Lehmann MH, Schwartz K, Coumel P, Bloise R (2001) Genotype-phenotype correlation in the long-QT syndrome: gene-specific triggers for life-threatening arrhythmias. Circulation 103: 89-95

97. Schwartz PJ, Stramba-Badiale M, Segantini A, Austoni P, Bosi G, Giorgetti R, Grancini F, Marni ED, Perticone F, Rosti D, Salice P (1998) Prolongation of the QT interval and the sudden infant death syndrome. N Engl J Med 338: 1709-1714

98. Shirasawa S, Arata A, Onimaru H, Roth KA, Brown GA, Horning S, Arata S, Okumura K, Sasazuki T, Korsmeyer SJ (2000) Rnx deficiency results in congenital central hypoventilation. Nat Genet 24: 287-290

99. Sun Z, Milos PM, Thompson JF, Lloyd DB, Mank-Seymour A, Richmond J, Cordes JS, Zhou J (2004) Role of a KCNH2 polymorphism (R1047 L) in dofetilide-induced Torsades de Pointes. J Mol Cell Cardiol 37: 1031-1039

100. Taylor BJ, Williams SM, Mitchell EA, Ford RP (1996) Symptoms, sweating and reactivity of infants who die of SIDS compared with community controls. New Zealand National Cot Death Study Group. J Paediatr Child Health 32: 316-322

101. Tester DJ, Ackerman MJ (2005) Sudden infant death syndrome: how significant are the cardiac channelopathies? Cardiovasc Res 67: 388-396

102. Tester DJ, Dura M, Carturan E, Reiken S, Wronska A, Marks AR, Ackerman MJ (2007) A mechanism for sudden infant death syndrome (SIDS): stress-induced leak via ryanodine receptors. Heart Rhythm, in press

103. Trochet, D, O'Brien LM, Gozal D, Trang H, Nordenskjold A, Laudier B, Svensson PJ, Uhrig S, Cole T, Niemann S, Munnich A, Gaultier C, Lyonnet L, Amiel J (2005) PHOX2B genotype allows for prediction of tumor risk in congenital central hypoventilation syndrome. Am J Hum Genet 76: 421-426

104. Vatta M, Ackerman MJ, Ye B, Makielski JC, Ughanze EE, Taylor EW, Tester DJ, Balijepalli RC, Foell JD, Li Z, Kamp TJ, Towbin JA (2006) Mutant caveolin-3 induces persistent late sodium current and is associated with long QT Syndrome. Circulation 114: 2104-2112

105. Wang DW, Desai RR, Crotti L, Arnestad M, Insolia R, Pedrazzini M, Ferrandi C, Vege A, Rognum T, Schwartz PJ, George AL (2007) Cardiac sodium channel dysfunction in sudden infant death syndrome. Circulation. 115: 368-376

106. Wang X, Zuckerman B, Pearson C, Kaufman G, Chen C, Wang G, Niu T, Wise PH, Bauchner H, Xu X (2002) Maternal cigarette smoking, metabolic gene polymorphism, and infant birth weight. JAMA 287: 195-202

107. Weese-Mayer DE, Berry-Kravis EM, Maher BS, Silvestri JM, Curran ME, Marazita ML (2003) Sudden infant death syndrome: association with a promoter polymorphism of the serotonin transporter gene. Am J Med Genet 117: 268-274

108. Weese-Mayer DE, Berry-Kravis EM, Zhou L, Maher BS, Curran ME, Silvestri JM, Marazita ML (2004) Sudden infant death syndrome: case-control frequency differences at genes pertinent to early autonomic nervous system embryologic development. Pediatr Res 56: 391-395

109. Weese-Mayer DE, Berry-Kravis EM, Zhou L, Maher BS, Silvestri JM, Curran ME, Marazita ML (2003) Idiopathic congenital central hypoventilation syndrome: analysis of genes pertinent to early autonomic nervous system embryologic development and identification of mutations in *PHOX2B*. Am J Med Genet 123: 267-278

110. Weese-Mayer DE, Shannon DC, Keens TG, Silvestri JM (1999) American Thoracic Society Statement on the diagnosis and management of idiopathic congenital central hypoventilation syndrome. Am J Respir Crit Care Med 160: 368-373

111. Weese-Mayer DE, Silvestri JM, Huffman AD, Smok-Pearsall SM, Kowal MH, Maher BS, Cooper ME, Marazita ML (2001) Case/control family study of autonomic nervous system dysfunction in idiopathic congenital central hypoventilation syndrome. Am J Med Genet 100: 237-245

112. Weese-Mayer DE, Silvestri JM, Marazita ML, Hoo JJ (1993) Congenital central hypoventilation syndrome: inheritance and relation to sudden infant death syndrome. Am J Med Genet 47: 360-367

113. Weese-Mayer DE, Zhou L, Berry-Kravis EM, Maher BS, Silvestri JM, Marazita ML (2003) Association of the serotonin transporter gene with sudden infant death syndrome: a haplotype analysis. Am J Med Genet 122: 238-245

114. Wilde AA, Jongbloed RJ, Doevendans PA, Duren DR, Hauer RN, van Langen IM, van Tintelen JP, Smeets HJ, Meyer H, Geelen JL (1999) Auditory stimuli as a trigger for arrhythmic events differentiate HERG-related (LQTS2) patients from KVLQT1-related patients (LQTS1). J Am Coll Cardiol 33: 327-332

115. Zhou FC, Sari Y, Zhang JK (2000) Expression of serotonin transporter protein in developing rat brain. Brain Res Dev Brain Res 119: 33-45

8. The genetic basis for obstructive sleep apnea: what role for variation in respiratory control?

Susan REDLINE and Sanjay R. PATEL

Case Western Reserve University, Cleveland OH 44106-6006, U.S.A.

8.1 Introduction

Obstructive sleep apnea (OSA) is a highly prevalent disorder with multiple co-morbidities. Its etiology is complex and multifactorial, with evidence that susceptibility is influenced by risk factors that include obesity and obesity-associated traits, craniofacial characteristics associated with reduced upper airway dimensions, as well as ventilatory deficits that predispose to pharyngeal collapsibility during sleep, when neuromuscular output is either reduced or relatively unstable. Although studies of the genetic etiology of the disorder are few, there are growing data that have quantified the heritability of OSA, described potential modes of transmission, and have identified suggestive and/or biologically plausible candidate genes. This chapter will review the evidence for a genetic basis for OSA, with special attention to the role of ventilatory phenotypes. In addition, candidate genes for ventilatory phenotypes potentially relevant to OSA will be reviewed based on existing animal and human candidate gene studies. Although this research is in its infancy, the importance of investigating the genetic basis of ventilatory gene variants that contribute to the pathogenesis to OSA is underscored by the high co-morbidity of the disease and the current limitations of treatment options. Genetic studies of relevant ventilatory phenotypes promise to identify targets suitable for pharmacological manipulation.

8.2 OSA: Definition and health impact

OSA is defined by the occurrence of recurrent episodes of apneas and hypopneas in sleep, associated with oxygen desaturation, sleep fragmentation, and with symptoms of disruptive snoring and daytime sleepiness. The prevalence of OSA varies according to the threshold levels of abnormality detected by polysomnography and the age of the population studied. Several epidemiological suggest that it affects approximately 2-4% of children [97], 4 to 15% of middle-aged adults [130], and more than 20% of the elderly [6]. Adverse behavioral and health effects are attributable to recurrent overnight exposures to profound adverse physiological stresses, including hypoxemia, hypercapnia and sleep fragmentation, which individually or interactively lead to chemoreflex activation, reduction in delta and REM sleep, arousal, and sympathetic nervous system activation [28; 70]. OSA is a risk factor for hypertension (HTN), cardiovascular disease, stroke, and mortality [61; 62; 72; 85]. This large public health burden of OSA underscores the need to better understand its causes, including genetic susceptibility, facilitating identification of susceptible individuals who are most likely to benefit from treatment, and better defining pathophysiological processes that may be amenable to specific therapies.

A challenge for epidemiological and genetic studies of OSA has been agreement as to the best phenotype by which to define the disease and quantify its severity. Most studies of OSA have defined the trait, or phenotype, by the apnea hypopnea index (AHI), a count of the number of complete or partially obstructed breaths per hour of sleep. Genetic analyses of this quantitative phenotype have used statistical transformations of the AHI to approximate a normal distribution, needed for many statistical tests, or have used age-specific thresholds for defining disease status. The advantages of the AHI include its relative simplicity, high night-to-night reproducibility, and wide-spread use clinically. It has been argued, however, that this "count" does not provide a measure of the full range of the severity of OSA, which may relate to the duration of individual events, degree of associated hypoxemia and sleep fragmentation, and associated functional and physiological consequences. A summary count which does not distinguish event characteristics also may limit its utility in studies that include elderly individuals, who because of underlying morbidity may experience a relatively higher number of central events, which may or may not be related to the phenotype of interest. Also, certain genetic influences may affect predominantly the duration rather than the number of respiratory events. Preliminary association studies suggest that duration of apneas, possibly a surrogate for arousability, may vary according to angiotensin converting enzyme (ACE) polymorphisms [126; 127]. More comprehensive physiological data may better capture features that define clearer and more specific phenotypes. In addition, combining polysomnographic data with other information, including symptoms, signs, and outcome data, to derive a multidimensional phenotype is another approach to improve phenotype characterization; this approach, however, has not yet been systematically applied to genetic studies of OSA.

8.3 OSA: Evidence for a familial basis

A familial basis for OSA was first proposed in 1978 by Strohl et al. who reported three brothers with this disorder [109]. Since that time, it has become increasingly clear that OSA commonly clusters within families [18; 20; 33; 38; 64; 87; 93-95]. A wide range of phenotype assessments have been employed, but significant familial aggregation of the AHI, or of symptoms of OSA, has been observed in studies from the US, Finland, Denmark, Iceland, the UK, and Israel [38; 47; 50; 64; 87; 94; 96; 106]. Studies have utilized a variety of designs, including studies of cohorts, small and large pedigrees, twins, and case-control studies. The literature includes reports of adults and/or children, and familial clustering has been reported from studies that have included [33; 94] or excluded [64] obese individuals. Despite study design and population differences, all studies have consistently shown familial aggregation of the AHI level and symptoms of OSA. These studies have provided clear evidence that a positive family history of OSA or snoring is an important risk factor for an elevated AHI and associated symptoms such as snoring, daytime sleepiness, and apneas. However, the estimated magnitude of effect has varied greatly.

Four large twin studies have shown that concordance rates for snoring, a principal symptom of OSA, are significantly higher in monozygotic than dizygotic twins [19; 27; 50]. A study of adult male twins showed significant genetic correlations for sleepiness as well as snoring, with models consistent with common genes underlying both symptoms [19]. A large Danish cohort study [47] showed that the age, BMI and co-morbidity-adjusted risk of snoring was increased 3-fold when one first degree relative was a snorer, and increased 4-fold when both parents were snorers.

The prevalence of OSA, as determined by an elevated AHI, among first degree relatives of OSA probands has been reported to vary from 22 to 84% [25; 38; 64; 87; 94]. Among the studies that included controls, the odds ratios (ORs), relating the odds of an individual having OSA in a family with affected relatives to that for someone without an affected relative, have varied from 2 to 46 [38; 64; 94]. Pedigree studies from both the US and Iceland have shown consistent associations, with the overall recurrent risk for OSA in a family member of an affected proband to be approximately 2 [33; 94], which is lower than that reported from case control studies. OSA has been reported to occur more commonly as a multiplex (affecting ≥ 2 members) in >50% of the families with at least one affected member than a simplex disorder (i.e., occurring in a single family member) [94]. Heritability estimates, which quantify the proportion of the variance in a trait attributable to common familial factors, for the AHI from both pedigree [18] and adult twin [20] studies are approximately 35 to 40%. Significant and similar parent-offspring ($r = 0.21$, $p = 0.002$) and sib-sib correlations ($p = 0.21$, $p = 0.003$) have been observed, which are greater than spousal correlations [94], consistent with genetic rather than purely familial environmental similarities accounting for familial clustering. Further evidence for a genetic basis for OSA is derived by the observation that the odds of sleep apnea syndrome, defined as AHI >15 and sleepiness, increases with increasing numbers of affected relatives; i.e., the odds for sleep apnea syndrome given one, two, or three affected relatives with these findings, as compared to individuals with no affected relatives, adjusted for age, gender, race and BMI range from 1.5 to 4.0, respectively [94].

8.3.1 Segregation analysis: modeling patterns of inheritance

Although family studies are useful for establishing a likely familial and genetic basis for a trait and for quantifying the magnitude of its heritability, specialized statistical tools are needed to characterize the mode of inheritance. The latter is addressed with the use of segregation analysis, a modeling technique whereby statistical models that make alternative assumptions about the underlying mode of inheritance and impact of environmental factors on a given trait are compared in an attempt to identify the model that best explains the observed distribution of traits. Segregation analyses of self-reported snoring in 584 pedigrees in the Tucson Epidemiological Study suggested a major gene effect; however, evidence for this weakened after adjustments were made for obesity and gender [42]. Segregation analysis was used to analyze data on the AHI from the Cleveland Family Study. In a sample of Caucasians (177 families) and African Americans (123 families), the observed distribution of the AHI was consistent with the segregation of major genetic factors within both sets of families [18]. The results suggested possible racial differences in the mode of inheritance. In Caucasians, analysis suggested recessive Mendelian inheritance of the AHI, accounting for 21-27% of the variance, with an additional 8-9% of the variation due to other familial factors, either environmental or polygenic. In African Americans, the BMI and age-adjusted AHI gave evidence of segregation of a co-dominant gene with an allele frequency of 0.14 accounting for 35% of the total variance. Adjustment of the AHI for BMI weakened the findings in the Caucasians and strengthened them in the African Americans. The analyses in Caucasians suggested that a major gene for OSA may be closely related to genes for obesity. However, the results from African Americans provide support for an underlying genetic basis for OSA independent of the contribution of BMI, and underscore the potential role of genetic risk factors for alternative intermediate traits for OSA in this race group, such as ventilatory control.

8.3.2 Genetic linkage studies in humans

Genome-wide linkage analysis, in which one attempts to identify genetic loci that co-segregate with the trait of interest, is a means for identifying chromosomal regions in which susceptibility genes may reside. The lod score, the logarithmic odds ratio of linkage to no linkage, is calculated to quantify the extent to which transmission of an allele at a particular genetic marker occurs in parallel with the trait of interest. To date, the only published linkage studies for OSA are from the Cleveland Family Study [64; 78; 80]. Although the preliminary findings did not report lod scores for age, sex and body mass index (BMI) adjusted AHI levels that achieved genome-wide significance (lod >2.8), several regions had intermediate lod scores [3]. Among Caucasians, the highest lod scores for AHI (adjusted for BMI) were reported at chromosomal regions 2p, 12p, and 19q with scores of 1.64, 1.43, and 1.40 respectively [78]. Further fine mapping of the Caucasian cohort at chromosome 19q resulted in increased evidence of linkage to OSA in this region with a peak lod score of 2.4 [55]. Among African-Americans, a region with suggestive evidence of linkage has been reported on chromosome 8q, with a lod score of 1.3 [80]. Several biologically plausible candidate genes are located within the most promising chromosomal regions in these analysis The chromosome 2p region contains acid phosphatase 1, apoprotein B precursor, proopiomelanocortin (POMC), and the alpha-2β-adrenergic receptor. A limitation of these analyses was the potential over-adjustment of the AHI for the trait BMI, which may be in the causal genetic pathway. Further preliminary analyses have attempted to dissect the potential independent and overlapping genetic etiologies of AHI and BMI [54; 92]. These findings suggest that there are distinct genetic factors underlying susceptibility to OSA, and that although the inter-relationship of OSA and obesity may be partially explained by a common causal pathway involving one or more genes regulating both traits, there is also evidence for other genetic loci that are linked with AHI, but not with BMI. Such factors may include genetic determinants for craniofacial structure and for ventilatory control.

8.4 Genetic etiology-risk factors and their use as intermediate phenotypes

As a complex trait, a number of risk factors likely interact to increase propensity for repetitive upper airway collapse occurring with sleep, causing the clinical entity, OSA. Well-recognized risk factors are obesity, male gender, and small upper airway size. Ventilatory control mechanisms likely contribute to OSA susceptibility. Our group has proposed three 'intermediate' pathogenic pathways through which genes might act to increase propensity to OSA: (1) obesity and related metabolic syndrome phenotypes; (2) craniofacial morphometry; and (3) ventilatory control. In addition, genetic variants that contribute to the control of sleep-wake states and circadian rhythm control may also be important in the pathogenesis of OSA [40; 48]. Since intermediate traits such as these may be more closely associated with specific gene products and may be less influenced by environmental modification than more complex phenotypes, they may facilitate the identification of genetic polymorphisms for OSA, if indeed such traits are on a causal pathway leading to OSA. The limitation of this approach is that the genes so identified may not be sufficient to describe the clinically important phenotype, which may only occur in the context of other genetic and environmental factors. Intermediate pathways also may interact to influence OSA susceptibility. For example, obesity may lead to fat deposition in the parapharyngeal regions narrowing the upper airway, which may only lead to OSA when neuromuscular activation of upper airway muscles also declines below critical levels. The role of genetic factors for obesity and craniofacial structure are reviewed elsewhere [83].

8.5 Ventilatory control as an intermediate OSA risk factor

In the population, there is a wide distribution of values quantifying various ventilatory phenotypes, including indices of central inspiratory drive and chemoreception. Evidence for heritability of ventilatory patterns and hypoxic sensitivity are reviewed in Chapter 2. In brief, heritability estimates for oxygen saturation or hypoxic chemoresponsiveness range from approximately 30 to 75% [14; 113], consistent with a substantial genetic contribution. In contrast, the response to hypercapnia has not consistently been found to display familial aggregation [10; 22; 52; 113].

Population variability in the respiratory neuromuscular responses to the influences of state (sleep-wake), chemical drive, load sensitivity, and/or arousal will result in different ventilatory responses to sleep-related stresses, shaping both the magnitude of ventilation and ventilatory pattern and the propensity for respiratory oscillations in sleep. Upper airway collapse may occur by preferential reduction in the level of activation of upper airway muscles as compared to chest wall muscles, by promoting ventilatory instability and, subsequently, periodic breathing, or by impairing the arousal response to chemical stimuli (hypoxemia or hypercapnia) or to an inspiratory load. Obstructive or central sleep apnea, or both, may result, depending on the changes in the relative innervations of upper airway dilator compared to diaphragmatic muscles.

Classic physiological studies have provided models for predicting airway patency [21; 51; 74; 129], which appears to be influenced dynamically by an array of complex processes associated with the control of respiratory neuromuscular function. However, adopting a cohesive model that defines the role of ventilatory control "deficits" as risk factors for OSA has been hindered by the apparent disparate pathways by which ventilatory control abnormalities might lead to OSA. It has been argued that ventilatory instability could result from blunted or augmented chemosensitivity, arousability, and/or load detection. With sleep onset, the central inspiratory drive to upper airway motorneurons, a major determinant of airway patency, is reduced or fluctuates [23; 43]. In the presence of a blunted ventilatory drive, neural stimulation of the upper airway dilator muscles may become insufficient to overcome increases in upper airway resistance (UAR) occurring during recumbency and sleep, which may be especially high in those with anatomically compromised airways. Conversely, an overly vigorous ventilatory drive may lead to instability of breathing with alternating cycles of hyperventilation and hypoventilation, most commonly recognized as "periodic breathing". During the nadir of this cycling, reduction in chemical drive due to hypocapnia may reduce neural output to upper airway muscles to a level below the threshold required to keep the airway open. The latter is also more likely to occur in individuals with increased UAR [44]. Thus, interpreting ventilatory phenotypes in light of anatomic phenotypes may be critical for defining the risk of each phenotype on OSA susceptibility. Furthermore, the influence of any given ventilatory control "abnormality" on OSA predisposition is likely to vary as a function of other ventilatory phenotypes. Much of the literature may be limited by its focus on studying individual ventilatory phenotypes. However, airway collapse and/or ventilatory stability is likely to be influenced by the interacting sensitivities to hypoxia and hypercapnia (and to peripheral and central chemoreception), each of which may be characterized by overall sensitivity (dose-response) as well as threshold sensitivities.

Anatomic factors are likely to be important in influencing whether deficits in ventilatory control will lead to upper airway collapse and OSA. In particular, any given reduction in central inspiratory drive will result in greater increases in UAR in individuals with anatomically compromised airways than in others [102]. Individuals with different degrees of UAR also will vary in their compensatory ventilatory requirements for maintaining airway patency [59; 120]. Thus, anatomy is an important determinant of the vulnerability for upper airway collapse in settings where background ventilatory drive fluctuates.

Despite the physiological plausibility of ventilatory deficits in the pathogenesis of OSA, there are relatively little data that quantify the contribution of ventilatory phenotypes to OSA predisposition. The literature is limited by difficulties in interpreting physiological studies in OSA patients, in whom chronic exposure to hypoxemia and sleep disruption may have contributed to any observed deficits in neural control, and the frequent co-morbidity with obesity or obstructive lung disease, which themselves may alter ventilation. Other limitations relate to the heterogeneity of OSA and the limited tools available for selecting relatively homogeneous subgroups for physiological assessment.

8.5.1 Evidence for deficits in ventilatory chemosensitivity in the pathogenesis of OSA

Studies that have attempted to gauge the magnitude of differences in respiratory chemosensitivity between samples of patients with OSA and controls are summarized in Table 1. These studies have generally assessed ventilatory responses to increasing levels of hypoxia and/or hypercapnia during challenges performed during wakefulness. Use of hypoxic challenge testing has attempted to identify the role of peripheral chemoreceptors, whereas studies of progressive hypercapnia have addressed the influences of central drive (with different background levels of hypoxia or hyperoxia used to further define peripheral and central contributions). Such chemosensitivity challenges are difficult to perform during sleep; however, limited research suggests that results obtained during wakefulness can be extrapolated to responses during sleep [53; 100]. Most studies included small and highly selected samples, and had variable control groups. Factors that may have influenced the magnitude and direction of the responses include subject heterogeneity (whether subjects were hypercapnic or eucapnic, the extent of co-morbidities such as chronic obstructive lung disease) and the testing protocol. In general, depressed ventilatory responses to progressive hypercapnia have been observed in hypercapnic OSA patients. However, less consistent findings have been observed for eucapnic patients, and the largest study [105], showed no differences in hyperoxic hypercapnic ventilatory responses in OSA patients compared to controls.

Table 1 : Evidence for variation in chemosensitivity in the pathogenesis of OSA
-: Not reported; OSA: obstructive sleep apnea; C: control; COPD: chronic obstructive pulmonary disease; HTN: hypertensive. HVR (hypoxic ventilartory response) and HCVR (hypercapnic ventilatory response) refer to studies using a variety of rebreathing or steady state exposures to progressive hypoxia or hypercapnia.

Samples	Hypoxic Responses	Hypercapnic Responses	Comments
13 OSA (7 hypercapnic) [31]	↓	↓	Blunted responses only in those with hypercapnia
8 OSA, 8 C [90]	-	No difference	Inspiratory load responses during hypercapnic breathing reduced
8 OSA; 10 C [102]	-	No difference	With decreased inspiratory drive, increased upper airway resistance in OSA
21 OSA [53]			HVR inversely associated with nadir desaturation
40 OSA (15 HTN); 20 C [111]	↓	-	Although HVR lower in OSA vs Cs; HVR higher among HTN OSA vs non-HTN OSA; also decreased ventilation with hyperoxia greatest in HTN-OSA.
49 OSA [123]	↑ (compared to reference values)	-	No association between HVR and AHI
20 obese OSA; 11 OSA/COPD; 13 non obese C [89]	-	↓	Lower HCVR seen only in OSA/COPD ;
27 OSA; 27 C [15]	-	No difference	Higher resting ventilation in OSA
25 Snorers; 52 OSA; 25 C [7]	↑	↑	Highest ventilatory responses in those with more severe OSA
16 OSA; 12 C [71]	↑	No difference	HVR responses positively associated with blood pressure, heart rate and sympathetic nervous responses to hypoxia
16 OSA; 9 C [77]	↓	↓	Degree of depression of HVR associated with nadir of oxygen desaturation with sleep; HCVR and peripheral chemosensivity increased after administration of a dopamine D2 receptor antagonist in OSA group.
104 OSA; 115 sleep referrals-non-OSA [105]	-	No difference	
12 OSA [69]	-	"Normal"	
19 OSA; 13 C [32]	↓	-	HVR varied directly with desaturation and arousal frequency; lower responses in OSA group also observed to hyperoxic challenge
30 OSA; 12 C [60] (AHI<10)	↑*	↑*	OSA group had higher increases in morning vs evening measurements; *Evening and morning measurements for hypoxic hypercapnia challenge testing tended to be higher in OSA

Hypoxic responses in OSA have been even more variable. Lower hypoxic ventilatory responses have been associated with more severe nocturnal desaturation [32; 53; 77], indicating that either: 1) blunted hypoxic sensitivity predisposes to more severe nocturnal desaturation; or 2) blunted ventilatory responses represent the effects of intermittent hypoxemia on causing reduced peripheral chemosensitivity. An intriguing study showed lower responses to both hypoxia and hypercapnia in OSA patients compared to controls, with evidence of augmented hypercapnic responses and improved peripheral chemosensitivity in the OSA subjects after administration of a dopamine D2 receptor antagonist [77]. The improved peripheral chemosensitivity after drug administration was interpreted as consistent with an inhibitory role of dopamine in carotid body neurotransmission, with abnormalities in dopaminergic activity providing a mechanistic basis for lowered hypoxic ventilatory responses in OSA. Other studies, however, suggest that more vigorous hypoxic ventilatory responses occur in OSA, especially in patients with HTN [71; 111]. Several studies also have shown that the magnitude of the ventilatory responses to hypoxia is positively correlated with the severity of OSA [7; 32; 123], suggesting that a brisk response to hypoxia contributes to ventilatory instability and severe OSA. The latter findings appear to be strongest among OSA patients with HTN. Since the strength of pressor and muscle sympathetic nervous system responses to acute hypoxic challenges is correlated with the magnitude of ventilatory responses [41; 71; 123], it is possible that altered chemoreceptor function and its association with OSA differs in OSA patients with and without HTN. Thus, HTN may be an additional phenotype that identifies a subgroup of OSA patients predisposed to both disorders because of enhanced peripheral chemoreceptor activity. Alternatively, HTN, via influences on baroreceptors or other factors, may modify ventilatory responses in OSA subjects. Additionally, higher levels of resting ventilation, unexplained by measured metabolic factors, have been demonstrated in OSA patients compared to controls, suggesting altered ventilatory control in OSA may occur unrelated to hypoxic sensitivity [15].

A deficit in ventilatory control may represent a primary risk factor (genetically determined) or be an acquired phenotype. Several human and animal studies have shown that exposures to recurrent hypoxemia, nocturnal acidosis and/or sleep fragmentation may alter ventilatory control, perhaps through altered peripheral or central neurotransmitter secretion or clearance, and/or effects on short or long term facilitation. Sleep deprivation may reduce hypercapnic and hypoxic ventilatory responsiveness [101; 122]. Animal studies have shown that sleep fragmentation reduced arousal and ventilatory responses to hypercapnia and hypoxia in sleep [86], and in normal humans, resulted in increased snoring and critical pressure levels (P_{crit}), indicative of increased airway collapsibility [103]. Assessment of the extent to which OSA may be a causal risk factor for altered ventilatory control has been addressed by evaluating changes in ventilatory sensitivity before and after sleep [30; 60]. Ventilatory responses to hyperoxic hypercapnic rebreathing were increased in the morning relative to the evening in OSA patients but not in controls [60], suggesting that increased central chemosensitivity occurred in response to overnight OSA stresses in vulnerable individuals.

One approach for determining the role of ventilatory control in OSA is to assess the extent to which ventilatory phenotypes change after OSA treatment. Studies that have assessed differences in ventilatory control in OSA patients before and after surgery or treatment with continuous positive airway pressure (CPAP) are summarized (Table 2). Although these studies have largely shown treatment-associated increased ventilatory responses to challenge tests with progressive hypercapnia, treatment-related changes were most consistent in subjects who were hypercapnic prior to OSA treatment. One study reported elevations in ventilatory responses to progressive hypoxemia in OSA patients studied prior to treatment compared to laboratory normative values [116]. After CPAP treatment, responses to hypoxia significantly declined while responses to hypercapnia increased, with similar trends seen in both eucpanic and hypercapnic subjects. In contrast, Lin reported that baseline ventilatory responses to both hypoxia and hypercapnia increased following CPAP treatment in hypercapnic but not eucapnic patients with OSA [59]. Thus, it seems likely that the recurrent nocturnal stresses imposed by OSA contribute to observed deficits in ventilatory phenotypes, but the extent may vary among OSA subgroups and according to the specific ventilatory phenotype examined.

Table 2: Changes in hypoxic (HVR) and hypercapnic (HCVR) ventilatory responses in OSA patients following OSA treatment

-: Not reported; OSA: obstructive sleep apnea; HVR and HCVR refer to studies using a variety of re-breathing or steady state exposures to progressive hypoxia or hypercapnia.

Sample	Intervention	HVR	HCVR	Comments
1 OSA; hyper-capnic [11]	Surgery	-	↑	
5 OSA [36]	3 months post-surgery	-	↑	
19 OSA [16]	CPAP	-	↑	Changes only in hypercapnic patients, not eucapnic
6 hypercapnic OSA; 24 eucapnic OSA [59]	2-4 weeks CPAP	↑	↑	Reduced responses in hyper-capnic group that improved with CPAP; no changes in eucapnic
28 OSA [116]	2 weeks-6 months CPAP	↓	↑	Changes in eucapnic and hy-percapnic patients
12 OSA [69]	3 nights CPAP	-	↑	

The inadequacy of approaches for evaluating variations of ventilation us-ing "steady-state" systems has prompted the development of a several elegant techniques for measuring dynamic responses believed to be related to ventilatory stability and propensity for respiratory oscillation [44; 121; 129]. A variety of data indicate the sensitivity of humans to relatively minor perturbations in gas tensions, which may precipitate oscillatory behavior, or periodic breathing. Exposure to hy-poxia, as with high altitude sojourns, precipitates periodic breathing in some but not all individuals. The relevance of ventilatory instability as a phenotype for OSA is supported by the observation that OSA patients may manifest highly oscillatory ventilatory rhythms, or periodic breathing, even while breathing through a by-passed unobstructed upper airway [75]. In some OSA patients addition of CO_2 to a breathing circuit stabilizes breathing pattern, and eliminates central and/or obstructive apneas [114], effects that presumably reflect dampening of chemical drive, reducing large swings in ventilation. Experimentally induced periodic breathing may lead to upper airway obstruction during the nadir of ventilatory output, with UAR increasing most in snorers [44; 75].

Measurement of loop gain (LG), the propensity of the respiratory system to develop an unstable or oscillatory behavior, has been gauged using mathematical formulae derived from engineering models that incorporate a series of negative feedback loops relating the gains of the plant (i.e., lungs) and controller (neuromuscular output), with time constants among the feedback loops (e.g., influenced by circulatory times). In these models, delayed information transfer (e.g., because of prolonged circulatory times), increased controller gain (e.g., ventilatory chemosensitivity), and plant gain (the lung) may dynamically interact and result in oscillatory patterns manifest as ventilatory under- and overshooting. Two general approaches have been used to quantify such dynamic behavior. In one approach, ventilatory output is experimentally amplified [121; 129], and in the other, pseudorandomly spaced exposures to brief pulses of hyperoxia or hypercapnia are presented, with measurement of immediate changes in ventilation [44]. These studies have shown increased LG in both obese and non-obese patients with OSA, and also have shown that increased LG is associated with severity of OSA [120; 129]. The strongest associations between LG and severity of OSA have been observed in individuals with pharyngeal closing pressures close to atmospheric pressure, representing a level of intermediate anatomic risk [120]. The relative complexity of these measurement strategies, however, may prevent their widespread use for general phenotyping of large populations at risk for OSA. Ibrahim et al. recently suggested that a simpler measure of ventilatory stability, a measurement of breathing variability at sleep onset obtained using routine polysomnography, may provide a feasible means for characterizing a dynamic ventilatory phenotype [45]. Their preliminary data suggest that increased ventilatory variability in the two minutes following sleep onset (when central inspiratory drive may fluctuate) is moderately correlated with the AHI. Additional work is needed to further validate this and other phenotypic measures of ventilation.

8.5.2 Evidence for deficits in load compensation in the pathogenesis of OSA

Abnormalities in load compensation, i.e., evidence of the generation of compensatory increased inspiratory effort in response to increased UAR, may explain the observed variability in the severity of OSA (or propensity to upper collapse) among individuals with comparable degrees of anatomic compromise. Several studies have measured the awake ventilatory responses to external inspiratory loading, comparing patients with OSA to controls [35; 66; 90; 115]. These have uniformly shown depressed load responses in OSA patients, with one study showing reversal following CPAP therapy [35]. It has been suggested that impaired load responses are related to reduced inspiratory effort sensation [115]. The latter, which also appears reversible following CPAP therapy, may indicate abnormalities in proprioception in chronically traumatized upper airway mucosal tissues, or to changes in cognition or neurotransmitter levels that may be caused by OSA-related hypoxemia or sleep fragmentation.

8.5.3 Inferences from pediatric studies

Pediatric OSA is most frequently recognized for its association with adeno-tonsillar hypertrophy. Uncontrolled studies have showed that OSA improves in approximately 70-80% of children after adeno-tonsillectomy [108; 110]. However, the persistence of OSA in 20 to 30% of patients who have undergone surgery [68; 104; 108; 110] suggests that there is a sizable subgroup of children with additional risk factors for OSA. This is supported by evidence that the association between OSA severity and the size of the tonsils and adenoids is only moderate, with considerable overlap in adenotonsillar size between children with OSA and asymptomatic children [8; 9]. In addition to the potential impact of other anatomic factors (e.g., soft palate enlargement), abnormalities or variations in upper airway neuromotor control may also contribute to individual susceptibility to OSA and responsiveness to surgery.

8.6 Evidence for genetically determined ventilatory control abnormalities in familial OSA

Several studies have performed various ventilatory challenges in family members of patients with OSA with the aim of determining whether deficits in ventilatory responses aggregate within families, and whether such deficits predict OSA severity. El Bayadi et al. reported the results of anatomic and physiological studies performed in 10 members from three generations of a family with an affected proband with OSA [26]. OSA was documented in 9 of the 10 family members, all of whom had a body mass index of <30. Blunted responses to ventilatory challenge tests to progressive hypoxia were demonstrated in all five subjects who underwent ventilatory challenge testing. Mild baseline hypercapnia was also present in these five subjects, three of whom also had blunted responses to hyperoxic hypercapnia rebreathing. Cephalometry also showed variable degrees of upper airway anatomic compromise. These findings supported a family basis for OSA, with evidence that inherited abnormalities in anatomic and physiological risk factors both contributed to the disease severity.

Significantly lower ventilatory responses to hypoxic challenge testing were demonstrated in a study that compared ventilatory control responses in 31 subjects from 12 families with two or more members with OSA, compared to responses in 9 age and gender matched controls [91]. The selection of subjects from families showing familial aggregation for OSA may have improved the ability to detect potentially inherited abnormalities in an intermediate phenotype. In addition, by studying relatively healthy relatives (including 12 without OSA), this study minimized the influence of confounding by other co-morbidities and limited the influence of any effect of recurrent sleep related stresses directly influencing the ventilatory outcomes.

Differences in responses to inspiratory resistive loading during sleep were examined in 10 apparently healthy adult offspring of OSA probands and in 14 control offspring of healthy parents [88]. Both groups had similar load responses during wakefulness, but during sleep with inspiratory loading, the offspring of OSA probands breathed at a lower tidal volume than controls, with development of hypopneas at lower levels of UAR, and were less likely to show EEG arousal with hypopneas than the controls. Similarly, nasal occlusion during sleep was demonstrated to precipitate more hypopneas in the healthy relatives of OSA probands than controls [56]. However, these studies could not differentiate the extent to which differences in ventilatory patterns between the groups was due to differences in collapsibility versus ventilatory drive.

Differences in chemoresponses to progressive hypercapnia and to hypoxia has been evaluated in the healthy relatives of hypercapnic OSA patients compared to eucapnic OSA patients [46]. Although the chemoresponses among relatives were correlated, the relatives of patients with and without hypercapnia did not differ. Thus, this study did not provide evidence for genetically determined differences in chemoresponsiveness influencing the propensity for hypercapnia in OSA.

Several studies have reported a co-aggregation of OSA with Sudden Infant Death Syndrome (SIDS) or Acute Life Threatening Events [33; 37; 63; 65; 115]. Members from families with both OSA and SIDS cases have been reported to have anatomic features, such as brachycephaly leading to upper airway narrowing, as well as reduced hypoxic ventilatory responsiveness [63; 116]. These observations suggest that the two disorders may have a common genetic predisposition acting via ventilatory control and/or craniofacial structure pathways. The recent description of widespread serotonergic brainstem abnormalities in SIDS victims [84], and the putative role of this pathway in respiratory drive, suggest an interesting biological basis for this potential genetic link (see Chapter 7).

8.7 Candidate genes for OSA that may operate through ventilatory control

Specific genetic variants that influence OSA susceptibility through effects on ventilatory control have not yet been identified. Nonetheless, animal studies and several human association studies suggest several candidate genes that may affect OSA susceptibility through effects on ventilatory drive. Genes of interest are those that influence pathways important for peripheral and central chemoreception, as well as those that are involved in autonomic regulation. The following descriptions are a brief list of potential candidates for which there are some data from human studies implicating their role in OSA; however, given the complexity of the neurobiological systems that control respiration and upper airway motor function, it is likely that there are important candidates not yet studied in relationship to OSA.

8.7.1 Serotonergic pathways

Serotonergic (5-hyroxytryptophan; 5HT) receptors are widespread, found in the carotid body and hypoglossal neurons, as well as in the brainstem near ventilatory control centers important for chemoreception. Animal work suggests that serotonergic neurotransmission, through peripheral actions at the level of the carotid body or hypoglossal nerve, or centrally, at medullary respiratory control centers, influences a wide range of functions relevant to OSA, including upper airway reflexes, ventilation, and arousal, as well as sleep-wake cycling through thalamic innervation. Although the pharmacology is complex [119], there is growing evidence implicating the importance of this pathway in the pathogenesis of SIDS (see Chapter 7), which, as discussed above, may share common genetically determined risk factors with OSA. Changes in serotonergic neuronal activity during sleep provide biological plausibility for the potential importance of this neurotransmitter in the pathogenesis of OSA. Polymorphisms in three genes-SLC6A4 (encoding a serotonin transporter protein which clears serotonin from the synaptic space), HTR2A (encoding the serotonin 2A receptor), and HTR2C (encoding the serotonin 2C receptor) each have been studied in relationship to OSA. An insertion/deletion polymorphism in the promoter region of SLC6A4, which is associated with variations in serotonergic reuptake activity, was weakly associated with OSA among men but not women in a small Turkish study, but this association was not confirmed in a Chinese study [128; 131]. The intronic polymorphism in SLC6A4, which has unclear functional significance, was weakly associated with OSA in both studies. A polymorphism for HTR2A was also reported to be associated with OSA among males, but not females, in a Turkish cohort [13], but this association was not replicated in a larger Japanese study [98]. That study also did not identify a role for a HTR2C polymorphism in OSA [98].

8.7.2 Apolipoprotein E (APOE)

An allele of the apolipoprotein E gene (ε4) which has been previously associated with increased risk for both cardiovascular disease and Alzheimer Disease [76], was reported to be associated with OSA in two cohort studies of predominantly Caucasians [34; 49]. This finding was not replicated in two other studies, including a study of an older Japanese American cohort [29]. The Cleveland Family Study reported preliminary evidence for linkage to AHI near the APOE locus on chromosome 19, with stronger evidence for linkage after inclusion of fine mapping markers [55]. However, the APOE genotype did not explain the linkage findings and was not associated with OSA status. These findings suggested that the susceptibility locus for OSA is not APOE but another locus close to it.

8.7.3 Angiotensin Converting Enzyme (ACE)

Angiotensin II, an important vasoconstrictor, also appears to modulate afferent activity from the carotid body chemoreceptor, and thus, may be a determinant of ventilatory drive [2]. Angiotensin II levels are regulated by the actions of ACE, which is encoded by the *ACE* gene. Experimental and observational studies indicate that an insertion polymorphism in this gene, which produces reduced serum and tissue ACE activity, is more common in individuals with greater endurance and better adaptation to high altitude, perhaps related to altered greater ventilatory response to exertional hypoxia [81; 124; 125]. The *ACE* insertion/deletion polymorphism also has been associated with muscle fiber composition [132]. Several studies of Chinese cohorts have reported an association between this polymorphism and OSA, including a study showing an association between apnea duration and *ACE* genotype [126; 127; 133]. However, more recent studies have been unable to confirm this finding [12; 58; 82; 134].

8.7.4 Leptin signalling

Animal studies suggest that leptin, an adipose-derived circulating hormone known to influence appetite regulation and energy expenditure, not only is important in the regulation of body weight, but also has important effects on ventilatory drive. Mice homozygous for a knockout mutation in leptin hypoventilate and have a blunted ventilatory response to hypercapnia [112]. Further, leptin replacement improves the ventilatory responses to hypercapnia in both wakefulness and sleep in leptin deficient mice [73]. Leptin's stimulatory effects on hypercapnic ventilatory response appear to be mediated through melanocortin, which is produced from a precursor polyprotein, POMC. As described earlier, the Cleveland Family Study reported suggestive evidence for linkage to an area on chromosome 2p that includes the POMC locus, an area also reported by others to be strongly linked to serum leptin levels [79]. Thus, hypothalamic and pituitary pathways involved in leptin signaling may be important in influencing ventilatory control as well as obesity phenotypes relevant to OSA.

8.8 Inferences from other conditions which involve ventilatory control deficits

The hereditary aspects of ventilatory control and its relevance to congenital central hypoventilation syndrome, Prader Willi syndrome, Rett syndrome, and sudden infant death syndrome, are detailed in other chapters (see Chapters 4, 7, 15, 16). Genetic variants for these conditions may also be relevant to OSA because of overlapping genetic mechanisms that involve ventilatory control and upper airway collapsibility. Thus, genes important in neural crest development or migration, such as *PHOX2B, EDN3, RET, GDNF, ASCL1* and *BDNF* [4; 17; 99]; genes suspected to be important in Prader Willi syndrome (*NDN* and *SNRPN*)

[135]; and those implicated in Rett (*MECP2*) [5], may be relevant to OSA. Other clues may be gleaned from genetic studies of Treacher Collins Syndrome (e.g., *TCOF1*) [24; 117], Charcot-Marie-Tooth (*CMT1A*; *PMP22*) [118], as well as Pierre Robin Syndrome, i.e., disorders where abnormalities of ventilatory control and/or pharyngeal development occur as clearly defined, and often dramatic phenotypes [107].

8.9 Studies from animals

The growing body of animal work that implicates specific genes and physiological pathways in ventilatory control is reviewed in other chapters (see Chapters 11-16). In addition to the genes discussed earlier, those that may be most relevant to OSA because of putative effects on chemoreception include genes in pathways that regulate: orexins (via effects on arousal and muscle tone); endothelins; nitric oxide synthesis, and those involved with carotid body cellular activity.

8.10 Pleiotropy

Pleiotropy refers to the actions of genes that affect multiple intermediate phenotypes (such as ventilatory control and/or craniofacial morphology and/or obesity). Genes affecting more than one pathway may result in more severe OSA than genes that influence airway patency through only a single mechanism. Studies of individuals with multiple OSA risk factors, although potentially introducing heterogeneity and requiring complex analytical approaches, may nonetheless help identify genes that result in multiple "hits" leading to OSA. Systems that may operate in this fashion include the orexin system, which regulates both the sleep/wake state and appetite [67], serotonergic systems, that influence hypothalamic satiety centers, ventilatory control, mood, and upper airway dilator muscle tone [39; 57]; leptin pathways, influencing obesity and ventilation; and endothelins, influencing both craniofacial structure and ventilation.

8.11 Conclusion

Despite the challenges in studying a complex trait, there is growing evidence that genetic factors influence the expression of OSA. Although such genetic factors likely increase OSA susceptibility through multiple pathways, including effects on body weight and fat distribution and craniofacial morphology, it is clear that the latter phenotypes do not explain the population variability in OSA. Physiological studies suggest that additional susceptibility is likely related to aspects of ventilatory control that influence airway patency through effects on the innervation of the upper airway muscles during sleep and with inspiratory loading, and/or through effects on ventilatory stability. Evaluating the role of the latter factors in the pathogenesis, as well as the genetic etiology of OSA, however, has been limited by the paucity of specific and sensitive techniques for measuring these properties in large numbers of individuals. Further work is needed to develop tools to permit reliable ventilatory phenotyping in large populations. Clues into mechanistic pathways and candidate gene systems are available from animal studies and family studies of individuals with profound clinical syndromes that affect ventilation. Promising areas of research include further assessment of the role of genetic variants in systems that influence serotonergic and leptin pathways, and in genetic systems that influence the development of ventilatory brainstem reflexes and the carotid body. Study of genetic systems that influence more than one phenotype implicated in the pathogenesis of OSA through pleiotropic pathways may provide insights into the interacting pathophysiologies leading OSA. Identification of such systems holds promise for better understanding the molecular mechanisms underlying OSA susceptibility that could be targets for pharmacological manipulation.

Acknowledgements

Supported in part by NIH HL 46380 and HL 81385

References

1.   Positional cloning of a gene involved in the pathogenesis of Treacher Collins syndrome. The Treacher Collins Syndrome Collaborative Group. Nat Genet 12: 130-136
2.   Allen AM (1998) Angiotensin AT1 receptor-mediated excitation of rat carotid body chemoreceptor afferent activity. J Physiol 510: 773-781
3.   Altmuller J, Palmer LJ, Fischer G, Scherb H, Wjst M (2001) Genomewide scans of complex human diseases: true linkage is hard to find. Am J Hum Genet 69: 936-950
4.   Amiel J, Salomon R, Attie T, Pelet A, Trang H, Mokhtari M, Laudier B, de Pontual L, Gener B, Trochet D, Etchevers H, Rau P, Simonneau M, Vekemans M, Munnich A, Gaultier C, Lyonnet S (1998) Mutations of the RET-GDNF signaling pathway in Ondine's curse. Am J Hum Genet 62: 715-717
5.   Amir RE, Van den Veyver IB, Wan M, Tran CQ, Francke U, Zoghbi HY (1999) Rett syndrome is caused by mutations in X-linked MECP2, encoding methyl-CpG-binding protein 2. Nat Genet 23: 185-188
6.   Ancoli-Israel S, Kripke DR, Klauber MR, Mason WJ, Fell R, Kaplan O (1991) Sleep-disordered breathing in community dwelling elderly. Sleep 14: 486-495
7.   Appelberg J, Sundstrom G (1997) Ventilatory response to CO_2 in patients with snoring, obstructive hypopnoea and obstructive apnoea. Clin Physiol 17: 497-507

8.   Arens R, McDonough JM, Corbin AM, Rubin NK, Carroll ME, Pack AI, Liu J, Udupa KJ (2003) Upper airway size analysis by magnetic resonance imaging of children with obstructive sleep apnea syndrome. Am J Respir Crit Care Med 167: 65-70

9.   Arens R, McDonough JM, Costarino AT, Mahboubi S, Tayag-Kier CE, Maislin G, Schwab RJ, Pack AI (2001) Magnetic resonance imaging of the upper airway structure of children with obstructive sleep apnea syndrome. Am J Respir Crit Care Med 164: 698-703

10.   Arkinstall WW, Nirmel K, Klissouras V, Milic-Emili J (1974) Genetic differences in the ventilatory response to inhaled CO_2. J Appl Physiol 36: 6-11

11.   Aubert-Tulkens G, Willems B, Veriter C, Coche E, Stanescu DC (1980) Increase in ventilatory response to CO_2 following tracheostomy in obstructive sleep apnea. Bull Europ Physiopath Resp 16: 587-593

12.   Barcelo A, Elorza MA, Barbe F, Santos C, Mayoralas LR, Agusti AG (2001) Angiotensin converting enzyme in patients with sleep apnoea syndrome: plasma activity and gene polymorphisms. Eur Respir J 17: 728-732

13.   Bayazit YA, Yilmaz M, Ciftci T, Erdal E, Kokturk O, Gokdogan T, Kemaloglu YK, Inal E (2006) Association of the -1438G/A polymorphism of the 5-HT2A receptor gene with obstructive sleep apnea syndrome. ORL J Otorhinolaryngol Relat Spec 68: 123-128

14.   Beall C, Strohl K, Blangero J, Williams-Blangero S, Brittenham G, Goldstein M (1997) Quantitative genetic analysis of arterial oxygen saturation in Tibetan highlanders. Hum Biol 69: 597-604

15.   Benlloch E, Cordero P, Morales P, Soler JJ, Macian V (1995) Ventilatory pattern at rest and response to hypercapnic stimulation in patients with obstructive sleep apnea syndrome. Respiration 62: 4-9

16.   Berthon-Jones M, Sullivan CE (1987) Time course of change in ventilatory response to CO_2 with long-term CPAP therapy for obstructive sleep apnea. Am Rev Respir Dis 135: 144-147

17.   Bolk S, Angrist M, Schwartz S, Silvestri J, Weese-Mayer D, Chakravarti A (1996) Congenital central hypoventilation syndrome mutation analysis of the receptor tyrosine kinase RET. Am J Med Genet 63: 603-609

18.   Buxbaum SG, Elston RC, Tishler PV, Redline S (2002) Genetics of the apnea hypopnea index in Caucasians and African Americans: I. Segregation analysis. Genet Epidemiol 22: 243-253

19.   Carmelli D, Bliwise DL, Swan GE, Reed T (2001) Genetic factors in self-reported snoring and excessive daytime sleepiness: a twin study. Am J Respir Crit Care Med 164: 949-952

20.   Carmelli D, Colrain IM, Swan GE, Bliwise DL (2004) Genetic and environmental influences in sleep disordered breathing in older male twins. Sleep 27: 917-922

21.   Cherniack NS (1981) Respiratory dysrhythmias during sleep. N Engl J Med 305: 325-330

22.   Collins DD, Scoggin CH, Zwillich CW, Weil JV (1978) Hereditary aspects of decreased hypoxic response. J Clin Invest 62: 105-110

23.   Dempsey JA, Skatrud JB (1986) A sleep-induced apneic threshold and its consequences. Am Rev Respir Dis 133: 1163-1170

24.   Dixon J, Hovanes K, Shiang R, Dixon MJ (1997) Sequence analysis, identification of evolutionary conserved motifs and expression analysis of murine tcof1 provide further evidence for a potential function for the gene and its human homologue, TCOF1. Hum Mol Genet 6: 727-737

25. Douglas NJ, Luke M, Mathur R (1993) Is the sleep apnoea/hypopnoea syndrome inherited? Thorax 48: 719-721

26. El Bayadi S, Millman RP, Tishler PV, Rosenberg C, Saliski W, Boucher MA, Redline S (1990) A family study of sleep apnea. Anatomic and physiologic interactions. Chest 98: 554-559

27. Ferini-Strambi L, Calori G, Oldani A, Della Marca G, Zucconi M, Castronovo V, Gallus G, Smirne S (1995) Snoring in twins. Respir Med 89: 337-340

28. Fletcher EC, Miller J, Schaaf JW, Fletcher JG (1987) Urinary catecholamines before and after tracheostomy in patients with obstructive sleep apnea and hypertension. Sleep 10: 35-44

29. Foley DJ, Masaki K, White L, Redline S (2001) Relationship between apolipoprotein E epsilon4 and sleep-disordered breathing at different ages. JAMA 286: 1447-1448

30. Fuse K, Satoh M, Yokota T, Ohdaira T, Muramatsu Y, Suzuki E, Arakawa M (1999) Regulation of ventilation before and after sleep in patients with obstructive sleep apnoea. Respirology 4: 125-130

31. Garay SM, Rapoport D, Sorkin B, Epstein H, Feinberg I, Goldring RM (1981) Regulation of ventilation in the obstructive sleep apnea syndrome. Am Rev Respir Dis 124: 451-457

32. Garcia-Rio F, Pino JM, Ramirez T, Alvaro D, Alonso A, Villasante C, Villamar J (2002) Inspiratory neural drive response to hypoxia adequately estimates peripheral chemosensitivity in OSAHS patients. Eur Respir J 20: 724-732

33. Gislason T, Johannsson JH, Haraldsson A, Olafsdottir BR, Jonsdottir H, Kong A, Frigge HL, Jonsdottir GM, Hakonarson H, Guleher J, Stefansson K (2002) Familial predisposition and cosegregation analysis of adult obstructive sleep apnea and the sudden infant death syndrome. Am J Respir Crit Care Med 166: 833-838

34. Gottlieb DJ, DeStefano AI, Foley DJ, Mignot E, Redline S, Givelber RJ, Young T (2004) APOE e4 is associated with obstructive sleep apnea/hypopnea: The Sleep Heart Health Study. Neurology 63: 664-668

35. Greenberg H, Scharf S (1993) Depressed ventilatory load compensation in sleep apnea: Reversal by nasal CPAP. Am Rev Respir Dis 148: 1610-1615

36. Guilleminault C, Cummiskey J (1982) Progressive improvement of apnea index and ventilatory response to CO_2 after tracheostomy in obstructive sleep apnea syndrome. Am Rev Respir Dis 126: 14-20

37. Guilleminault C, Heldt G, Powell N, Riley R (1986) Small upper airway in near-miss sudden infant death syndrome infants and their families. Lancet 1: 402-407

38. Guilleminault C, Partinen M, Hollman K, Powell N, Stoohs R (1995) Familial aggregates in obstructive sleep apnea syndrome. Chest 107: 1545-1551

39. Haxhiu MA, Mack SO, Wilson CG, Feng P, Strohl KP (2003) Sleep networks and the anatomic and physiologic connections with respiratory control. Front Biosci 8: 946-962

40. Heath AC, Kendler KS, Eaves LJ, Martin N (1990) Evidence for genetic influences on sleep disturbance and sleep pattern in twins. Sleep 13: 318-335

41. Hedner JA, Wilcox I, Laks L, Grunstein RR, Sullivan CE (1992) A specific and potent pressor effect of hypoxia in patients with sleep apnea. Am Rev Respir Dis 146: 1240-1245

42. Holberg CJ, Natrajan S, Cline MG, Quan SF (2000) Familial aggregation and segregation analysis of snoring and symptoms of obstructive sleep apnea. Sleep Breath 4: 21-30

43. Horner RL, Kozar LF, Kimoff RJ, Phillipson EA (1994) Effects of sleep on the tonic drive to respiratory muscle and the threshold for rhythm generation in the dog. J Physiol 474: 525-537

44.   Hudgel DW, Gordon A, Thanakitcharu S, Bruce Eugene N (1998) Instability of ventilatory control in patients with obstructive sleep apnea. Am J Respir Crit Care Med 158: 1142-1149

45.   Ibrahim L, Modarres-Zadeh M, Johnson NL, Mehra R, Redline S (2006) Novel measure of breathing variability at wake-sleep transition predicts sleep-disordered breathing severity. Am J Respir Crit Care Med: A567

46.   Javaheri S, Colangelo G, Corser B, Zahedpour MR (1992) Familial respiratory chemosensitivity does not predict hypercapnia of patients with sleep apnea-hypopnea syndrome. Am Rev Respir Dis 145: 837-840

47.   Jennum P, Hein HO, Suadicani P, Sorensen H, Gyntelberg F (1995) Snoring, family history, and genetic markers in men: the Copenhagen Male Study. Chest 107: 1289-1293

48.   Jones CR, Campbell SS, Zone SE, Cooper F, DeSano A, Murphy PJ, Jones B, Czajkowski L, Ptacek LJ (1999) Familial advanced sleep-phase syndrome: a short-period circadian rhythm variant in humans. Nat Med 5: 1062-1065

49.   Kadotani H, Kadotani T, Young T, Peppard PE, Finn L, Colrain IM, Murphy JM, Mignot E (2001) Association between Apolipoprotein E e4 and sleep-disordered breathing in adults. JAMA 285: 2888-2890

50.   Kaprio J, Koskenvuo M, Partinen M, Telakivi I (1988) A twin study of snoring. Sleep Res 17: 365

51.   Khoo MC, Berry RB (1996) Modeling the interaction between arousal and chemical drive in sleep-disordered breathing. Sleep 19: S167-169

52.   Kobayashi S, Nishimura M, Yamamoto M, Akiyama Y, Kishi F, Kawakami Y (1993) Dyspnea sensation and chemical control of breathing in adult twins. Am Rev Respir Dis 147: 1192-1198

53.   Kunitomo F, Kimura H, Tatsumi K, Okita S, Tojima H, Kuriyama T, Honda Y (1989) Abnormal breathing during sleep and chemical control of breathing during wakefulness in patients with sleep apnea syndrome. Am Rev Respir Dis 139: 164-169

54.   Larkin EK, Patel SR, Elston RC, Redline S (2006a) Linkage analyses of sleep disordered breathing using alternative measures of the Apnea Hypopnea Index (AHI). Am J Respir Crit Care Med: A568

55.   Larkin EK, Patel SR, Redline S, Mignot E, Elston RC, Hallmayer J (2006b) Apolipoprotein E and obstructive sleep apnea: evaluating whether a candidate gene explains a linkage peak. Genet Epidemiol 30: 101-110

56.   Lavie P, Rubin AE (1984) Effects of nasal occlusion on respiration in sleep: evidence of inheritability of sleep apnea prognosis. Acta Otolaryngol 97: 127-130

57.   Leibowitz SF, Hoebel BG (1998) Behavioral neuroscience of obesity. In: Bray GA, Bouchard C, James WPT editors. Handbook of Obesity. New York: Marcel Dekker, Inc., pp 313-358

58.   Li Y, Zhang W, Wang T, Lu H, Wang X, Wang Y (2004) Study on the polymorphism of angiotensin converting enzyme genes and serum angiotensin II level in patients with obstructive sleep apnea hypopnea syndrome accompanied hypertension. Lin Chuang Er Bi Yan Hou Ke Za Zhi 18: 456-459

59.   Lin CC (1994) Effect of nasal CPAP on ventilatory drive in normocapnic and hypercapnic patients with obstructive sleep apnoea syndrome. Eur Respir J 7: 2005-2010

60.   Mahamed S, Hanly PJ, Gabor J, Beecroft J, Duffin J (2005) Overnight changes of chemoreflex control in obstructive sleep apnoea patients. Respir Physiol Neurobiol 146: 279-290

61.   Marin JM, Carrizo SJ, Vicente E, Agusti AG (2005) Long-term cardiovascular outcomes in men with obstructive sleep apnoea-hypopnoea with or without treatment with continuous positive airway pressure: an observational study. Lancet 365: 1046-1053

62.   Marti S, Sampol G, Munoz X, Torres F, Roca A, Lloberes P, Sagales T, Quesada P, Morell F (2002) Mortality in severe sleep apnoea/hypopnoea syndrome patients: impact of treatment. Eur Respir J 20: 1511-1518

63.   Mathur R, Douglas NJ (1994) Relation between sudden infant death syndrome and adult-sleep apnoea/hypopnoea syndrome. Lancet 344: 819-820

64.   Mathur R, Douglas NJ (1995) Family studies in patients with the sleep apnea-hypopnea syndrome. Ann Intern Med 122: 174-178

65.   McNamara F, Sullivan CE (2000) Obstructive sleep apnea in infants: relation to family history of sudden infant death syndrome, apparent life-threatening events, and obstructive sleep apnea. J Pediatr 136: 318-323

66.   McNicholas W, Bowes G, Zamel N, Phillipson E (1984) Impaired detection of added inspiratory resistance in patients with obstructive sleep apnea. Am Rev Respir Dis 129: 45-48

67.   Mignot E (2001) A commentary on the neurobiology of the hypocretin/orexin system. Neuropsychopharmacol 25: S5-13

68.   Mitchell RB, Kelly J (2004) Adenotonsillectomy for obstructive sleep apnea in obese children. Otolaryngol Head Neck Surg 131: 104-108

69.   Moura SM, Bittencourt LR, Bagnato MC, Lucas SR, Tufik S, Nery LE (2001) Acute effect of nasal continuous positive airway pressure on the ventilatory control of patients with obstructive sleep apnea. Respiration 68: 240-241

70.   Narkiewicz K, Somers VK (1997) The sympathetic nervous system and obstructive sleep apnea: implications for hypertension. J Hypertens 15: 1613-1619

71.   Narkiewicz K, van de Borne PJ, Pesek CA, Dyken ME, Montano N, Somers VK (1999) Selective potentiation of peripheral chemoreflex sensitivity in obstructive sleep apnea. Circulation 99: 1183-1189

72.   Nieto FJ, Young TB, Lind BK, Shahar E, Samet JM, Redline S, et al. (2000) Association of sleep-disordered breathing, sleep apnea, and hypertension in a large community-based study. Sleep Heart Health Study. JAMA 283: 1829-1836

73.   O'Donnell CP, Schaub CD, Haines AS, Berkowitz DE, Tankersley CG, Schwartz AR, Smith PL, Robotham JL, O'Donnell CP (1999) Leptin prevents respiratory depression in obesity. Am J Respir Crit Care Med 159: 1477-1484

74.   Onal E, Burrows DL, Hart RH, Lopata M (1986) Induction of periodic breathing during sleep causes upper airway obstruction in humans. J Appl Physiol 61: 1438-1443

75.   Onal E, Lopata M (1982) Periodic breathing and the pathogenesis of occlusive sleep apneas. Am Rev Respir Dis 126: 676-680

76.   Ordovas JM, Schaefer EJ (2000) Genetic determinants of plasma lipid response to dietary intervention: the role of the APOA1/C3/A4 gene cluster and the APOE gene. Br J Nutr 83: S127-136

77.   Osanai S, Akiba Y, Fujiuchi S, Nakano H, Matsumoto H, Ohsaki Y, Kikuchi K (1999) Depression of peripheral chemosensitivity by a dopaminergic mechanism in patients with obstructive sleep apnoea syndrome. Eur Respir J 13: 418-423

78.   Palmer LJ, Buxbaum SG, Larkin E, Patel SR, Elston RC, Tishler PV, Redline S (2003) A whole-genome scan for obstructive sleep apnea and obesity. Am J Hum Genet 72: 340-350

79.   Palmer LJ, Buxbaum SG, Larkin EK, Elston RC, Tishler PV, Redline S (2002) A genome-wide search for quantitative trait loci underlying obstructive sleep apnea. Am J Respir Crit Care Med 165: A419

80.   Palmer LJ, Buxbaum SG, Larkin EK, Patel SR, Elston RC, Tishler PV, Redline S (2004) Whole genome scan for obstructive sleep apnea and obesity in African-American families. Am J Respir Crit Care Med 169: 1314-1321

81.   Patel S, Woods DR, Macleod NJ, Brown A, Patel KR, Montgomery HE, Peacock AJ (2003) Angiotensin-converting enzyme genotype and the ventilatory response to exertional hypoxia. Eur Respir J 22:755-760

82.   Patel SR, Larkin EK, Mignot E, Lin L (2007) The association of Angiotensin Converting Enzyme (ACE) polymorphisms with sleep apnea and hypertension sleep. Sleep 30: 531-533

83.   Patel SR, Tishler PV (2007) Familial and genetic factors. In: Kushida CA editor. Obstructive Sleep Apnea: Pathophysiology, Comorbidities, and Consequences New York: Informa Healthcare, in press

84.   Paterson DS, Trachtenberg FL, Thompson EG, Belliveau RA, Beggs AH, Darnall R, Chadwick AE, Krous HF, Kinney HC (2006) Multiple serotonergic brainstem abnormalities in Sudden Infant Death Syndrome. JAMA 296: 2124-2132

85.   Peppard PE, Young T, Palta M, Skatrud J (2000) Prospective study of the association between sleep-disordered breathing and hypertension. N Engl J Med 342: 1378-1384

86.   Phillipson EA, Bowes G, Sullivan CE, Woolf GM (1980) The influence of sleep fragmentation on arousal and ventilatory responses to respiratory stimuli. Sleep 3: 281-288

87.   Pillar G, Lavie P (1995) Assessment of the role of inheritance in sleep apnea syndrome. Am J Respir Crit Care Med 151: 688-691

88.   Pillar G, Schnall RP, Peled N, Oliven A, Lavie P (1997) Impaired respiratory response to resistive loading during sleep in healthy offspring of patients with obstructive sleep apnea. Am J Respir Crit Care Med 155: 1602-1608

89.   Radwan L, Maszczyk Z, Koziorowski A, Koziej M, Cieslicki J, Sliwinski P, Zielinski J (1995) Control of breathing in obstructive sleep apnoea and in patients with the overlap syndrome. Eur Respir J 8: 542-545

90.   Rajagopal K, Abbrecht P, Tellis C (1984) Control of breathing in obstructive sleep apnea. Chest 85: 174-180

91.   Redline S, Leitner J, Arnold J, Tishler PV, Altose MD (1997) Ventilatory-control abnormalities in familial sleep apnea. Am J Respir Crit Care Med 156: 155-160

92.   Redline S, Palmer LJ, Elston RC (2004) Genetics of obstructive sleep apnea and related phenotypes. Am J Respir Cell Biol 3: S35-39

93.   Redline S, Tishler PV, Strohl KP (2002) The genetics of the obstructive sleep apnea hypopnea syndrome. In: Pack AI ed. Sleep Apnea: Pathogenesis, Diagnosis, and Treatment. New York: Marcel Dekker, Inc., pp 235-264

94.   Redline S, Tishler PV, Tosteson TD, Williamson J, Kump K, Browner I, Ferrette V, Krejci P (1995) The familial aggregation of obstructive sleep apnea. Am J Respir Crit Care Med 151: 682-687.

95.   Redline S, Tosteson T, Tishler PV, Carskadon MA, Millman RP (1992a) Studies in the genetics of obstructive sleep apnea. Familial aggregation of symptoms associated with sleep-related breathing disturbances. Am Rev Respir Dis 145: 440-444

96.   Redline S, Tosteson T, Tishler PV, Carskadon MA, Millman RP (1992b) Studies in the genetics of obstructive sleep apnea. Familial aggregation of symptoms associated with sleep-related breathing disturbances. Am Rev Respir Dis 145: 440-444

97.   Rosen CL, Larkin EK, Kirchner HL, Emancipator JL, Bivins SF, Surovec SA, Martin RJ, Redline S (2003) Prevalence and risk factors for sleep-disordered breathing in 8- to 11-year-old children: association with race and prematurity. J Pediatr 142: 383-389

98.   Sakai K, Takada T, Nakayama H, Kubota Y, Nakamata M, Satoh M, Otaki S, Kuizumi T, Kusuda S, Ogawa Y, Tuchiya K, Yamanoto W, Nakamura T, Hayasaka K (2005) Serotonin-2A and 2C receptor gene polymorphisms in Japanese patients with obstructive sleep apnea. Intern Med 44: 928-933

99.   Sasaki A, Kanai M, Kijima K, Akaba K, Hashimoto M, Hasegawa H, Susuki E, Akazawa K, Gejyu F (2003) Molecular analysis of congenital central hypoventilation syndrome. Hum Genet 114: 22-26

100.   Satoh M, Hida W, Chonan T, Miki H, Iwase N, Taguchi O, Kituchi Y, Takishima I (1991) Role of hypoxic drive in regulation of postapneic ventilation during sleep in patients with obstructive sleep apnea. Am Rev Resp Dis 143: 481-485

101.   Schiffman PL, Trontell MC, Mazar MF, Edelman NH (1983) Sleep deprivation decreases ventilatory response to CO_2 but not load compensation. Chest 84: 695-698

102.   Series F, Cormier Y, Desmeules M, La Forge J (1989) Effects of respiratory drive on upper airways in sleep apnea patients and normal subjects. J Appl Physiol 67: 973-979

103.   Series F, Roy N, Marc I (1994) Effects of sleep deprivation and sleep fragmentation on upper airway collapsibility in normal subjects. Am J Respir Crit Care Med 150: 481-485

104.   Shintani T, Asakura K, Kataura A (1998) The effect of adenotonsillectomy in children with OSA. Int J Pediatr Otorhinolaryngol 44: 51-58

105.   Sin DD, Jones RL, Man GC (2000) Hypercapnic ventilatory response in patients with and without obstructive sleep apnea: do age, gender, obesity, and daytime $PaCO_2$ matter? Chest 117: 454-459

106.   Smart D, Haynes AC, Williams G, Arch JR (2002) Orexins and the treatment of obesity. Eur J Pharmacol 440: 199-212

107.   Spier S, Rivlin J, Rowe RD (1986) Sleep in Pierre Robin syndrome. Chest 90:711-715

108.   Stradling JR, Thomas G, Warley ARH, Williams P, Freeland A (1990) Effect of adenotonsillectomy on nocturnal hypoxaemia, sleep disturbance, and symptoms in snoring children. Lancet 335: 249-253

109.   Strohl KP, Saunders NA, Feldman NT, Hallett M (1978) Obstructive sleep apnea in family members. N Engl J Med 299: 969-973

110.   Suen JS, Arnold JE, Brooks LJ (1995) Adenotonsillectomy for treatment of obstructive sleep apnea in children. Arch Otolaryngol Head Neck Surg 121: 525-530

111.   Tafil-Klawe M, Thiele A, E., Raschke F, Mayer J, Peter JH, von Wichert W (1991) Peripheral chemoreceptor reflex in obstructive sleep apnea patients: a relationship between ventilatory response to hypoxia and nocturnal bradycardia during apnea events. Pneumologie 45: 309-311

112.   Tankersley CG, Kleeberger S, Russ B, Schwartz A, Smith P (1996) Modified control of breathing in genetically obese (ob/ob) mice. J Appl Physiol 81: 716-723

113.   Thomas DA, Swaminathan S, Beardsmore CS, McArdle FK, MacFadyen UM, Goodenough PC, Carpenter R, Simpson H (1993) Comparison of peripheral chemoreceptor responses in monozygotic and dizygotic twin infants. Am Rev Respir Dis 148: 1605-1609

114.   Thomas RJ, Daly RW, Weiss JWS (2005) Low-concentration carbon dioxide is an effective adjunct to positive airway pressure in the treatment of refractory mixed central and obstructive sleep-disordered breathing. Sleep 28: 69-77

115.   Tishler PV, Redline S, Ferrette V, Hans MG, Altose MD (1996) The association of sudden unexpected infant death with obstructive sleep apnea. Am J Respir Crit Care Med 153: 1857-1863

116.   Tun Y, Hida W, Okabe S, Kikuchi Y, Kurosawa H, Tabata M, Shirato K (2000) Inspiratory effort sensation to added resistive loading in patients with obstructive sleep apnea. Chest 118: 1332-1338

117.   Valdez BC, Henning D, So RB, Dixon J, Dixon MJ (2004) The Treacher Collins syndrome (TCOF1) gene product is involved in ribosomal DNA gene transcription by interacting with upstream binding factor. Proc Natl Acad Sci USA 101: 10709-10714

118.   Valentijn LJ, Baas F, Wolterman RA, Hoogendijk JE, van den Bosch NH, Zorn I,
Gabreels-Festen AW, de Visser M, Bolhuis PA (1992) Identical point mutations of
PMP-22 in Trembler-J mouse and Charcot-Marie-Tooth disease type 1A. Nat Genet 2:
288-291

119.   Veasey SC (2003) Serotonin agonists and antagonists in obstructive sleep apnea: therapeutic potential. Am J Respir Med 2: 21-29

120.   Wellman A, Jordan AS, Malhotra A, Fogel RB, Katz ES, Schory K, Edwards JK, White
DP (2004) Ventilatory control and airway anatomy in obstructive sleep apnea. Am J
Respir Crit Care Med 170: 1225-1232

121.   Wellman A, Malhotra A, Fogel RB, Edwards JK, Schory K, White DP (2003) Respiratory system loop gain in normal men and women measured with proportional-assist ventilation. J Appl Physiol 94: 205-212

122.   White DP, Douglas NJ, Pickett CK, Zwillich CW, Weil JV (1983) Sleep deprivation and
the control of ventilation. Am Rev Respir Dis 128: 984-986

123.   Wilcox I, Collins FL, Grunstein RR, Hedner J, Kelly DT, Sullivan CE (1994) Relationship between chemosensitivity, obesity and blood pressure in obstructive sleep apnoea.
Blood Pressure 3: 47-54

124.   Woods DR, Montgomery HE (2001) Angiotesin-converting enzyme and genetics at
high altitude. High Alt Med Biol 2: 201-210

125.   Woods DR, Pollard AJ, Collier DJ, Jamshidi Y, Vassiliou V, Hawe E, Humphries SE,
Montgomery HE (2002) Insertion/deletion polymorphism of the angiotensin I-
converting enzyme gene and arterial oxygen saturation at high altitude. Am J Respir Crit
Care Med 166: 362-366

126.   Xiao Y, Huang X, Qiu C (1998) Angiotension I converting enzyme gene polymorphism
in Chinese patients with obstructive sleep apnea syndrome. Zhonghua Jie He He Hu Xi
Za Zhi 21: 489-491

127.   Xiao Y, Huang X, Qiu C, Zhu X, Liu Y (1999) Angiotensin I-converting enzyme gene
polymorphism in Chinese patients with obstructive sleep apnea syndrome. Chin Med J
(Engl) 112: 701-704

128.   Ylmaz M, Bayazit YA, Ciftci TU, Erdal ME, Urhan M, Kokturk O, Kemaloglu YK, Inal
E (2005) Association of serotonin transporter gene polymorphism with obstructive sleep
apnea syndrome. Laryngoscope 115: 832-836

129.   Younes M, Ostrowski M, Thompson W, Leslie C, Shewchuk W (2001) Chemical control stability in patients with obstructive sleep apnea. Am J Respir Crit Care Med 163:
1181-1190

130.   Young T, Peppard PE, Gottlieb DJ (2002) Epidemiology of obstructive sleep apnea: a
population health perspective. Am J Respir Crit Care Med 165: 1217-1239

131.   Yue WH, Liu PZ, Hao W, Zhang XH, Wang XP, Zhang JS, Zhou XH, Xie YB, Ni M
(2005) Association study of sleep apnea syndrome and polymorphisms in the serotonin
transporter gene. Zhonghua Yi Xue Yi Chuan Xue Za Zhi 22: 533-536

132.   Zhang B, Tanaka H, Shono N, Miura S, Kiyonaga A, Shindo M, Saku K (2003) The I allele of the angiotensin-converting enzyme gene is associated with an increased percentage of slow-twitch type I fibers in human skeletal muscle. Clin Genet 63: 139-144

133.   Zhang J, Zhao B, Gesongluobu, Sun Y, Wu Y, Pei W, Ye J, Hui R, Liu L (2000) Angiotensin-converting enzyme gene insertion/deletion (I/D) polymorphism in hypertensive
patients with different degrees of obstructive sleep apnea. Hypertension Res 23: 407-411

134. Zhang LQ, Yao WZ, He QY, Wang YZ, Ren B, Lin YP (2004) Association of polymorphisms in the angiotensin system genes with obstructive sleep apnea-hypopnea syndrome. Zhonghua Jie He He Hu Xi Za Zhi 27: 507-510
135. Zoghbi HY, Percy AK, Glaze DG, Butler IJ, Riccardi VM (1985) Reduction of biogenic amine levels in the Rett syndrome. N Engl J Med 313: 921-924

9. Apnea and irregular breathing in animal models: a physiogenomic approach

Motoo YAMAUCHI[1] , Fang HAN [2] and Kingman P. STROHL [3]

[1] Visiting Scientist, Department of Medicine, Louis Stokes DVA Medical Center, Case Western Reserve University, Cleveland OH .[2]Associate Professor Peoples Hospital, Beijing Medical University, Beijing PRC; currently, Visiting Professor, Department of Medicine, Louis Stokes DVA Medical Center, Case Western Reserve University, Cleveland OH. [3]Professor of Medicine, Department of Medicine, Louis Stokes DVA Medical Center, Case Western Reserve University, Cleveland OH, U.S.A.

9.1 Introduction

The respiratory control consists of the controller (brain), controlled system (lungs and chest wall), and sensors (e.g. carotid body). This system will maintain pH and to a lesser extent oxygenation within a fairly narrow range despite changes in metabolism and environmental conditions with growth, development, and activities of daily living. Ventilatory behavior (the expression of frequency, tidal volume, and ventilation) is one measurable output in this respiratory system. Many combinations of tidal volume and frequency are found among species [1] and depend only in part on the mechanical properties of the controlled system. In the rat or the mouse there is a broad range of frequency values that are energetically similar [5; 10]. Thus, both genes and environment can modify ventilatory behavior in a global manner without extracting significant energy cost in regard to gas exchange.

Ventilatory behavior is described by the timing and respective volumes of inspiratory and expiratory phases, and is traditionally reported as mean and variance values. The variations in the patterns of breathing over time and the appearance of apneas and unstable breathing is less well described in rodents, in contrast to reports from human studies where descriptions of breathing irregularities, apneas, and hypopneas are commonly reported in analyses of sleep-disordered breathing [37]. Lessons from clinical medicine indicate the importance of such patterning over time [38]. Historically, the condition of recurrent central apneas in the setting of congestive heart failure (Cheyne-Stokes respiration) is representative of irregular, albeit regularly irregular, breathing in humans [39]. The waxing and waning features of ventilation in this condition may occur during wakefulness but are most apparent during sleep, and these cycles of hyperpnea and apnea indicate wide swings in gas exchange associated with arousals from sleep resulting in a waketime neuroexcitatory state and excessive fatigue/sleepiness. These patterns of irregular breathing over time are amendable to fundamental analysis and can be modeled as a dynamic response initiated and sustained by the neural components of the respiratory control system [24]. Ventilatory irregularity is also apparent in apnea of prematurity, and is a feature in inherited conditions such as congenital hypoventilation syndrome, Rett syndrome, and Prader-Willi syndrome (see Chapters 4, 15 and 16).

Animal models of irregular breathing exist in large mammals. English bulldogs [23] and elephant seals [9] provide examples of sleep apnea useful for comparative physiology. Rodent models, however, are more amenable to physiologic and genetic dissection of environmental exposures, consequences, and intermediate phenotypes [3]. In the case of transgenic animals or in common strains animals are examined for intermediate traits, like obesity or ventilatory control during wakefulness and sleep, which may lead to ventilatory instability and relate to sleep apnea expression or progression. Utilizing inbred strains permits analyses of age, environmental exposures, and pedigree, etc. that can confound studies in human populations.

This chapter will review evidence that reproducible differences in ventilatory behavior occur as the consequence of an interaction between physiologic features of the respiratory control system and genetic background in unanesthetized rodent models. In particular the question being addressed is whether it is likely that genes contribute to non-steady state events such as the appearance of apneas, post-sigh-pauses, ventilatory irregularity, and periodic breathing. Examination of this literature offers support for an argument that genetic setpoints exist for many of the time-domain features in the regularity of breathing, especially in response to hypoxia and re-oxygenation. These observations along with the genetic homologies among rats, mice and humans offer opportunity for what others have called a "physiogenomic" approach [36] to not only steady-state but irregular breathing patterns.

9.2 Measuring ventilatory behavior

The measurement of ventilatory behavior in unanaesthetized animals is usually made by plethysmography, of which there are two types. The first is a head-out plethysmography method, where a seal around the neck permits the measurement of breathing efforts and airflow. In this method the animal is restrained by the neck collar. The second, chamber plethysmography, is the measurement of breathing while the unrestrained animal is in a closed chamber. Measures can be made while the box is briefly sealed or when there is a balanced flow through the box. While there are differences in the values among these methods depending on the size and frequency response of the chamber and the system, the keys to identifying ventilatory behavior in rodent models are accuracy and reproducibility.

For studies in the unanesthetized rodent, the most common instrument is the chamber or barometric measurement of breathing. With this method, frequency is a rather precise value and tidal volume is semi-quantitative under the best circumstances [14]. In regard to reproducibility, it is important to standardize as much as possible the setting of the measurements, room environment, personnel, etc. (Table 1).

Table 1. Some features to optimize phenotype collection

- Appropriate size and shape of barometric chambers to minimize dispersal of pressure and convection effects
- Consistent testing facilities in regard to quiet surroundings, odor suppression, and consistency in technical assistance over time
- Standard calibration of equipment, testing steps, and time of testing
- Standardized cleaning of chambers between testing of different animals
- Quality control over time using animals from in-bred strains

Reproducibility is affected by light, smell, noise, etc., and these "environmental" factors can be reduced by careful attention to the testing chamber protocol [12]. Time-of-day effects are important to consider, examine, and/or control [17; 42]. Effects resulting from variation in testing conditions enter into estimates of environmental rather than genetic variance. If an environmental effect is large or uncontrolled, it will obscure genetic contributions. Bias can be introduced at every level of collection of phenotype data and will also enter into analysis as an environmental factor, obscuring subtle physiologic or genetic effects. While environmental effects are conceivably quite large, a number of studies indicate that the technique can uncover underlying genetic influences, ie. strain or knock-out effects, and even sub-chromosomal locations, acting on traits of breathing frequency and depth (see Chapter 10).

Ideally, one would select a phenotype trait or set of traits that are sensitive to a single feature of the respiratory control system-controller, controlled system, or sensor. However, plethysmographic measures do not permit such mechanistic insight [3]. Other methods to consider might include direct measures of respiratory muscle activity using implanted electrodes, measures of respiratory neural drive in anesthetized or reduced preparations, and/or in vitro models. With these approaches there is the possibility of introducing variation through the direct or indirect effects of anesthesia or surgery, or the process of reducing proprioceptive or reflex influences. However, one can use a reduced preparation to disclose brainstem or genetic mechanisms at a cellular level once the effect is shown in the whole animal.

9.3 Defining apnea and ventilatory irregularity

Ventilatory behavior has several time-domains and is dynamically configured over time. Respiratory control in rodents, as in humans, is not only used for breathing but also for sighing, gasping, eating, defecation, exercise, etc. Fig. 1 shows a number of examples of "ventilatory behavior" at rest. For the purposes of studying ventilatory behavior, one challenge is to distinguish among these conditions, and in regard to apnea define the event in relation to non-respiratory events. In humans, a cessation of airflow at the nose and mouth of >10 seconds defines an apnea, and distinctions are made between apneas with (obstructive apnea) and without (central apnea) the presence of respiratory efforts [37]. In the rodent, obstructive apneas are rarely observed, owing to the lack of collapsibility of upper airway structures, and to the limitations of barometric testing. Pauses in breathing, i.e. central apneas, do occur. These central-type apneas have been operationally defined as either as a pause in breathing >1 second, or as a pause of >2 respiratory cycles. One can further make the distinction that such pauses or apneas are either

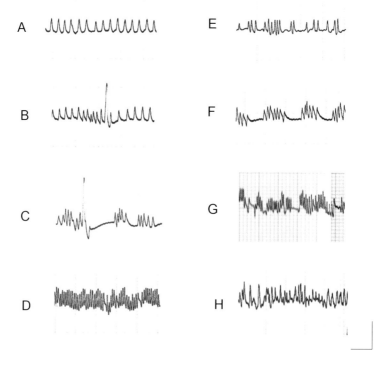

Fig. 1. Examples of different breathing patterns at rest in plethysmographyu from adult mice. Calibrations : horizontal axis 1.2 seconds, and vertical axis ~ 110 microliters. Panels are representative of breathing over time for : A : A/J mouse-quiet wakefulness ; B : A/J mouse-augmented breath without pause ; C : B6 mouse-augmented breath with a pause ; D : B6 mouse-sniffing ; E : B6 mouse-unstable breathing at rest ; F : B6 mouse-periodic breathing post-hypoxia; G : B6 mouse-rearing; H : B6 mouse-scratching and grooming.

preceded by a sigh (also called an augmented breath) or not. An augmented breath is a tidal volume that exceeds by at least 125% a "normal breath". Using such a distinction permits one to count isolated pauses in breathing as compared to pauses occurring after a larger lung inflation, termed "post-sigh apneas" or "post-sigh pauses". Not all augmented breaths produce a post-sigh apnea. Such differences may be mechanistically relevant. As will be discussed, some interventions modify spontaneous pauses without altering the appearance of augmented breaths and subsequent pauses.

On exposure to hypoxia there is an early peak response followed by a decline in ventilation. The peak response originates initially from sensing in the carotid body [18] and is often followed by a progressive decline as hypoxia continues [10]. The magnitude of both the acute response and "roll-off" differs among species and among strains of rats and is the subject of another chapter (see Chapter 10). Prolonged hypoxemia of many minutes to hours appears to enhance a tendency towards variations in tidal volume and frequency including patterning of breathing [11; 27]. Intermittent hypoxia may also evoke changes in the operating point for respiratory rhythm generation [32].

Ventilatory behavior potentially relevant to human sleep apnea may occur upon reoxygenation after exposure to hypoxia. In this transient there are two broad classes of response: short-term potentiation where ventilation remains elevated above baseline (pre-hypoxic) values and another called post-hypoxic (frequency or ventilatory) decline where ventilation and/or frequency fall below baseline. Short-term potentiation (STP) of ventilation would promote persistence of ventilation [27], and protect against further reductions in drive leading to apnea [46]. Patients with obstructive sleep apnea hypopnea syndrome [19] and those with Cheyne-Stokes respiration [12] lack STP during wakefulness. Ventilatory responses to hypoxia (and re-oxygenation) involve interconnected neuronal pools in the medulla and pons [22; 28; 31]. Central neurotransmitters implicated in re-oxygenation responses include serotonin and adenosine [6; 40] and nitric oxide [33]. The pontine A5 region is involved in coordinating this response [13].

Inheritance operates on these non-steady state or acute responses. Fig. 2 illustrated the differences in the post-hypoxic ventilatory behavior (frequency, tidal volume and minute ventilation) between A/J and B6 mouse strains. Post-hypoxic ventilatory decline (PHVD) seen in the B6 in response to re-oxygenation 100% oxygen after five minutes of exposure to hypoxic gas is not present in the A/J strain [20]. Differences in this response are also present between Sprague Dawley and Brown Norway strains of rat [41]. Such responses could be important intermediate traits relevant to the termination vs. the reinitiation of an apnea.

There is no common language in reporting breathing over time in rodent models. Most publications are content with a picture showing an example of slow breathing, apneas or pauses in breathing, or an irregular breathing pattern, without formal quantification. The intent is to illustrate visibly apparent differences in ventilatory behavior between control and test conditions or among different strains of animals or conditions. For example, in a brain-derived neurotrophic factor or *Bdnf*

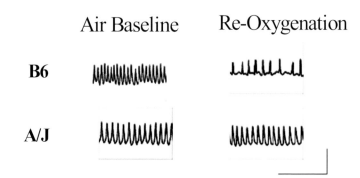

Fig. 2. Differences in B6 and A/J mice occur with re-oxygenation.These strip chart re-cordings are representative of breathing on room air (Air Baseline) and then 15-30 seconds after re-oxygenation with 100% oxygen after five minutes of hypoxia (10% oxygen/balance nitrogen). Calibra-tons: horizontal 5 is seconds and vertical axis is ~200 microliters.

knock-out model, loss of one or both *Bdnf* alleles resulted in depression of central respiratory frequency and reduced animal survival through post-natal day 15 [4; 6]. The example from this paper is presented in Fig. 3. The value of this and other knock-out models is that the irregular breathing appears to be linked to an adverse outcome (perinatal death). The issue is that irregular breathing can be a common feature in the neonate. Even in this example the irregular breathing seen at birth appears improved as the animal gets older. Furthermore, this irregular breathing pattern can be seen in the adult (see Fig. 1).

The study of adult rodents, especially unanesthetized adults, can provide many examples of ventilatory instability, in part caused by behavioral factors, but often unexpected. Metrics are needed for these observations, as illustrated in Fig. 1 and 3. Occasional articles have described variations in ventilatory behavior us-ing metrics like the coefficient of variation or CV, as defined by the ratio of the standard deviation over the mean value [25]. In these instances a breathing set is examined for temporal variations in frequency or tidal volume, and higher values of the CV imply greater irregularity in ventilatory behavior.

Another way to display the data in regard to the potential for recurring patterns or lack thereof, is in a Poincaré plot (Fig. 4). The variation between values over time is examined in regard to short- and long-term tendencies to return to a particular value. These plots can be compared between control and test conditions. Certain conditions should be met regarding the number of observations. There is no consensus as to which site in the respiratory control system these patterns can be attributed.

In humans, recurrent apneas are captured as number of apneas/hour, >3 cycles of apneas followed by hyperpnea, and, less often, the measures capturing the cycle length or cycle strength in a string of recurrent apneas. This definition has been used to describe post-hypoxic instability in a mouse model [21]. Different measures are reported for instances where there are recurrent non-apneic rather than apneic pauses. These approaches have some value for offering mechanistic insight into the influence of respiratory control system as do the analyses of human patterns [24].

There are a number of other linear and non-linear approaches to describe ventilatory regularity over time. A discussion of these approaches is beyond the scope of this review, and there is at present no demonstration as to their utility, for instance in the forecasting of ventilatory irregularity, apneas, or adverse outcomes from ventilatory disturbances in rodent models.

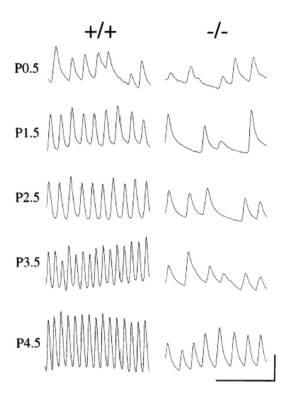

Fig. 3. Example from Erikson et al. 1996. Plethysmograph records of resting ventilation in room air from one *Bdnf*[+/+] and a second *Bdnf*[-/-] littermates during the first 4 days of postnatal life (P 0.5–P4.5). These are examples of the differences in the pattern of resting ventilation between wild-type and knockout animals, which were apparent as early as 6–12 hr after birth. Scale bars: vertical, 20 microliters; horizontal, 2 sec. This Figure is from [15] (copyright 1996 J Neuroscience).

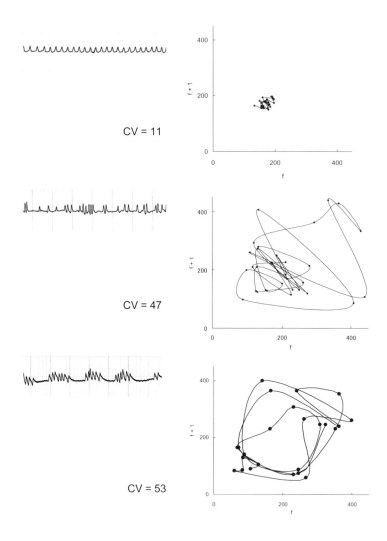

Fig. 4. Three examples of breathing patterns. Shown are fragments of records that illustrate regular breathing at rest, irregular breathing at rest (middle), and periodic breathing (bottom tracing). Each tracing is from the B6 strain. What is illustrated are the strip chart recording, the value of the co-efficient of variation (CV) for frequency, and a Poincaré for each condition. The Poincaré chart has as the horizontal axis the frequency estimation of the first breath and as the vertical axis the frequency estimation of the next breath. These are attached by a line. Note that the patterns seen in the frequency plots are not reflected in the CV values.

9.4 Models of apnea

Central-type apneas, defined as a pause in breathing of >2 respiratory cycles (>1 sec), occur in rats and mice. One may also further make the distinction between central apneas preceded by a sigh (post-sigh apneas) or those not preceded by a sigh (a respiratory pause). Using such a distinction permits one to distinguish between those pauses that are spontaneous and those operating after lung inflation.

In the early 1990's three different laboratories reported the presence of apneas in adult rats. A conditioning paradigm of intermittent sleep interruption during the first 4 weeks of life in Sprague Dawley rats will increase the number of spontaneous respiratory pauses some 10-12 weeks later [43]; however, the number of post-sigh apneas is unchanged. Furthermore, in adult conditioned animals, apneic pauses, as opposed to post-sigh pauses, were dramatically reduced by a counter-conditioning stimulus (white noise) [44]. These observations indicate that the respiratory control system in the brainstem of the neonate contains a degree of developmental plasticity which is experience-dependent and is a potential substrate for adult patterns of sleep-disordered breathing. In models where the intervention is REM deprivation in the adult, differences occur between the numbers of spontaneous and post-sigh in recovery sleep [7]. In this study there was a decrease in post-sigh apneas without a change in spontaneous apneas. All these results suggest the existence of distinct mechanisms for the two types of apneas.

There exist group differences among strains in the expression of spontaneous and post-sigh sleep apneas in adult animals. Spontaneously hypertensive (SHR) rats exhibit more spontaneous apneas as well as post-sigh apneas/hour than normotensive Wistar-Kyoto (WKY) rats during slow wave sleep [8]. Administration of an anti-hypertensive agent to the SHR rats will equalize blood pressure to that of the WKY strain and but will also reduce apnea number to that of WKY normotensive rats. These results suggest interactions among hypertensive traits, anti-hypertensive agents, and sleep-related apneas in adult rats. To further explore this relationship this group examined genetically hypertensive SHR strain and the SHR strain treated neonatally with antihypertensive agents to produce a phenotypically normotensive SHR rat. Apneas were elevated more than 15-fold during NREM sleep in both animal groups carrying hypertension-related genes versus normotensive WKY rats; during REM sleep, a genetic background for hypertension was associated with an even higher apnea index. Overall mean respiratory rate, minute ventilation, and sleep architecture, however, were equivalent among all animal groups. Blood pressure and heart rate were similar in both normotensive groups but elevated in the hypertensive animals. This result connects the trait of spontaneous apnea during sleep to a genetically determined phenotype not secondary to hypertension.

Apneas during sleep have been recorded in adult mice [30]. Cessation of plethysmographic signals for longer than two respiratory cycles in 129/Sv mice were observed during sleep but not in quiet awake periods. Sleep apneas were further classified into two types. Post-sigh apneas occurred exclusively during slow-wave sleep (SWS), whereas spontaneous apneas arose during both SWS and REM sleep. Post-sigh apneas were increased with hypoxia but decreased with both hyperoxia and hypercapnia, while spontaneous pauses were not. Comparable studies have not been done in other strains of mice.

Experimentally induced lesions of the pre-Bötzinger complex in adult animals will produce apneas in wakefulness and in sleep [26; 45]. Unilateral lesions also result in a rather stable model showing REM related pauses during sleep [16]. The observation that cholinergic transmission in pre-Bötzinger complex will modulate the appearance of apneas [35] makes it likely that both genetic variation and disease might influence the expression of REM-related apneas in humans.

Taken together, in rodent models spontaneous apneas can be modulated physiologically without affecting post-sigh apneas, an event that is more likely to be the result of stretch or chemoreceptor feedback. In addition the occurrence of spontaneous apneas varies among strains and is influenced by neonatal experiences and by genetic background. The finding of apneas produced in sleep by intentional ablation of the pre-Bötzinger complex further illustrates an important role of state in the generation of ventilatory behavior.

9.5 Model of recurrent apneas

As discussed in the sections above, ventilatory instability can take several forms but this section will emphasize our studies of recurrent apneas, characterized as a form of periodic apnea, present in the inbred C57BL/6J (B6). In the B6 strain the occurrence of unstable breathing is largely independent of degree of the previous hypoxia with or without added carbon dioxide, or oxygen level of the re-oxygenation gases [20; 21]. Using administration of a relatively selective NOS blocker (7-nitroindazole (NI), 60 mg/kg), both B6 and A/J strains of mice exhibit PHFD and the absence of STP. The B6 strain however will exhibit both post-hypoxic frequency decline and periodic breathing with longer cycles and apnea length. This is in contrast to A/J mice, which do not show periodic breathing before or after 7-NI. These findings support the observation that PHFD is not closely coupled to the occurrence of post-hypoxic recurrent apneas.

Finding of recurrent apneas in some but not all recombinant inbred mice (B6XA/J) strains confirmed an inherited basis for this trait in a non-genetically engineered adult mouse [21]. In unpublished studies we have found that first generation animals (F1; n= 4 males and 8 females) of A/J and B6 parental strains, purchased from Jackson Laboratories, Bar Harbor, ME, exhibit no recurrent central apnea (a > 3 successive cycles) in the post hypoxic-period when re-oxygenated with either air or 100% oxygen. The strength of oscillation over the 20-60 seconds following re-oxygenation was negligible, more like the A/J than the B6 parental animals. While a small number of observations, it appears that post-hypoxic breathing instability is not modeled as a Mendelian dominant trait or one with high penetrance.

A methodology for the mapping of complex traits in mice utilizes chromosome substitution strains (CSSs) as an alternative mapping resource [29]. CSSs are a panel of strains in which each A/J chromosome is placed through selective breeding onto the B6 background in a homozygous fashion, effectively replacing the B6 chromosome. The B6 CSSs were constructed because the two strains differ in a number of physiological traits, the genome sequence for each strain is largely completed, and complete bacterial artificial chromosome libraries exist for each strain (bacpac.chori.org).

We have undertaken a survey of baseline values, hypoxic responses, and post-hypoxic ventilatory behavior in 94 animals (n= 4-10 animals in each strain) from the CSS panel of A/J and B6 animals. In this data panel that included study of chromosomal substitution strains of all 20 mouse chromosomes as well as the Y chromosome, only the b6a1 strain failed to produce periodic breathing before as well as after 7-NI administration, suggesting that chromosome 1 might be the initial regions of interest. Table 2 indicates values for the apnea time (Ta, secs) and cycle length (Tc, secs) for the A/J and B6 strains and four selected CCS strains.

Table 2. Apnea time and cycle length for the A/J and CSSs

* Ta = Apnea length (sec) Tc = cycle length (sec)   Mean (SD)

** number with apnea before 7-NI/Total

$p < 0.01$ among strains before and after 7-NI (ANOVA)

Strain (N)**	Ta*		Tc*	
	Before	After 7-NI	Before	After
A/J Jax (0/20)	0	0	0	0
b6 (17 of 20)	3.5 (1.5)	6 (3)	7 (1)	10 (1.7)
b6a1 (0/10)	0	0	0	0
b6a11 (5 of 6)	2.5 (2)	5.3 (1)	5 (2)	9 (2)
b6a14 (4 of 6)	3.5 (3)	6.8 (4)	5.5 (3)	6.2 (3)
b6a16 (4 of 5)	4 (1.7)	7 (4)	6 (1.1)	9 (4)

In this panel there was an independence of the trait of recurrent apneas from post-hypoxic events, called short-term potentiation or STP or post-hypoxic frequency decline (PHFD). In CSS b6a1 (n =10), b6a11 (n =6), ba14 (n =6), and b6a16 (n–5), either the PHFD observed in the B6 parental strain is attenuated or the A/J phenotype of STP partially reappears despite the presence of one A/J chromosome. This result indicates to us that several chromosomes and perhaps multiple alleles influence the responses to reoxygenation. Reversal of the B6 recurrent apnea phenotype by substitution of an A/J chromosome was independent of frequency, minute ventilation, and the ventilatory response to hypoxia at 1 minute or 5 minutes of exposure (data not shown), consistent with our published observations in recombinant inbred strains [21]. Precedent for multiple chromosomal influences on a trait is not new, being evident in quantitative traits such as blood pressure [36], where alleles on different chromosomes can act and interact to increase or decrease trait values.

We also observed a variance among the four chromosomal substitution strains in response to 7-NI (60mg/kg) administration. With 7-NI, the A/J strain is known to exhibit PHFD, as does the B6, and a tendency for this to happen occurs in CCS strain b6a14 as well. However, CSS strains b6a1 and b6a16 do not produce PHFD and CSS b6a16, which exhibited PHFD before, now showed a tendency for STP with 7-NI. These drug-by-strain interactions may be used in identifying mechanisms for pharmacogenomic influences of respiratory stimulants.

As a result of these observations in CCSs and in the drug-by-strain studies using 7-NI, we hypothesize that the time-domain features in the response to reoxygenation are distinct at a genetic level, and that the tendency for PHFD or STP is not directly related to the occurrence of recurrent apneas.

9.6 Conclusion

Identification of gene regions influencing ventilatory behavior will be relevant to understanding human diseases characterized by abnormal respiratory patterns. Knowing such genes would lead to greater understanding of the heterogeneity in responses to environmental influences and, potentially, to develop a set of markers that help address breathing patterns in human health and disease. Despite a lack of consensus among laboratories on how to describe breathing patterns over time, there are a number of rodent studies that identify breathing patterns, including apneas, breathing irregularity, and transient patterns resembling periodic breathing, amenable to physiologic and pharmacological study. Variability in breathing depends on strain, be inherited, and be modified in neurophysiological or drug-by-strain studies, indicating the influences of both genes and environment on the expression of irregular breathing patterns. Incorporation of these patterning traits into maps of functional connections among genes that act on ventilatory behavior might directly relate to understanding of the biology of human disorders of respiratory control and homeostasis.

References

1.   Agostoni E, Thimm F, Fenn WO (1959) Comparative features of mechanics of breathing. J Appl Physiol 14: 679-683
2.   Ahmed M, Serrette C, Kryger MH, Anthonisen NR (1994) Ventilatory instability in patients with congestive heart failure and nocturnal Cheyne-Stokes breathing. Sleep 17: 527-534
3.   American Thoracic Society (1999) Finding genetic mechanisms in syndromes of sleep disordered breathing

4.   Balkowiec A, Katz DM (1998) Brain-derived neurotrophic factor is required for normal development of the central respiratory rhythm in mice. J Physiol 510: 527-533

5.   Bennett FM, Tenney SM (1982) Comparative mechanics of mammalian respiratory system. Respir Physiol 49: 131-140

6.   Bonham AC (1995) Neurotransmitters in the CNS control of breathing. Respir Physiol 101: 219-230

7.   Carley DW, Trbovic S, Radulovacki M (1996a) Effect of REM sleep deprivation on sleep apneas in rats. Exp Neurol 137: 291-293

8.   Carley DW, Trbovic S, Radulovacki M (1996b) Sleep apnea in normal and REM sleep-deprived normotensive Wistar-Kyoto and spontaneously hypertensive (SHR) rats. Physiol Behav 59: 827-831

9.   Castellini MA, Milsom WK, Berger RJ, Costa DP, Jones DR, Castellini JM, Rea LD, Bharma S, Harris M (1994) Patterns of respiration and heart rate during wakefulness and sleep in elephant seal pups. Am J Physiol 266: R863-869

10. Crossfill ML, Widdicombe J (1961) Comparative mechanics of the mammalian respiratory system. Respir Physiol 49: 131-140

11. Dahan A, Berkenbosch A, DeGoede J, van den Elsen M, Olievier I, van Kleef J (1995) Influence of hypoxic duration and posthypoxic inspired O_2 concentration on short term potentiation of breathing in humans. J Physiol 488: 803-813

12. DeLorme MP, Moss OR (2002) Pulmonary function assessment by whole-body plethysmography in restrained versus unrestrained mice. J Pharmacol Toxicol 47: 1-10

13. Dick TE, Coles SK (2000) Ventrolateral pons mediates short-term depression of respiratory frequency after brief hypoxia. Respir Physiol 121: 87-100

14. Enhorning G, van Schaik S, Lundgren C, Vargas I (1998) Whole-body plethysmography, does it measure tidal volume of small animals? Can J Physiol Pharmacol 76: 945-951

15. Erickson JT, Conover JC, Borday V, Champagnat J, Barbacid M, Yancopoulos G, Katz DM (1996) Mice lacking brain-derived neurotrophic factor exhibit visceral sensory neuron losses distinct from mice lacking NT4 and display a severe developmental deficit in control of breathing. J Neurosci 16: 5361-5371

16. Feldman JL, Del Negro CA (2006) Looking for inspiration: new perspectives on respiratory rhythm. Nat Rev Neurosci 7: 232-242

17. Fenelon K, Seifert EL, Mortola JP (2000) Hypoxic depression of circadian oscillations in sino-aortic denervated rats. Respir Physiol 122: 61-69

18. Gauda EB (2002) Gene expression in peripheral arterial chemoreceptors. Microsc Res Tech 59: 153-167

19. Georgopoulus D, Giannouli E, Tsara V, Argiropoulou P, Patakas D, Anthonisen NR (1992) Respiratory short-term poststimulus potentiation (after-discharge) in patients with obstructive sleep apnea. Am Rev Respir Dis 146: 1250-1255

20. Han F, Subramanian S, Dick TE, Dreshaj IA, Strohl KP (2001) Ventilatory behavior after hypoxia in C57BL/6J and A/J mice. J Appl Physiol 91: 1962-1970

21. Han F, Subramanian S, Price ER, Nadeau J, Strohl KP (2002) Periodic breathing in the mouse. J Appl Physiol 92: 1133-1140

22. Haxhiu MA, Mack SO, Wilson CG, Feng P, Strohl KP (2003) Sleep networks and the anatomic and physiologic connections with respiratory control. Front Biosci 8: d946-962

23. Hendricks JC, Kline LR, Kovalski RJ, O'Brien JA, Morrison AR, Pack AI (1987) The English bulldog: a natural model of sleep-disordered breathing. J Appl Physiol 63: 1344-1350

24. Khoo MC, Kronauer RE, Strohl KP, Slutsky AS (1982) Factors inducing periodic breathing in humans: a general model. J Appl Physiol 53: 644-659

25. Li A, Nattie E (2006) Catecholamine neurones in rats modulate sleep, breathing, central chemoreception and breathing variability. J Physiol 570: 385-396

26. McKay LC, Janczewski WA, Feldman JL (2005) Sleep-disordered breathing after targeted ablation of preBötzinger complex neurons. Nat Neurosci 8: 1142-1144

27. Menendez AA, Nuckton TJ, Torres JE, Gozal D (1999) Short-term potentiation of ventilation after different levels of hypoxia. J Appl Physiol 86: 1478-1482

28. Mironov SL, Richter DW (1998) L-type Ca2+ channels in inspiratory neurones of mice and their modulation by hypoxia. J Physiol 512: 75-87

29. Nadeau JH, Singer JB, Matin A, Lander ES (2000) Analysing complex genetic traits with chromosome substitution strains. Nat Genet 24: 221-225

30. Nakamura A, Fukuda Y, Kuwaki T (2003) Sleep apnea and effect of chemo-stimulation on breathing instability in mice. J Appl Physiol 94: 525-532

31. Nattie EE (2001) Central chemosensitivity, sleep, and wakefulness. Respir Physiol 129: 257-268

32. Powell FL, Milsom WK, Mitchell GS (1998) Time domains of the hypoxic ventilatory response. Respir Physiol 112: 123-134

33. Price ER, Han F, Dick TE, Strohl KP (2003) 7-nitroindazole and posthypoxic ventilatory behavior in the A/J and C57BL/6J mouse strains. J Appl Physiol 95: 1097-1104

34. Richter DW, Ballanyi K, Schwarzacher S (1992) Mechanisms of respiratory rhythm generation. Curr Opin Neurobiol 2: 788-793

35. Shao XM, Feldman JL (2005) Cholinergic neurotransmission in the preBötzinger complex modulates excitability of inspiratory neurons and regulates respiratory rhythm. Neuroscience 130: 1069-1081

36. Stoll M, Cowley Jr. AW, Tonellato PJ, Greene AS, Kaldunski ML, Roman RJ (2001) A genomic-systems biology map for cardiovascular function. Science 294: 1723-1726

37. Strohl K, Cherniack N, Gothe B (1986) Physiological basis of therapy for sleep apnea. Am Rev Respir Dis 134: 791-802

38. Strohl K, Redline S (1996) State-of-the-Art: Recognition of obstructive sleep apnea. Am J Respir Crit Care Med 154: 279-289

39. Strohl KP (2003) Periodic breathing and genetics. Respir Physiol Neurobiol 135: 179-185

40. Subramanian S, Erokwu B, Han F, Dick TE, Strohl KP (2002) L-NAME differentially alters ventilatory behavior in Sprague-Dawley and Brown Norway rats. J Appl Physiol 93: 984-989

41. Subramanian S, Han F, Erokwu BO, Dick TE, Strohl KP (2001) Do genetic factors influence the Dejours phenomenon? Adv Exp Med Biol 499: 209-214

42. Tankersley CG, Irizarry R, Flanders S, Rabold R (2002) Circadian rhythm variation in activity, body temperature, and heart rate between C3H/HeJ and C57BL/6J inbred strains. J Appl Physiol 92: 870-877.

43. Thomas A, Austin W, Friedman L, Strohl K (1992) A model of ventilatory instability induced in the unrestrained rat. J Appl Physiol 73: 1530-1536

44. Thomas A, Friedman L, MacKenzie C, Strohl K (1995) Modification of conditioned apneas in rats: evidence for cortical involvement. J Appl Physiol 78: 1215-1218

45.   Wenninger JM, Pan LG, Klum L, Leekley T, Bastastic J, Hodges MR (2004) Small reduction of neurokinin-1 receptor-expressing neurons in the pre-Botzinger complex area induces abnormal breathing periods in awake goats. J Appl Physiol 97: 1620-1628
46.   Younes M (1989) The physiologic basis of central apnea and periodic breathing. Curr Pulmonol 10: 265-326

10. Genetic determinants of respiratory phenotypes in mice

Clarke G. TANKERSLEY

Johns Hopkins Bloomberg School of Public Health, Department of Environmental Health Sciences, Baltimore, MD 21205, U.S.A.

10.1 Introduction

Individual variation in breathing pattern has been recognized for over a century and a half, but only recently have we been able to assign specific genetic determinants that regulate this variation. An understanding of these potential genetic determinants is on the horizon with the eventual sequencing of genomes from different mammalian species, including mouse and human genomes. Given that breathing characteristics are obviously complex traits, the interactions between various genetic determinants are numerous and evolve from different regulatory pathways. In a simple model system of mammalian respiration, genes that regulate neural pathways must interact with genes that regulate lung function. These gene interactions are then subject to serving the maintenance of metabolic homeostasis. The purpose of this chapter is to explore different breathing characteristics in inbred mice, and to query the role of lung mechanics in regulating variation in the magnitude and pattern of breathing. Specifically, we address the hypothesis implicating genetic determinants that regulate differential breathing phenotypes in certain mouse strains may evolve from factors controlling variation in lung mechanics.

10.1.1 Coincident phenotypes in lung disease

Individuals predisposed to abnormal respiratory control mechanisms are susceptible to increased morbidity and mortality rates. This association is underscored in patients with coincidental and significant lung pathophysiology (see Chapter 2). In asthma, for example, patients characterized with hypoventilatory phenotypes are at greater risk of near-fatal attacks [15; 16; 21; 27]. Likewise, hypoventilatory phenotypes are shown to significantly affect disease exacerbation in patients with other forms of chronic obstructive pulmonary disease [2; 13; 17; 26; 28; 34; 39]. Similar associations are demonstrated in morbidly obese patients [18; 51] and in patients susceptible to sleep-disordered breathing [7; 10-12]. Taken together, the role familial factors and genetic predisposition play in conferring hypoxic and hypercapnic hypoventilatory phenotypes represents a critical risk factor; one which interacts with lung pathophysiological mechanisms in susceptible individuals [6; 9; 15; 19; 20; 29; 31].

The genetic basis for many of the disease processes associated with asthma and other chronic obstructive pulmonary disease states have been frequently studied in transgenic mouse models. Inbred mouse strains serve as a powerful tool for studying genetic determinants that regulate heterogeneity observed in lung distension and elasticity. Robust phenotype-genotype relationships are identified using Mendelian strategies (i.e. forward genetics), which are instrumental to assigning unique variation to specific genes or chromosomal locations. Since physiological heterogeneity frequently results from multifactorial mechanisms and polygenic determinants, trait variation is often complex. Trait variation is derived from single nucleotide polymorphisms (SNPs; i.e., DNA sequence variation) at various chromosomal loci, which merge in quantitative interactions between two or more genetic determinants. Identifying quantitative trait loci (QTL) that regulate heterogeneity in lung structure and function is the basis for understanding disease susceptibility and advancing genetic therapeutic interventions as these relate to the lung [8]. Since hypoventilatory phenotypes are frequently inseparable from lung remodeling events associated with various disease processes, genetic determinants of hypoventilatory syndromes might derive from adverse changes in lung structure and mechanics.

10.1.2 Mouse model of breathing control and lung mechanics

Our group and other laboratories have been studying the genetic variability in lung structure and function using phenotypic differences observed among standard inbred strains of mice [32; 33; 37; 50]. Our laboratory has uniquely perfected methods to measure a wide spectrum of lung volume and mechanical property differences among a variety of inbred strains. The most outstanding between-strain differences implicate the role of robust genetic determinants that modulate variation in lung mechanics through volume-dependent and volume-independent mechanisms. For example, our lab has shown that lung volume measurements at airway pressures between 0 and 30 cmH_2O are 50% greater in C3H/HeJ (C3) compared with C57Bl/6J (B6) mice [50]. While lung compliance is also greater in C3 than B6 mice, as anticipated, specific compliances also differ between these strains suggesting that separate genetic determinants modulate lung elastic recoil by volume-dependent and volume-independent mechanisms. A similar distinction surrounds strain variation in lung resistance. In this case, the A/J strain demonstrates airway resistances at baseline and during bronchoprovocation that dramatically exceed both C3 and B6 mice [4; 5; 8; 24; 25], yet lung volumes of A/J mice resemble the B6 strain [50]. With respect to strain-specific phenotypes in lung mechanics, other labs have replicated our original findings [31; 37]. Therefore, we believe the variability in lung mechanics among strains is consistently regulated by robust genetic determinants, which are conserved within strains but vary between strains. In this chapter, we augment this work by including strain differences between C3 and B6 mice in lung impedance measurements. These novel approaches are intended to further explore the lung variability between strains, and focus on lung parenchymal characteristics and alveolar structural proteins, so as to better define the origin of functional differences in breathing between these strains.

The following flow chart (Fig. 1) outlines a simplified model of the essential genetic determinants that play a role in regulating breathing differences between C3 and B6 mice and their first-generation offspring (F1). It is well-established that C3 mice have a slow, deep baseline breathing pattern compared to the rapid, shallow breathing pattern of B6 mice. The timing of inspiration is a breathing trait that largely contributes to this strain difference. The F1 offspring resembles the B6 progenitor, and we believe that QTL for variation in inspiratory timing at baseline is located on a region on mouse chromosome 3 [44; 47; 49].

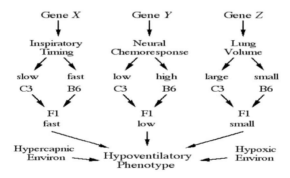

Fig. 1. A genetic model of heritable ventilatory traits is depicted for C3, B6 and F1 mice. The model incorporates at least three genes that influence baseline inspiratory time, neural chemoresponsiveness, and lung mechanical variation. The F1 mice inherit alleles from the C3 progenitor, which confers low chemical ventilatory response. In addition, F1 mice inherit alleles from the B6 progenitor, which confers fast inspiratory time and a relatively small lung. The environment can interact with these genetic determinants to influence the expression of different phenotypes. Our goal in this chapter is to review these genetic determinants and evaluate their interaction.

Breathing differences between C3 and B6 mice also include variation in the magnitude of breathing for responses to chemical challenge, such as hypercapnia and hypoxia. In this case, the magnitude of breathing can be characterized by the minute ventilation (V_E) or mean inspiratory flow (V_T/T_I), and C3 mice show a relatively low response compared with B6 mice. The F1 offspring resembles the C3 progenitor in this case, and we believe QTL on mouse chromosomes 1, 5 and 9 regulate variation in neurochemoresponse between the progenitors [36; 42; 43; 45].

Finally, our model of genetic control of breathing considers the importance of two additional and related mechanisms, including variation in lung structure and mechanics and strain differences in metabolic rate. As shown in the flow chart above, C3 mice have a relatively large lung volume (V_{30}, volume at 30 cm H_2O of airway pressure) and compliance compared to B6 mice. The F1 offspring resembles the B6 progenitor in both V_{30} and lung compliance. The work of Reinhard et al. has shown that QTL on mouse chromosomes 5, 15, and 17 are important in regulating the strain differences in lung volume and compliance [33].

10.1.3 Genetic hypothesis

With this background, the central focus of this chapter is to carefully consider variation in respiratory phenotypes in mice using a well-characterized model of breathing control. The hypothesis that lung mechanics and metabolism play a significant role in affecting the variation in breathing phenotypes has not been explored. Therefore, a primary aim is to dissect a set of very complex breathing traits into components that interact with lung mechanical properties or metabolic rate. For example, timing of breathing is determined by the elastic properties of the lung. Also, the tidal volume achieved during hypercapnic breathing is dependent on the total lung capacity and compliant properties of the lung. Therefore, the genes that control variation in breathing characteristics may ultimately regulate the development of the lung. While the central nervous system must be involved in this integrated process, both the brain and the lung must develop in concert with one another. Likewise, the integration of the brain and the lung must maintain metabolic homeostasis delivering oxygen and eliminating carbon dioxide. In addition, the genes that are expressed in the brain are potentially modulated by the physical properties of the lung.

10.2 Experimental methods and design

10.2.1 Animals

Male and female C3 and B6 mice (n = 6 mice/gender/strain) were purchased from Jackson Laboratories. The animals were housed in a facility at the Johns Hopkins University, Bloomberg School of Public Health. The temperature of the facility was maintained at ~21.5°C, and the light-dark cycle varied every 12h, beginning at 0700. The animals were housed in microisolation cages, and were fed ad libitum with a pelleted stock diet. The study was approved by the Johns Hopkins University Animal Care and Use Committee, and complied with the American Physiological Society Guidelines.

10.2.2 Whole-body plethysmography

Breathing characteristics were assessed at baseline and during hypercapnic, normoxic (5% CO_2 in 21% O_2) inspirate challenge protocols using whole-body plethysmography. Using unanesthetized and unrestrained conditions, each animal was permitted to acclimate in the chamber at least 30 min before ventilation measurements were obtained. Chamber temperature was maintained within the thermoneutral zone for mice (i.e. 26-28°C), and was recorded with each ventilatory measurement using a Type-T thermocouple. Compressed air was humidified (90% relative humidity) and directed through the chamber at a flow rate ~300 ml·min^{-1}. At a constant chamber volume, changes in pressure due to inspiratory and expiratory temperature fluctuations were measured using a differential pressure transducer (Model 8510B-2, Endevco Co.). Additional details regarding the plethysmographic method and the computation of other ventilatory traits have been reported elsewhere [46; 47].

10.2.3 Measurements of metabolic rate

Oxygen consumption (VO_2) and carbon dioxide production (VCO_2) were measured with a commercially available indirect open circuit calorimetric system (Oxymax Deluxe, Columbus Instruments, Inc., Columbus, OH) in-line with 200 ml cylindrical Plexiglass chambers. Unhumidified compressed air was delivered through the chamber, under the control of a calibrated flow meter. The flow was adjusted to maintain a small difference between chamber inflow (21% O_2 in N_2) and outflow oxygen concentrations. The airflow out of the chamber was dried using a column of anhydrous $CaSO_4$, and was sampled for 30 s for fractional concentrations of O_2 and CO_2 using a limited diffusion O_2 sensor and nondispersive infrared CO_2 sensor (Columbus Instruments, Columbus, OH). Sensor output was transmitted to a dedicated computer operated by data-acquisition software (Oxymax v 5.3, Columbus Instruments, Columbus, OH) for on-line computation and display of metabolic parameters. Additional details regarding indirect calorimetric methods have been reported elsewhere (40).

10.2.4 Respiratory impedance measurements

Animals were anesthetized with intraperitoneal injections of pentobarbital sodium at a dose of 85 mg/kg body weight. After the trachea was cannulated, the mouse was connected via the tracheal cannula to a computer-controlled small animal ventilator (FlexiVent, Montreal, PQ) while in a supine position. All mice were mechanically ventilated with 100% O_2 at 150 breaths/min and a tidal volume of 10 ml/kg at a positive end-expiratory pressure (PEEP) of 2 cmH$_2$O. After 10 min of mechanical ventilation, each animal was paralyzed with an intraperitoneal injection of 0.05 ml succinylcholine (9 mg/ml). A computer-generated volume signal composed of 19 mutually primed sinusoids ranging from 0.25 to 19.625 Hz was applied to the airway opening. Five minutes after a deep inspiration at an airway pressure of 30 cmH$_2$O, the impedance of the respiratory system (Z_{rs}) was measured, and then fitted by Flexivent software to a constant phase model. Additional details regarding the respiratory impedance method and the computation of other lung mechanics traits, including tissue elastance (H) and tissue damping (G) have been reported elsewhere [14].

10.2.5 Pressure-Volume (PV) curves

After the impedance measurements, mice were mechanically ventilated with 100% O_2 at a PEEP of 2 cmH_2O for 1 min, and the cannula was sealed with a stopcock for 3 min to degas the lung. Quasistatic PV curves were then performed in situ. The rate of inflation and deflation was standardized by a dual infusion-withdraw pump (model 900-610, Harvard Apparatus, Dover, MA), and the airway pressure was measured by using a differential pressure transducer (model 8510B-2, Endevco). The initial inflation rate was controlled at ~0.75 ml/min to ensure that all lung regions opened before being switched to a rate of 2 ml/min for the remaining inflation-deflation maneuvers. The pressure limits of the inflation and deflation airway pressures were 30 and -10 cmH_2O, respectively. The volume on deflation 30 cmH_2O (V_{30}) was considered to represent total lung capacity, respectively. Additional details regarding the PV maneuver have been reported elsewhere [33; 50].

10.3 Observations and results

As shown in Fig. 2, the breathing pattern at baseline differs substantially ($P < 0.05$) between B6 and C3 mice, which is consistent with our previous studies. That is, B6 mice show a rapid shallow breathing pattern relative to the slow deep pattern of C3 mice. Here, the results also suggest that these strain differences are evident for both males and females. The total time of the breathing cycle is more rapid in B6 mice during hypercapnic stimulation for both genders (right panel). Also, the hypercapnic tidal breath is notably ($P < 0.05$) reduced in volume for both B6 and C3 females compared with their male counterparts (left panel).

As shown in Fig. 3, the oxygen consumption and carbon dioxide production is significantly different in B6 females compared with both B6 males and C3 females (left panel). However, when we consider the baseline ventilation as referenced to the VO_2 or VCO_2, or the ventilatory equivalent (right panel), it is evident that the magnitude of ventilation in B6 mice is significantly ($P < 0.05$) greater than C3 mice for a given level of metabolism in both males and females. This suggests that C3 mice demonstrate a hypoventilatory response relative to B6 mice under normal unstimulated conditions.

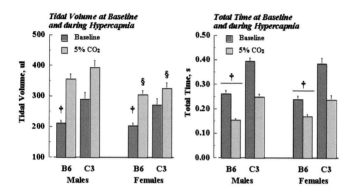

Fig. 2. B6 and C3 mice demonstrate substantially different pattern of breathing; that is, B6 mice are characterized by rapid, shallow breathing pattern at baseline relative to the slower, deeper pattern in C3 mice. During mild hypercapnic challenge, B6 mice achieve a similar tidal volume as C3 mice, while reducing the total time of each breath. [†] P <0.05; B6 vs. C3. [§] P <0.05; females vs. males.

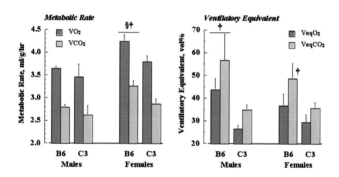

Fig. 3. Female B6 mice show substantially higher oxygen consumption (VO₂) and carbon dioxide production (VCO₂) compared with other groups; male B6 and C3 mice demonstrate similar VO₂ and VCO₂. The ventilatory equivalent (V_{EQ}; V_E normalized to VO₂ or VCO₂) is significantly greater in B6 compared with C3 mice. [†] P <0.05; B6 vs. C3. [§] P <0.05; females vs. males.

As shown in Fig. 4, there is a significantly greater percent increase in tidal volume and minute ventilation from baseline in response to a 5% CO_2 challenge for B6 mice (left panel). That is, the increase in V_T with hypercapnia is consistently greater in both male and female B6 mice. However, B6 males show a substantially (P <0.05) greater increase relative to B6 females.

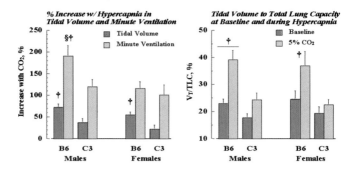

Fig. 4. The percent increase in hypercapnic V_T from baseline is notably greater in B6 relative to C3 mice. However, a significantly greater hypercapnic V_E occurs in only B6 males. When the V_T response is normalized for strain differences in total lung capacity (TLC), B6 mice show a substantially higher proportion of TLC in achieving hypercapnic V_T compared with C3 mice. [†] P <0.05; B6 vs. C3. [§] P <0.05; females vs. males.

If we refer the volume of the tidal breath to the total lung capacity, then we can better understand how lung reserve is being utilized. In Fig. 4 (right panel), there is a markedly (P <0.05) greater V_T/TLC ratio for B6 compared with C3 mice. This is especially true for the V_T achieved during hypercapnic challenge, and is consistently greater in both B6 male and female mice compared with their C3 counterparts. These results suggest that B6 mice are utilizing a substantially greater proportion of lung reserve at baseline and in response to hypercapnic stimulation.

From cosegregation analysis as shown in Fig. 4 and 5, the total time of the breathing cycle at baseline is significantly (P <0.05) related to lung mechanical properties, including lung elastance (left panel, r-value = - 0.71) and tissue damping (right panel, r-value = - 0.68). These results suggest that approximately 50% of the variation in respiratory timing between B6 and C3 mice can be accounted for by variation in lung elastance and tissue damping. Moreover, the cosegregation plots suggest that genetic determinants regulate strain variation in respiratory timing by modulating function by variation in lung mechanical properties A second cosegregation analysis is depicted in Fig. 6, where inspiratory timing is also strongly associated (e.g. r-value = - 0.72 for lung elastance; P <0.01) with lung mechanical properties in both male and female B6 and C3 mice. Since our laboratory has shown that variation in inspiratory timing between B6 and C3 mice is significantly modulated by genetic determinants on mouse chromosome 3, the possibility exists that these determinants ultimately act to control variation in lung volume-dependent lung mechanics.

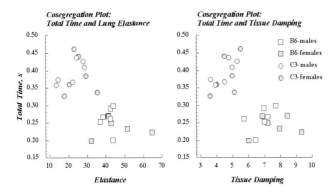

Fig. 5. The average total time of each breath at baseline is significantly correlated with lung elastance and tissue damping. In addition, the strain-dependent breathing (i.e. total timing) phenotypes between B6 and C3 mice cluster or cosegregate with the lung mechanical phenotypes (i.e. lung elastance and tissue damping).

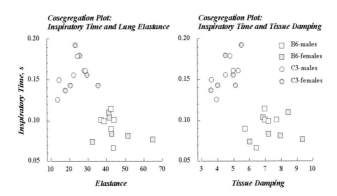

Fig. 6. The average inspiratory time of each breath at baseline is also significantly correlated with lung elastance and tissue damping. The strain-dependent breathing (i.e. inspiratory timing) phenotypes between B6 and C3 mice cosegregate with the lung mechanical phenotypes (i.e. lung elastance and tissue damping).

Likewise, the variation in V_T among male and female B6 and C3 mice is strongly associated (e.g. r-value = - 0.74 for lung elastance; p <0.01) lung mechanical properties as shown in Fig. 7. However, the association appears to be more robust for male (r-value = - 0.83) than female mice (r-value = - 0.67); that is, 69% of the variation in V_T in males was accounted for by variation in lung elastance, while only 45% in females.

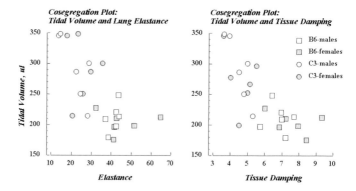

Fig. 7. The average tidal volume of each breath at baseline is also significantly correlated with lung elastance and tissue damping. In this case, the strain-dependent breathing (i.e. tidal volume) phenotypes between B6 and C3 mice do not appear to cosegregate as discretely with the lung mechanical phenotypes (i.e. lung elastance and tissue damping); however, the association remains robust.

The cosegregation results between breathing characteristics and lung mechanics extend to include the respiratory timing during hypercapnic stimulation as seen in Fig. 8. Whereas the total time of the breathing cycle during hypercapnic challenge is lesser than at baseline, the association is robust. That is, both total time (left panel, r-value = - 0.78) and inspiratory time (right panel, r-value = - 0.72) are strongly correlated with lung elastance. The association between inspiratory timing and lung elastance is particular robust in male mice (r-value = - 0.92; P <0.01).

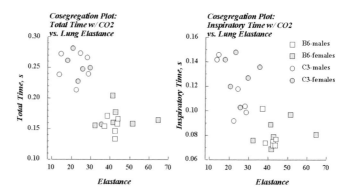

Fig. 8. The average total and inspiratory timing in response to hypercapnic challenge is also significantly correlated with lung elastance. The strain-dependent hypercapnic breathing (i.e. total and inspiratory timing) phenotypes between B6 and C3 mice cosegregate with the lung mechanical phenotypes (i.e. lung elastance).

10.4 Significance of genetic determinants

The current results demonstrate several different ways to evaluate respiratory phenotypes in inbred mouse strains. The measure of V_{EQ} for O_2 and CO_2 is used in the current analysis to normalize the magnitude of V_E at baseline between B6 and C3. While the measurement of V_{EQ} for O_2 and CO_2 suggests that B6 mice have a greater V_E for a given level of metabolism (Fig. 2), measurements of V_E, VO_2 and VCO_2 are not different between strains when considered separately. A second unique measurement of breathing in mice shown in the current study is referencing strain and gender variation in V_T to TLC. In this regard, B6 mice demonstrate a larger utilization of lung reserve to achieve hypercapnic ventilation compared with C3 mice (right panel, Fig. 4). Although the percent increase in V_T from baseline to hypercapnic challenge for both male and female B6 mice exceeds C3 counterparts, the percent increase in V_E with hypercapnia does not differ between female B6 and C3 mice (left panel, Fig. 4). Finally, several breathing phenotypes in this mouse model appear to correlate with fundamental lung mechanical properties suggesting that ~50% of the variation in breathing traits among individual mice is accounted for by variation in lung elastance and tissue damping. With several traits, the clustering of breathing and lung mechanical phenotypes suggests that the gene regulation of breathing is determined by the lung.

Changes in V_{EQ} for O_2 and CO_2 have been used during exercise testing in human subjects to monitor changes in V_E with increases in VO_2 and VCO_2, serving as an index to understanding the incremental rise in V_E with increased work. Variation in V_{EQ} for O_2 and CO_2 has also been measured during breathing at rest. In this context, a rise in V_E in the face of an unchanged VO_2 and VCO_2 has been used to indicate increased ratio between dead-space volume (V_D) and V_T [23]. More recently, this same concept has been used in mechanically ventilated patients with acute lung injury [3].

There are only a select few papers that have shown V_{EQ} for O_2 and CO_2 in mice [35]. More recently, our lab has used V_{EQ} for O_2 and CO_2 to better understand changes in V_E in the face of variation in metabolic rate due to the SOD1[G93A] mutation. The SOD1[G93A] mutant is frequently used as a model of amyotrophic lateral sclerosis, and the gene mutation effects mitochondrial function [48]. In this model, resting V_E is coupled to metabolism during 10-14 wks of age when the SOD1[G93A] mutants appear healthy. As mutant mice demonstrate progress neurodegeneration and greater paralysis between 16-18 wks, a rise in V_{EQ} for O_2 and CO_2 occurs indicating an uncoupling between ventilation and metabolism. It is possible in the current study that the greater V_{EQ} for O_2 and CO_2 at baseline in B6 mice indicates that there is greater V_D compared with C3 mice. Alternatively, B6 may be responding to a given level of O_2 demand or CO_2 production with a greater V_E due to a higher strain-specific chemosensitivity. Although the interpretation surrounding variation in V_{EQ} for O_2 and CO_2 is very complex, breathing phenotypes are better understood when referenced to metabolic rate. This is particularly appropriate in transgenic models or other interventions where mitochondrial function is altered [30].

While there are clear limitations in the use of whole-body plethysmography to measure V_T in mice, the strength of the measurement focuses on capturing breathing characteristics non-invasively in a conscious, unrestrained animal. In the present study, we consider the measurement of V_T as normalized to the total lung capacity. The results suggest that B6 and C3 mice utilize 18-25% of TLC to achieve V_T as one component of a strain-specific breathing pattern at baseline. Within the relatively narrow range, baseline V_T/TLC for male B6 mice is greater than C3 mice. One explanation is that this greater proportion of TLC utilized in B6 mice is part of a compensatory mechanism to overcome the greater V_D due to a higher strain-specific baseline breathing frequency. Therefore, if the timing of baseline breathing is regulated by the lung elastic properties, then a greater proportion of TLC is required to achieve a given V_E to maintain metabolic homeostasis.

The V_T response to hypercapnic challenge is also referenced to TLC in the current analysis to better assess ventilatory chemosensitivity. The V_T/TLC response of B6 mice to hypercapnia is substantially greater than C3 mice (~40% for B6 vs. 25% for C3). In two recent studies, investigators have explored breathing consequences of adverse lung development attributable to specific gene mutations [1; 22]. In both studies, lung morphological changes were the focus of potential mechanisms that led to altered breathing characteristics at baseline or during chemical challenge; however, lung volumes and mechanical properties were not reported. The results of the current study also involve two inbred strains with notably different lung morphometric characteristics. Alveoli of C3 mice are significantly enlarged compared to B6, suggesting that alveolar surface area is reduced in C3 mice [41]. Although it is unclear how alveolar enlargement directly impacts altered breathing phenotypes, the current study suggests that the link may involve the association between the timing of breathing and the lung elastic properties. One might anticipate that the alveolar enlargement may result in a loss of elastic recoil.

10.5 Conclusion

The clustering of breathing phenotypes with lung elastance in B6 and C3 mice is likely very complex. In general, cosegregation of phenotypes might lead to the conclusion that the alleles regulating the two outcomes are linked or physically located in proximity with one another. For example, the gene for elastin (i.e. *Eln*) is located on mouse chromosome 5, in proximity to a region where a QTL for the ratio of inspiratory-to-total time (i.e. T_I/T_{TOT}) during mild hypercapnia is located [36]. Therefore, the cosegregation of the timing of breathing, particularly during mild hypercapnia (Fig. 8), may implicate the *Eln* gene as a candidate gene for regulating variation in breathing in this model. Likewise, Reinhard et al. [33] identified a QTL on mouse chromosome 5 for differential lung compliance using the C3 progenitor. This QTL is in close proximity to a QTL for mean inspiratory flow (i.e. V_T/T_I), particularly related to the interaction between hypercapnic and hypoxic ventilation [43]. While these results represent circumstantial evidence, they may suggest that genetic control of breathing may be linked to determinants of lung mechanical propertics.

In conclusion, this chapter focused on the hypothesis that genetic determinants of breathing are very complex involving the interaction of genes expressed in the brain and lung. The use of inbred mouse strains and numerous transgenic models have shown the role of altered genes on a variety of breathing characteristics. The development of the respiratory system must integrate the mechanisms of neural control and lung mechanical properties. Therefore, a complex interplay of expression between gene determinants that program the respiratory center of the brain and the lung structure seems to be a prerequisite. Shea and Guz [38] proposed, in a review honoring the contributions of Pierre Dejours, that individual variation in breathing pattern at baseline and the magnitude of ventilation during chemical challenge is likely due to a complex interaction between lung mechanics and chemosensitivity. Here, we propose that individual genetic variation facilitates this interchange to occur.

References

1. Bonora M, Bernaudin JF, Guernier C, Brahimi-Horn MC (2004) Ventilatory responses to hypercapnia and hypoxia in conscious cystic fibrosis knockout mice Cftr$^{-/-}$. Pediatr Res 55: 738-746

2. Chen JC, Mannino DM (1999) Worldwide epidemiology of chronic obstructive pulmonary disease. Curr Opin Pulm Med 5: 93-99

3. Coss-Bu JA, Walding DL, David YB, Jefferson LS (2003) Dead space ventilation in critically ill children with lung injury. Chest 123: 2050-2056

4. De Sanctis GT, Merchant M, Beier DR, Dredge RD, Grobholz JK, Martin TR, Lander ES, Drazen JM (1995) Quantitative locus analysis of airway hyperresponsiveness in A/J and C57BL/6J mice. Nat Genet 11: 150-154

5. De Sanctis GT, Singer JB, Jiao A, Yandava CN, Lee YH, Haynes TC, Lander ES, Beier DR, Drazen JM (1999) Quantitative trait locus mapping of airway responsiveness to chromosomes 6 and 7 in inbred mice. Am J Physiol 277: L1118-L1123

6. Douglas NJ, Luke M, Mathur R (1993) Is the sleep apnoea/hypopnoea syndrome inherited? Thorax 48: 719-721

7. Douglas NJ, White DP, Weil JV, Pickett CK, Zwillich CW (1982) Hypercapnic ventilatory response in sleeping adults. Am Rev Respir Dis 126: 758-762

8. Ewart SL, Mitzner W, DiSilvestre DA, Meyers DA, Levitt RC (1996) Airway hyperresponsiveness to acetylcholine: segregation analysis and evidence for linkage to murine chromosome 6. Am J Respir Cell Mol Biol 14: 487-495

9. Fleetham JA, Arnup ME, Anthonisen NR (1984) Familial aspects of ventilatory control in patients with chronic obstructive pulmonary disease. Am Rev Respir Dis 129: 3-7

10. Gold AR, Schwartz AR, Wise RA, Smith PL (1993) Pulmonary function and respiratory chemosensitivity in moderately obese patients with sleep apnea. Chest 103: 1325-1329

11.   Gothe B, Altose MD, Goldman MD, Cherniack NS (1981) Effect of quiet sleep on resting and CO_2-stimulated breathing in humans. J Appl Physiol 50: 724-730

12.   Gould GA, Whyte KF, Rhind GB, Airlie MA, Catterall JR, Shapiro CM, Douglas NJ. The sleep hypopnea syndrome (1988) Am Rev Respir Dis 137: 895-898

13.   Higgins M. Risk factors associated with chronic obstructive lung disease (1991) Ann N Y Acad Sci 624: 7-17

14.   Huang K, Rabold R, Schofield B, Mitzner W, Tankersley CG (2007) Age-dependent changes of airway and lung parenchyma in C57Bl/6J mice. J Appl Physiol 102: 200-206

15.   Hudgel DW, Weil JV (1974) Asthma associated with decreased hypoxic ventilatory drive. A family study. Ann Intern Med 80: 623-625

16.   Hudgel DW, Weil JV (1975) Depression of hypoxic and hypercapnic ventilatory drives in severe asthma. Chest 68: 493-497

17.   Jeffrey AA, Warren PM, Flenley DC (1992) Acute hypercapnic respiratory failure in patients with chronic obstructive lung disease: risk factors and use of guidelines for management. Thorax 47: 34-40

18.   Kaufman BJ, Ferguson MH, Cherniack RM (1959) Hypoventilation in obesity. J Clin Invest 38: 500-507

19.   Kawakami Y, Irie T, Kishi F, Asanuma Y, Shida A, Yoshikawa T, Kamishima K, Hasegawa H, Murao M (1981) Familial aggregation of abnormal ventilatory control and pulmonary function in chronic obstructive pulmonary disease. Eur J Respir Dis 62: 56-64

20.   Kawakami Y, Irie T, Shida A, Yoshikawa T (1982) Familial factors affecting arterial blood gas values and respiratory chemosensitivity in chronic obstructive pulmonary disease. Am Rev Respir Dis 125: 420-425

21.   Kikuchi Y, Okabe S, Tamura G, Hida W, Homma M, Shirato K, Takishima T (1994) Chemosensitivity and perception of dyspnea in patients with a history of near-fatal asthma. N Engl J Med 330: 1329-1334

22.   Kinkead R, LeBlanc M, Gulemetova R, Lalancette-Hebert M, Lemieux M, Mandeville I, Jeannotte L (2004) Respiratory adaptations to lung morphological defects in adult mice lacking Hoxa5 gene function. Pediatr Res 56: 553-562

23.   Kinney JM, Askanazi J, Gump FE, Foster RJ, Hyman AI (1980) Use of the ventilatory equivalent to separate hypermetabolism from increased dead space ventilation in the injured or septic patient. J Trauma 20: 111-119

24.   Levitt RC, Mitzner W (1989) Autosomal recessive inheritance of airway hyperreactivity to 5-hydroxytryptamine. J Appl Physiol 67: 1125-1132

25.   Levitt RC, Mitzner W, Kleeberger SR (1990) A genetic approach to the study of lung physiology: understanding biological variability in airway responsiveness. Am J Physiol 258: L157-L164

26.   Marin JM, Montes de Oca M, Rassulo J, Celli BR (1999) Ventilatory drive at rest and perception of exertional dyspnea in severe COPD. Chest 115: 1293-1300.

27.   McFadden ERJ (1991) Fatal and near-fatal asthma. N Engl J Med 324: 409-411

28.   Montes de Oca M, Celli BR (1998) Mouth occlusion pressure, CO_2 response and hypercapnia in severe chronic obstructive pulmonary disease. Eur Respir J 12: 666-671

29.   Mountain RC, Zwillich CW, Weil JV (1978) Hypoventilation in obstructive lung disease; the role of familial factors. N Engl J Med 298: 512-525

30.   Prabhakar NR, Mitra J, Adams EM, Cherniack NS (1989) Involvement of ventral medullary surface in respiratory responses induced by 2,4-dinitrophenol. J Appl Physiol 66: 598-605

31.   Redline S, Leitner J, Arnold J, Tishler PV, Altose MD (1997) Ventilatory-control abnormalities in familial sleep apnea. Am J Respir Crit Care Med 156: 155-160

32.   Reinhard C, Eder G, Fuchs H, Ziesenis A, Heyder J, Schulz H (2002) Inbred strain variation in lung function. Mamm Genome 13: 429-437

33.   Reinhard C, Meyer B, Fuchs H, Stoeger T, Eder G, Ruschendorf F, Heyder J, Nurnberg P, de Angelis MH, Schulz H (2005) Genomewide linkage analysis identifies novel genetic Loci for lung function in mice. Am J Respir Crit Care Med 171: 880-888

34.   Rosen RL (1986) Acute respiratory failure and chronic obstructive lung disease. Med Clin North Am 70: 895-907.

35.   Schlenker EH, Hansen SN, Pfaff DW (2002) Gender comparisons of control of breathing and metabolism in conscious mice exposed to cold. Neuroendocrinology 76: 381-389

36.   Schneider H, Patil SP, Canisius S, Gladmon EA, Schwartz AR, O'donnell CP, Smith PL, Tankersley CG (2003) Hypercapnic duty cycle is an intermediate physiological phenotype linked to mouse chromosome 5. J Appl Physiol 95: 11-19

37.   Schulz H, Johner C, Eder G, Ziesenis A, Reitmeier P, Heyder J, Balling R (2002) Respiratory mechanics in mice: strain and sex specific differences. Acta Physiol Scand 174: 367-375

38.   Shea SA, Guz A (1992) Personnalité ventilatoire : an overview. Respir Physiol 87: 275-291

39.   Sorli J, Grassino A, Lorange G, Milic-Emili J (1978) Control of breathing in patients with chronic obstructive lung disease. Clin Sci Mol Med 54: 295-304

40.   Soutiere SE, Tankersley CG (2001) Challenges implicit to gene discovery research in the control of ventilation during hypoxia. High Alt Med Biol 2: 191-200

41.   Soutiere SE, Tankersley CG, Mitzner W (2004) Differences in alveolar size in inbred mouse strains. Respir Physiol Neurobiol 140: 283-291

42.   Tankersley CG (2001) Selected Contribution: Variation in acute hypoxic ventilatory response is linked to mouse chromosome 9. J Appl Physiol 90: 1615-1622

43.   Tankersley CG, Broman KW (2004) Interactions in hypoxic and hypercapnic breathing are genetically linked to mouse chromosomes 1 and 5. J Appl Physiol 97: 77-84

44.   Tankersley CG, DiSilvestre DA, Jedlicka AE, Wilkins HM, Zhang L (1998) Differential inspiratory timing is genetically linked to mouse chromosome 3. J Appl Physiol 85: 360-365

45.   Tankersley CG, Elston RC, Schnell AH (2000) Genetic determinants of acute hypoxic ventilation: patterns of inheritance in mice. J Appl Physiol 88: 2310-2318

46.   Tankersley CG, Fitzgerald RS, Kleeberger SR (1994) Differential control of ventilation among inbred strains of mice. Am J Physiol 267: R1371-R1377

47.   Tankersley CG, Fitzgerald RS, Levitt RC, Mitzner WA, Ewart SL, Kleeberger SR (1997) Genetic control of differential baseline breathing pattern. J Appl Physiol 82: 874-881

48.   Tankersley CG, Haenggeli C, Rothstein JD (2007) Respiratory impairment in a mouse model of amyotrophic lateral sclerosis. J Appl Physiol 102: 926-932

49.   Tankersley CG, Kulaga H, Wang MM (2001) Inspiratory timing differences and regulation of Gria2 gene variation: a candidate gene hypothesis. Adv Exp Med Biol 499: 477-482

50.   Tankersley CG, Rabold R, Mitzner W (1999) Differential lung mechanics are genetically determined in inbred murine strains. J Appl Physiol 86: 1764-1769
51.   Zwillich CW, Sutton FD, Pierson DJ, Greagh EM, Weil JV (1975) Decreased hypoxic ventilatory drive in the obesity-hypoventilation syndrome. Am J Med 59: 343-348

11. Genes and development of respiratory rhythm generation

Jean CHAMPAGNAT, Gilles FORTIN and Muriel THOBY-BRISSON

UPR 2216 Neurobiologie Génétique et Intégrative, Institut de Neurobiologie Alfred-Fessard, C.N.R.S., 1, avenue de la terrasse, Gif sur Yvette 91198 Cedex, France

11.1 Introduction

How research program will most effectively translate basic scientific findings into potential treatment for respiratory disorders is a major concern of clinicians neurobiologists. Extensive discussion at the last Oxford Conference on Modeling and Control of Breathing (Lake Louise Alberta, Canada, September 19-24 2006) revealed the need for more adequate definition of respiratory phenotypic traits in animal models and human pathologies. This might greatly help relating specific syndromes to structural and/or neurochemical abnormalities in the brainstem. Starting with the idea that recent data on hereditary respiratory-related syndromes such as Joubert, Rett, and CCHS (see Chapters 4, 16) point to a major implication of developmental genes, the present chapter considers developmental steps from early specifications in the neural tube to the function of respiratory rhythm generators and controllers. During development, the assembly of neural circuits that encode animal behavior, results from several mechanisms contributing to generate distinct neuronal cell types in appropriate number and position. Regionalization of the neural tube controls proliferation and specifications of progenitors, until they exit from the cell cycle, at precise location according to antero-posterior and dorso-ventral axis of the neural tube [70; 106]. Neurogenesis refers to a variety of processes by which neuronal types differentiate, migrate and form nuclei to eventually produce neuronal populations and their synaptic interconnections [106]. Once activity starts in primordial neurons, neuronal circuitry is refined in a use-dependent manner [57] and the size of neuronal populations depends on the activity of neurotrophic factors interacting with apoptotic processes [82].

In all vertebrates, regionalization of the brainstem (rhombencephalon) along the antero-posterior axis, leads the partitioning of the neuroepithelium into a series of cellular compartments, called rhombomeres (r1-r8), [69]. This process takes place between embryonic days E8 and E12 in mice, between Hamburger and Hamilton [47] stages HH9 and HH24 in chicks, and influences later differentiation and spatial distribution of neuronal patterns [69]. The branchiomotor nuclei conform to the rhombomeric pattern with a two-segment periodicity. Trigeminal motoneurons originate from r2 and r3 and send their axons to an exit point in r2. In mammals, the facial branchial nucleus originates in r4. The ventral and dorsal motor nuclei of glossopharyngeal-vagal nerves derive from r6-r8.

Strong regulatory constraints couple *Hox* gene expression to the progression of embryogenesis. Therefore, the chromosomal organization of *Hox* genes into four clusters is highly conserved in vertebrates. In birds and mammals, the formation of territories patterned by *Hox* genes is accompanied by a sequential activation of these genes from 3' to 5' in the clusters. As a result, early structures are given an anterior identity with 3'*Hox* genes as key determinants, while progressively later structures start expressing more 5' *Hox* genes and acquire a more posterior identity. Thus, *Hox* genes play a key role in anteroposterior patterning of the rhombencephalon at pre- and early-segmental stages of development. The most 3' *Hox* genes, *Hoxa1* and *Hoxb1* are expressed up to the rhombomeric r3/r4 boundary and they are directly down regulated in r3 by Krox20, a zinc-finger transcription factor also known as Egr-2. This signalling is required for the proper development of r3, r4 and of the boundary between r3 r4 [39; 70; 97]. Neurogenesis starts within rhombomeres at end segmental stages, shortly before hindbrain segmentation disappears. It is therefore important to consider whether this early developmental regulation of progenitors influences the function of mature brainstem circuits at later stages.

Rhombomeres are polyclonal developmental compartments containing all types of progenitors that are required to produce the large variety of cellular elements forming neuronal circuits and reflex arches at a given level of the neural tube. Developmental studies in the embryonic spinal cord have provided a genetic framework for analyzing how the individual components of circuits are configured into a network for controlling behaviors [42; 54]. Eleven classes of neurons and there respective progenitor domain have been identified along the dorso-ventral axis of the neural tube. This general organization of the spinal cord is maintain in the hindbrain where selective progenitor domains are added (e.g. giving rise to branchiomotor motoneurones described above), and dorso-ventral signaling interacts with the antero-posterior segmental organization. For example, serotonergic neurons are born in r2r3 and r5r7 where they derive from the same dorso-ventral progenitor domain as branchiomotor motoneurons [16; 87; 88]. Downregulation of the homeobox gene *Phox2b* is required for the progenitors to switch from a motor to a serotonergic fate. However, in the murine r4, the generation of facial branchial motor neurons carries on and consequently, no serotonergic differentiation occurs because the switch is prevented by the maintenance of *Phox2b* expression by *Hoxb1* [87]. Although the dorso-ventral progenitors of respiratory neurons are still under scrutiny, it is therefore important to keep in mind that cell fates in the hindbrain are determined by the interaction between distinct antero-posterior and dorso-ventral programs of specification.

In rodents, respiratory groups have been located with the help of preparations isolated *in vitro*. A correlation can be suggested with the rhombomere related organization of the branchio-motor nuclei [20]. Caudally, at the vagal glosso-pharyngeal level, the inspiratory rhythm is generated in the pre-Bötzinger complex (pre-BötC, see [31]), and persists in coronal brainstem slices of newborn rodents, isolated *in vitro* [44; 66; 101]. Central chemosensitivity results from the specific responsiveness to CO_2 of different structures, among which the retro-trapezoid nucleus [78], ATP-releasing [42] and serotonergic structures [99]. At the facial level, another rhythm generator, called the parafacial respiratory group (pFRG, [31; 81]), has been recently delineated by activity-dependent imaging. In the hindbrain isolated *in vitro*, permanent rhythm generation requires a balanced interaction between the pFRG and the pre-BötC, which can be disrupted by potent respiratory depressants such as opioïds [27; 72; 75-77]. At the trigeminal level, ventral pontine controls of the rhythm [13] include noradrenergic neurons of the A5 group exerting a depressant effect upon the more caudal respiratory group [28]. The most rostral hindbrain, close to the hindbrain/midbrain boundary contains the pontine respiratory group as well as several structures controlling breathing (reviewed in [73]).

The respiratory rhythm is not the first permanent rhythm produced in the brainstem. Intrinsic neuronal activity starts in the neural tube with a low frequency rhythm (LF, range min^{-1}) called "primordial activity" [20] that can be recorded from branchiomotor nerves in isolated hindbrain preparation [33]. Understanding how the faster respiratory rhythm emerges within the slowly active hindbrain is a major challenge that interests respiratory neurobiologists and clinicians. In the present chapter, we first review embryological observations investigating in the chick, the evolution of the primordial rhythm producing an episodic pattern. In this species, maturation of embryonic central pattern generators of the hindbrain is manifest in the course of embryonic day 6 (E6) when each LF burst becomes suddenly followed by cycles of additional high frequency burst discharges (range sec^{-1}) that form an episode [33]. The analysis in the chick embryo reveals importance of the rhombomere-related expression of *Krox-20*, also required in mice to prevent neonatal apneas. Developmental processes and modulatory controls involved in the set up of the two interconnected respiratory generators are developped in the second part of this chapter.

11.2 Primordial embryonic rhythm in the neural tube

In the murine hindbrain, primordial rhythmic activity is widely propagated from midline areas involving rostral serotonergic neurons [51] and dorso-medial structures in the caudal medulla [110]. In fact, spontaneous activity occurring at low frequency and invading large territories has been described in various species and a wide variety of developing structures in the CNS such as the spinal cord [15; 24; 49; 79; 80; 95; 117], the cortex [37; 89], the hippocampus [7], the retina [36; 57], and the brainstem [1; 33; 47]. Therefore, the LF activity observed in the embryonic chick and mouse brainstem probably corresponds to an immature, non-respiratory primordial activity that is required for the maturation of neuronal circuits.

Later maturation ensues in the chick leading to the episodic pattern. Acute transversal sections of the rhombencephalon show that in chick, in contrast to the widely distributed primordial rhythm, this episodic activity is first generated at a precise, parafacial, level of the neural tube [34; 35]. Embryological analysis reveals the involvement of genes that are expressed in a segmental-specific manner.

11.2.1 Embryological approach to understanding development of rhythm generators

To investigate development of circuits responsible for generation of episodic activities in chick embryos, we isolated segments of the neural tube and inverted the anterior/posterior polarity of rhombomeres just prior to the beginning of neurogenesis. The neural tube, dissected free of adherent tissue was transected at the level of rhombomere boundaries, making sure that the boundary and the immediately adjacent cells were excised; the anterior-posterior orientation of the excised rhombomere was marked by leaving tags of dorsal ectoderm on one-half of the explant. For transplantation, stage-matched host eggs were windowed and a gap in the neural tube was created at the desired site of insertion. The graft was then inserted, the orientation of the ectoderm being maintained or inverted. The egg was resealed and incubated for either 24 hours, for whole-mount *in situ* hybridization, or 5-6 days (until the embryo reached E7) for electrophysiological recording. Preparations for electrophysiology consisted of either whole hindbrains or isolated rhombomere islands (particularly r3r4 segments), made by dissecting out the tissues and transferring them to a recording chamber. Results suggest that a rhombomeric code is required for the assembly of a specific neuronal circuit within the hindbrain. Interestingly, assembling the parafacial rhythm-promoting episodic network appears to require two-segment functional units (r3r4) in the chick embryo [35].

11.2.2 The inductive role of Krox20 on rhythm generators

Starting with this r3r4 pair necessary for episodic rhythm generation, we have replaced homotopically r3 by two other odd rhombomeres either r1 or r5 [26]. An episodic rhythm resulted in r5r4 segment pairs, but not in r1r4 pairs that maintained an immature LF pattern. Thus, to identify a possible role of *Krox20*, which is expressed in r3 and r5 but not in r1 and which is known to specify aspects of r3 and r5 character [39] we used a gain of function strategy. We electroporated a *Krox20* expression vector at the r1 level in donor embryos. Electroporated r1 were then engrafted and isolated together with r4 to form r1r4 pairs, where *Krox20* expression is restricted to the odd rhombomere r1 ("r1[Krox20]").

An episodic rhythm resulted in r1[Krox20]r4 pairs indicating the instructive role of *Krox20* in the interaction with r4. Hence the expression of *Krox20* is sufficient to endow other rhombomeres, r4 in the normal hindbrain, with the capacity to induce rhythmic activity [11; 26].

11.2.3 Transrhombomeric interactions between r3 and r4

An interesting aspect of this induction is the apparent anterior-posterior polarity requirement whereby the *Krox20* expressing rhombomere can signal to even rhombomeres when these occupy a posterior (e.g. r3r4) but not an anterior (e.g. r2r3 or r4r5) position in the pair. Anterior-posterior (AP) reversal of r4 dramatically altered the development of the rhythm because the orientation of target growing neurons is altered. In contrast, the source of signaling should not be affected by the reversal. Therefore, to identify which of the two rhombomeres in the r3r4 pair exerts directional control on the other we inverted the AP polarity of each rhombomere independently. AP inverted (flipped, f) orientation of r3, in the pair f(r3)r4, did not change the episodic rhythm. However, AP reversal of r4 in the pair r3f(r4) resulted in both absence of episodic rhythm generation and an unexpected loss of regularity in the LF rhythm. This experiment suggests that r3, not r4, appears to be the source of the inter-rhombomeric influence and that the developing rhythmogenic network originates in r4, not r3 [26]. Therefore, the signaling initiated by *Krox20* expression influences the wiring of parafacial (r4) rhythm generators in the chick.

11.3 Parafacial rhythm generators : induction requires *Krox20* in r3 and *Hoxa1* in r4

Investigation of the respiratory behaviour of $Krox20^{-/-}$, $Hoxa1^{-/-}$ and $Kreisler^{-/-}$ mutants [21; 29; 52] has established links between the transient rhombomere-related expression of these genes and deletions, neoformations and reconfiguration of the respiratory network. These three mutations helped in unveiling a vital role of the rhombomeric r3r4 domain for the specification of an adapted respiratory rhythm at birth [10].

11.3.1 Transgenic mice models for neonatal apneas

Inactivation of *Krox20* and elimination of r4 in *Hoxa1* null mutants lead to neonatal death due to central respiratory deficits and probable collapse of the cardiovascular system. The $Hoxa1^{-/-}$ neonates show profound hypoplasia in the ventral pons including the facial branchiomotor nucleus (r4-derived) and of the underlying adjacent area where the pFRG is located. These mutants revealed the "anti-apneic" function of the pFRG at birth. In contrast to *Hoxa1* and *Krox20* null mice, homozygous *Kreisler* mice lacking r5, show neither life-threatening apneic respiratory patterns nor dramatic reduction of the ventral area of the pons containing the facial branchiomotor nucleus, but rather a hypoplasia of the ventral reticular formation causing an abnormal rostral positioning of the ambiguus nucleus [21]. Therefore, the ventral reticular neurons in r4 and r5 greatly differ with respect to their respiratory fates, and lethality in *Hoxa1* mutants involves the elimination of parafacial structures originating from r4 rather than from r5.

In vivo and *in vitro* analysis show that the caudal medulla including the pre-BötC is not affected by the *Krox20*[-/-] and *Hoxa1*[-/-] mutations [22; 29; 52]. Opioid peptides, such as enkephalins, acting through μ and δ receptors of the pre-BötC induce a powerful depressant modulatory control of respiratory rhythm [8; 72; 75; 81]. This system remains functional after the impairment of parafacial structures. Subcutaneous administration of the opioid antagonist naloxone to *Krox-20*[-/-] and *Hoxa1*[-/-] animals increased the respiratory frequency for several hours and definitively reversed the occurrence of apneas and lethality [22; 29; 52]. These observations indicate that neuromodulation by endogenous opioids contributes to the generation of life-threatening apneas that develop after impairment of the parafacial system and that antagonizing the enkephalinergic transmission might provide an effective pharmacological tool to be used when the neonatal anti-apneic function is disrupted.

11.3.2 Role of Hoxa1

Interestingly, progenitor cells mis-specified in *Hoxa1*[-/-] mutants were not necessarily eliminated during later developmental stages and their developmental fate mutants could be identified by examining cell migration patterns [29]. In wild-type E11.5 embryos, facial motoneurons migrate caudally to form the facial motor nucleus, whereas trigeminal motoneurons migrate dorsally. In the mutated r4 region, a greatly reduced number of facial motoneurons migrate caudally and an abnormal trigeminal-like lateral migration of r4 motoneurons is detected at E11.5 and completed at about E12.5. Thus, neurons with an anteriorized trigeminal-like phenotype appear to survive at the parafacial level of *Hoxa1* mutant hindbrains. Using thick horizontal brainstem slices, the effects of pharmacological stimulation of the r4-derived dorsal pons on the respiratory activity recorded from motor cranial nerve were compared between control and *Hoxa1* mutants [29]. In contrast to wild type preparations in which such stimulations had no effect on the rhythm frequency, stimulation in the mutant induced a robust increase in frequency, followed by a transient inhibition of the rhythm. In the mutants, labelling the pre-BötC with DiI revealed an axonal pathway and running laterally in the mutant pons that is not present in the wild-type. These results strongly suggest the presence of supernumerary functional efferent connections of the dorsal pons to the rhythm generator [22; 29]. These observations give insights into the crucial role of *Krox20/Hoxa1* on early phenotypic choices that affect parafacial neuronal progeny, leaving the pre-BötC functionally intact.

11.3.3 Segmentation and neurogenesis may interact at late segmental stages

Krox20 and *Hox* expression, boundary formation and neurogenesis do not take place at the same time in adjacent rhombomeres. Between E9.5 and E10.5 in mice, and between HH10 and HH15-20 in chick, r3 displays a marked delay compared with even rhombomeres in the timing of neuronal differentiation and axonal outgrowth [39; 69; 97]. Therefore, heterochrony of neurogenic processes at the parafacial level of the neuraxis allows neuronal differentiation in r4 and continuing expression of segmental genes such as *Krox20* in r3 at the same developmental stage. Given that rhombomeric heterochrony is found in both chick and mouse, we hypothesize that there are likely to be conserved signaling interactions by which the expression of *Krox20* in r3 may influence neural circuits developing in r4.

For example, all growing axons of r4-derived neurons avoid territories expressing *Krox20* or avoid the ectopic patches of *Krox20* expression after electroporation [39]. The gene encoding EphA4, a member of the Eph family of the tyrosine kinase receptor might be interesting to investigate [26]. It is a known target of the Krox20 transcription factor. It is expressed in odd rhombomeres while their ephrin ligands are expressed in alternating even rhombomeres and participate in inter-rhombomeric cell sorting. Expression of Eph receptors and their ephrin ligands have been shown to affect neuronal wiring at several levels of the central nervous system, particularly in the spinal cord, where EphA4-positive neurons are excitatory components of the locomotor central pattern generator [63]. In the hindbrain, the heterochronic development of the r3r4 segment provides a precise window of time, of about one day, during which Krox20-initiated signaling persists in r3 while future rhythmic circuits in r4 start to differentiate.

11.4 Onset of the respiratory rhythm generation

In rodents, the pre-BötC, a region immunoreactive for the neurokinin receptor NK1R [44; 85; 102; 114], is thought to contain all of the necessary elements for respiratory rhythm generation in rodents [60; 66; 93; 101]. This group is readily identifiable and amenable to analysis *in vitro* in preparations in which the pFRG is eliminated by a transection caudal to the facial motor nucleus : pFRG and pre-BötC are therefore distinct and interacting respiratory generators (as reviewed by [31]). The pre-BötC neural network contains rhythmic neurons connected by glutamatergic synapses, some of them exhibiting pacemaker burst-generating properties [55; 61; 90; 107; 108]. The destruction of neurons expressing the NK1R in the pre-BötC leading to ataxic respiration [45] and sleep apneas [74] supports the view that the pre-BötC plays a primary role in respiratory rhythmogenesis in adults.

11.4.1 Emergence of the pre-Bötzinger rhythm generator

Using immunohistochemical, electrophysiological and calcium imaging techniques on transverse slices of the brainstem, recent evidence showed that the pre-BötC is anatomically and functionally defined in the mouse embryo at E15 [110]. Rhythm generation depends on glutamatergic transmission through AMPA/kainate receptors, as in older preparations [44; 61; 101]. This glutamatergic transmission is concomitant to establishment of NK1R immunoreactivity in this region. At this stage, substance P and the opioid agonist DAMGO exert respective excitatory and depressing neuromodulatory influences on the rhythm [44; 53; 72; 75]. Moreover the onset of respiratory activity in the mouse embryo during E15 corresponds to the developmental stage at which fetal movements become detectable [1; 105]. In keeping with the shorter gestational period of mice compared to rats, mouse developmental time E15 probably corresponds to rat E17 [111], when formation of the respiratory rhythm generator is completed and begins to drive respiratory-like activity and fetal breathing movements [28; 46; 58; 85].

11.4.2 Synchronization of burst-generating neurons in the pre-Bötzinger complex

Recent results provide insights on the possible mechanisms that turn on pre-BötC activity [113]. Glutamate excitatory transmission is dependent on the release from glutamate-filled presynaptic vesicles loaded by three members of the solute carrier family, Slc17a6-8, that function as vesicular transporters (VGLUTs). VGLUT2 (Slc17a6) was found to be required for the function of the respiratory neuronal network. While VGLUT1 is dispensable for life, *Vglut2* null mutant pups die immediately after birth due to complete loss of rhythm generation in the pre-Bötz. Observations in embryos at the stage when the pre-BötC activity normally emerges shows that in these mutants respiratory rhythm is not turned on, although burst-generating mechanisms and excitatory transmission through e.g. immature excitatory GABAergic synapses are available. The entire set of pre-BötC characteristics being differentiated at E15, functional VGLUT2 glutamatergic synapses underlying neuronal synchronization seems to constitute the rate limiting step for establishment of the pre-BötC as a rhythm generator.

11.4.3 The pre-Bötzinger complex is distinct from the primordial generator

The spatial organization of neuronal circuits supporting primordial embryonic activity (LF) was also examined at the onset of pre-BötC activity [110]. LF bursts arose unilaterally from a dorso-medial area flanking the 4th ventricle. The LF events were initiated from either the right or the left sides of the slice and often alternated in time from one side to the other. Generation of this LF activity in the embryonic mouse brainstem does not seem to rely on single anatomical structures or specific circuits since it was observed in both facial and medullary slices. In contrast, at later stages, the bilateral pre-BötC rhythm generation site was spatially restricted to a region ventral to the nucleus ambiguous whose activity is preserved in a transverse slice [110]. Therefore, calcium imaging of the embryonic pre-BötC establishes that the HF generator in mice does not derive from the pre-existing LF generating network. However, none of the characteristics tested so far have allowed identification of a candidate population of migrating neural cells destined to form this HF generator prior to E15, at a stage when the LF generator is already active. In a previous study in the rat embryo, Pagliardini et al. [85] similarly reported the lack of NK1 positive cells in the pre-BötC location before the onset of HF activity. There seems to be no neuron exhibiting the pre-BötC-like phenotype that would be located between the germinative zone and the final ventrolateral location of the pre-BötC. This strongly supports the view that the entire set of pre-BöTC characteristics differentiates at E15 in ventro-lateral post-migratory neurons, distant from the dorso-medial sites where LF is generated.

11.4.4 Mouse/chick chimeras reveal the post-otic origin of the pre-Bötzinger complex

Embryological experiments confirm that the post-otic territory (r6-r8, caudal to the otic vesicle) from which the pre-BötC develop, is endowed with a selective capacity to generate HF. In these experiments [12], pre-otic (r3r4), post-otic (r6r8) and spinal segments of the mouse neural tube were isolated and grafted into the same position in stage-matched chick hosts (heterospecific grafting). Three days later, cranial nerves exiting the graft and the host indicate that HF is generated in post-otic segments, while more anterior pre-otic and more posterior spinal territories generate LF. These experiments suggest that a selective developmental program enforcing HF rhythm generation is already set at end segmental stages in post-otic segments. Development of the pre-BötC burst generators ensues, leading eventually to the onset of fetal breathing when glutamatergic synchronizing relationships are active.

11.4.5 Clinical importance of the dual oscillator hypothesis after birth

There is emerging evidence that the pre-BötC and the pFRG are distinct and play a critical role in generating respiratory rhythm. Ablation of the pre-BötC in adult rats resulted in a progressive, increasingly severe disruption of the respiratory pattern initially during sleep and then also during wakefulness. Therefore, sleep-disordered breathing and apneas may result from the loss of pre-BötC neurons in elderly humans and patients with neurodegenerative disease [74]. Studies in neonatal $Krox20^{-/-}$ and $Hoxa1^{-/-}$ rodents show that apneas can also be induced by disrupting the development of the pFRG. In an attempt to improve survival by stimulating the respiratory rhythm, a deficit of the retro-trapezoid CO_2-sensitive nucleus, located close to the pFRG was first excluded. Respiration was stimulated by CO_2, only for the duration of the CO_2 exposure, and all animal died during the first day of life [52]. In contrast, administration of naloxone to block opioid receptors in the pre-BötC [8; 72; 75; 81] definitely suppressed the occurrence of apneas and all animals survived past 18h of age [52]. These observations revealed a postnatal stage of development after which the breathing rhythm looses postnatal irregularity in wild type and apneas are no more detrimental to survival after the elimination of the pFRG in mutants. Therefore, the novel hypothesis on a dual respiratory rhythm generation has important implications in respiratory pathology, with the pFRG playing a major role to prevent sudden infant death while the pre-BötC becomes prevalent with age to alleviate sleep apneas in adults.

11.5 Neurotrophic control of breathing

During a restricted period of development, most neuronal types are dependent on trophic support thought to be mediated by selective membrane receptors interacting with diffusible growth factors such as neurotrophins [82]. Both clinical evidence and observations on transgenic mice indicate that genetic disruption of the neurotrophic support during development of the respiratory network, may cause aberrant respiratory patterns after birth, particularly in the human Rett syndrome [2; 3; 30]. There is consensus to consider that neurotrophins control the size of several neuronal populations at a given location in the neural tube. The role of the neurotrophin brain-derived neurotrophic factor (Bdnf) acting through TrkB receptor, was first identified on motoneurons [83; 98] and later on sensory neurons in $Bdnf^{-/-}$ mutant mice [30; 67]. This neurotrophin, probably involved in the Rett syndrome (see Chapter 16) is active until adulthood, particularly on the serotonergic spinal plasticity of breathing following intermittent hypoxia [5] (see Chapter 17). At early stages of development, Bdnf is also important in establishing connections between hindbrain pre-motor reticular neurons and motoneurons [56]. These initial observations prompted the analysis of Bdnf function in the development of central respiratory circuitry. Interestingly, the entire respiratory reflex arc in the vagal-glossopharyngeal domain seems to be affected by Bdnf. The neurotrophin has been implicated in the chemoafferent control of breathing [30] while the Bdnf receptor, TrkB, is strongly expressed in the nTS and the pre-BötC [109], suggesting its involvement in the pre-BötC control of the respiratory frequency.

11.5.1 Bdnf is involved in the development of chemoafferent neurons controlling breathing

The inactivation of genes encoding Bdnf or TrkB leads to abnormal patterns of quiet breathing and chemosensory responses. In $Bdnf^{-/-}$ mice $in\ vivo$, ventilatory responses to hyperoxia and hypoxia are depressed, but sensitivity to hypercapnia is intact. Both respiratory frequency and tidal volume per body mass are smaller than normal, leading to chronic hypoventilation that contributes to lethality during the first 3 weeks following birth [30]. Lack of Bdnf has been shown to influence survival in cranial sensory neurons [14] and function of central respiratory circuits [6]. Hence, $Bdnf^{-/-}$ mice show a dramatic hypoplasia of cranial sensory ganglia while the brainstem anatomy appears unaffected. The dopaminergic chemoafferent sub-population of the glossopharyngeal sensory neurons is particularly affected in petrosal cranial sensory ganglia [30]. Therefore, Bdnf like other neuronal growth factors is believed to regulate survival by modulating genetically programmed cell death *in vivo* thereby quantitatively matching neuronal populations to the size of their target field [82].

11.5.2 Bdnf controls activity of the pre-Bötzinger complex

In vitro observations indicate that $Bdnf^{-/-}$ mutants also exhibit abnormal central rhythm generation [6] suggesting that deficit in chemafferent inputs affects development of the rhythm generator. Alternatively, Bdnf may act centrally because in addition to its importance as a neuronal survival factor, Bdnf also regulates synaptic plasticity [19; 62; 68] and neuronal membrane conductances [9; 84; 96; 100]. In the vagal/glossopharyngeal hindbrain, expression of the Bdnf receptor, TrkB, and the effects of Bdnf application in rhythmic pre-BötC neurons showed that Bdnf plays an acute role in the modulation of respiratory rhythmogenesis after birth [109]. The Bdnf receptor, TrkB, strongly expressed in the pre-BötC, is co-localized with the substance P receptor NK1, a marker of a subtype of respiratory neurons in this region [44; 114]. In fact, single cell RT-PCR analysis demonstrated that individual pre-BötC neurons express TrkB, consistent with a direct action of Bdnf on these cells. Given that more than one third of TrkB positive pre-BötC neurons co-express NK1, and that NK1 neurons are required for normal breathing [45; 74], this population of cells is a likely target at which Bdnf acts to modulate respiratory rhythmogenesis *in vitro* [109]. In fact, a large subset of pre-BötC neurons, including cells involved in generating or regulating the respiratory rhythm, responds to Bdnf, and bath application of Bdnf to spontaneously active brainstem slices significantly decreases respiratory frequency generated in the pre-BötC. This effect is associated with an enhancement of excitatory synaptic currents in pre-BötC neurons and modifications of the properties of the hyperpolarization-activated cationic current Ih [109]which is involved in respiratory control *in vitro* [94; 108].

Altogether, experiments on the function of Bdnf at the vagal/glossopharyngeal level of the hindbrain demonstrate a tight coupling between development and function of rhythm generating circuits. On the one hand, neurotrophins act at specific stages of prenatal development when the size of neuronal subpopulation is controlled. But on the other hand, the same neurotrophin is also involved as a neuromodulator of neuronal activity and preserves this function at postnatal stages of development until adulthood [5]. Therefore, Bdnf acting through TrkB receptors in specific sub-populations of respiratory control cells, initiates the activation of signalling pathways exerting a wide range of modulatory influences, including the non-genomic control of membrane excitability and the genomic regulation of cell survival in specific neuronal sub-populations.

11.6 Brainstem modulatory controls of breathing

11.6.1 Central chemosensitivity

In embryos and neonates, the paired-like homeobox transcription factor, Phox2b, provide the best documented case of matching gene expression with function in the autonomic reflex arc during development [18; 41]. Mutations of the human homolog *PHOX2B* have been identified in patients with congenital hypoventilation syndrome, CCHS [4], a rare congenital syndrome which is characterized by a decreased sensitivity to hypercapnia and hypoxia, frequently associated with anomalies of the autonomic nervous system including Hirschsprung disease (see Chapter 4).

The majority of Phox2b dependent neuronal precursors, irrespective of their neurotransmitter phenotype, or of any aspect of their cellular phenotype or developmental history, are fated to partake in the medullary reflex circuits of the visceral nervous system (see Chapter 3). The homozygous inactivation of *Phox2b* affects glossopharyngeal and vagal afferents, the nTS and the nucleus ambiguus. Thus trophic support by Bdnf probably interferes with developmental processes by which cascades of transcription factors such as Phox2b eventually lead to the differentiation of specific neuronal cell types. Along this line, it is worth noting that one of the neuronal groups expressing *Phox2b* is the sub-population of dopaminergic chemoafferent neurons [17] that is selectively reduced by 50% in *Bdnf* mutants [30]. Clinical implications are important [116].

Alternatively, *Phox2b* mutation might interact with the differentiation of serotonergic neurons implicated in the plasticity of the respiratory response to episodic hypoxia [5], in the regulation of the pre-BötC function [72] and central chemosensitivity [99]. Transcriptional determinants of 5HT differentiation in r2r3 and r5r7 have been identified. A transition was found from branchiomotor/visceral motor to 5-HT neuron production, ventral progenitors switching its homeobox gene code from $Nkx2.2^+/Nkx2.9^+/Phox2b^+$ (motor) to $Nkx2.2^+/Nkx2.9^-/Phox2b^-$ (serotonergic). Loss of function experiments have demonstrated that *Nkx2.2* expression and downregulation of *Phox2b* are actually required for 5HT neuron production [16; 87; 88]. In normal r4, the switch is prevented by the maintenance of *Phox2b* expression by *Hoxb1*, whose expression is in turn maintained by *Nkx6.1/6.2* and *Hoxb2* [87]. Abnormal *Phox2b* expression may therefore affect development of the serotonergic control of breathing.

Recently, one of the central CO_2-sensitive structure of the ventral medulla, the retrotrapezoid nucleus [78] has been found to selectively express *Phox2b* [103] thereby identifying a potential site of altered chemosensitivity in *Phox-2b* deficient animal models (see Chapter 14) and human disease (see Chapter 4). Interestingly, this *Phox2b*-expressing site appears anatomically and developmentally different from the *Krox20*-dependent pFRG oscillator [112]. Altogether, these results suggest that breathing abnormalities in CCHS might result from abnormal chemosensitivity (see Chapter 4) and are probably distinct from neonatal apneas and sudden infant death resulting from the impairment of the pFRG control.

11.6.2 Pontine controls

Little is known on the early development of the rostral pontine control of breathing (reviewed in [73]). Gain-of-function experiments discussed above suggest that segmental signaling at end-segmental stages (E9.5) is sufficient to induce development of rhythm generators. However, segmentation of the hindbrain starts earlier, at E7.5 in mice [70]. Retinoic acid (RA) treatment at E7.5 with low (sub-teratogenic) doses was found convenient in mice to further investigate the importance of this stage for the breathing behavior. RA at this stage is known to affect the spread of *Hox* and *Gbx2* expression domains into the anterior hindbrain, followed by a retraction, leaving behind a posteriorised expression pattern [38; 65; 71]. Retinoic treatment induced pontine abnormalities associated with an hyperpnoeic episodic breathing, a clinical trait widely reported in pre-term neonates, in patients with Joubert syndrome, and in adults (Cheyne-Stokes respiration) with congestive heart failure and brainstem infarction. Respiratory and anatomical anomalies of treated mice resembled several traits reported in the Joubert syndrome, a genetically heterogeneous syndrome, with three known loci, on 9q34.3 (JBTS1), 11p11-q12 (CORS2/ JBTS2,[64]) and JBTS3, on chromosome 6q23.2-q23.3 [32]. Defects might involve the pontine respiratory group [59; 73; 91] as well as the cholinergic control of breathing by the pediculo-pontine tegmentum. Impairment of this latter cholinergic control is known to induce an abnormal breathing pattern in adult mice, as seen by homozygous inactivation of the gene encoding the enzyme acetylcholinesterase [23]. In contrast, in these mice, rhythm generation appears normal and no apneas were seen. This is probably because the pre-BötC adapts to hypercholinergy by decreasing the synaptic efficacy of nicotinic and muscarinic agonists [22]. Thus, different phenotypic traits may reflect the involvement of different brainstem regions. Given that developmentally distinct brainstem systems are involved in the breathing control, distinct pathologies reviewed in the present volume might result from pontine deficits (Joubert syndrome), abnormal chemosensitivity (CCHS) (see Chapter 4), neurotrophic control (Rett syndrome) (see Chapter 16), or the impairment of the rhythm generators (apneas).

11.7 Conclusion

Respiration is a rhythmic motor behavior that appears in the fetus and acquires a vital importance at birth. It is generated centrally, within neuronal networks of the hindbrain. This region of the brain is of particular interest since it is the best understood with respect to the cellular and molecular mechanisms that underlie its development. Examination of hindbrain activities in the chick and mouse embryo has revealed distinct central rhythm generators that are active before foetal maturation and conforms to the rhombomeric organization of the embryonic hindbrain. Functional properties of neurones are probably specified at the progenitor stage, by the combinatorial genetic patterns that are typical of their location in specific rhombomere territories. Therefore, inactivation of genes required for the normal formation of rhombomeres in mice leads to postnatal pathology; for example, an anomaly in the *Krox20/Hox* pathways may cause a neonatal apneic syndrome that affect respiration after birth and compromise the probability of survival, causing sudden infant death.

From a neurobiological point of view, these observations have been important to demonstrate that, although transient during early development, the rhombomeric pattern of the hindbrain segmentation strongly and irreversibly influences the establishment of the neuronal circuitry of the mature brainstem. Reconfiguration of neurons and synapses during foetal and postnatal stages is unable to restore viable neural functions when the basic scaffold is altered during early development. Thus, from studies of hindbrain development we may gain an understanding of how genes govern the early embryonic development of neuronal networks, how this might specify patterns of motor activities operating normally throughout life and how early developmental dysfunction might cause hereditary syndromes including postnatal breathing abnormalities in humans.

Acknowledgements

This work was supported by CNRS, ACI BDPI#57, E.C. grant "Brainstem Genetics" QLG2-CT-2001-01467.

References

1. Abadie V, Champagnat J, Fortin F (2000) Branchiomotor activities in mouse embryo. Neuroreport 11: 141-145
2. Angrist M, Bolk S, Halushka M, Lapchak PA, Chakravarti A (1996) Germline mutations in glial cell line-derived neurotrophic factor (GDNF) and RET in a Hirschsprung disease patient. Nat Genet 14 : 341-344
3. Amiel J, Salomon R, Attie T, Pelet A, Trang H, Mokhtari M, Gaultier C, Munnich A, Lyonnet S (1998) Mutations of the RET-GDNF signaling pathway in Ondine's curse. Am J Hum Genet 62: 715-717
4. Amiel J, Laudier B, Attie-Bitach T, Trang H, de Pontual L, Gener B, Trochet D, Etchevers H, Ray P, Simmoneau M, Vekemans M, Munnich A, Gaultier C, Lyonnet S (2003) Polyalanine expansion and frameshift mutations of the paired-like moeobox gene *PHOX2B* in congenital central hypoventilation syndrome. Nat Genet 33: 459-461
5. Baker-Herman TL, Fuller DD, Bavis RW, Zabka AG, Golder FJ, Doperalski NJ, Johnson RA, Watters JJ, Mitchell GS (2004) BDNF is necessary and sufficient for spinal respiratory plasticity following intermittent hypoxia. Nat Neurosci 7: 48-55

6. Balkowiec A, Katz DM (1998) Brain-derived neurotrophic factor is required for normal development of the central respiratory rhythm in mice. J Physiol 510: 527-533
7. Ben-Ari Y (2001) Developing networks play a similar melody. Trends Neurosci 24: 353-360
8. Bianchi AL, Denavit-Saubié M, Champagnat J (1995) Central control of breathing in mammals: neuronal circuitry, membrane properties and neurotransmitters. Physiol Rev 75: 1-45
9. Blum R, Kafitz KW, Konnerth A (2002) Neurotrophin-evoked depolarization requires the sodium channel Nav1.9. Nature 419: 687-692
10. Borday C, Abadie V, Chatonnet F, Thoby-Brisson M, Champagnat J, Fortin G (2003) Developmental molecular switches regulating breathing patterns in CNS. Respir Physiol Neurobiol 135: 121-132
11. Borday C, Chatonnet F, Thoby-Brisson M, Champagnat J, Fortin G (2005) Neural tube patterning by Krox20 and emergence of a respiratory control. Respir Physiol Neurobiol 149: 63-72
12. Borday C, Coutinho A, Gernon I, Champagnat J, Fortin G (2006) Pre-/post-otic rhombomeric interactions control emergence of a fetal-like respiratory rhythm in the mouse embryo. J Neurobiol 66: 1285-1301
13. Borday V, Kato F, Champagnat J (1997) A ventral pontine pathway promotes rhythmic activity in the medulla of neonate mice. Neuroreport 8: 3679-3683
14. Brady R, Zaidi SI, Mayer C, Katz D.M (1999) BDNF is a target-derived survival factor for arterial baroreceptor and chemoafferent primary sensory neurons. J Neurosci 19: 2131-2142
15. Branchereau P, Morin D, Bonnot A, Ballion B, Chapron J, Viala D (2000) Development of lumbar rhythmic networks: from embryonic to neonate locomotor-like patterns in the mouse. Brain Res Bull 53: 711-718
16. Briscoe J, Sussel L, Serup P, Hartigan-O'Connor D, Jessel TM, Rubenstein JL, Ericson J (1999) Homeobox gene Nkx2.2 and specification of neuronal identity by graded Sonic hedgehog signalling. Nature 398: 622-627
17. Brosenitsch TA, Katz DM (2002) Expression of Phox2 transcription factors and induction of the dopaminergic phenotype in primary sensory neurons. Mol Cell Neurosci 20: 447-457
18. Brunet JF, Pattyn A (2002) Phox2 genes – from patterning to connectivity. Curr Opin Genet Dev 12: 435-440
19. Carter AR, Chen C, Schwartz PM, Segal RA (2002) Brain-derived neurotrophic factor modulates cerebellar plasticity and synaptic ultrastructure. J Neurosci 22: 1316-1327
20. Champagnat J, Fortin G (1997) Primordial respiratory-like rhythm generation in the vertebrate embryo. Trends Neurosci 3: 119-124
21. Chatonnet F, Domínguez del Toro E, Voiculescu O, Charnay P, Champagnat J (2002) Different respiratory control systems are affected in homozygous and heterozygous *kreisler* mutant mice. Eur J Neurosci 15: 684-692
22. Chatonnet F, Dominguez del Toro E, Thoby-Brisson M, Champagnat J, Fortin G, Rijli FM, Thaeron-Antono C (2003a) From hindbrain segmentation to breathing: developmental pattern in rhombomeres 3 and 4. Mol Neurobiol 28: 277-294
23. Chatonnet F, Boudinot E, Chatonnet A, Taysse L, Daulon S, Champagnat J, Foutz AS (2003b) Respiratory survival mechanisms in *acetylcholinesterase* knockout mouse. Eur J Neurosci 18: 1419-1427

24.   Chub N, O'Donovan MJ (1998) Blockade and recovery of spontaneous rhythmic activity after application of neurotransmitter antagonists to spinal networks of the chick embryo. J Neurosci 18: 294-306

25.   Clarke J D, Lumsden A (1993) Segmental repetition of neuronal phenotype sets in the chick embryo hindbrain. Development 118: 151-162

26.   Coutinho AP, Borday C, Gilthorpe J, Jungbuth S, Champagnat J, Lumsden A, Fortin G (2004) Induction of a parafacial rhythm generator by rhombomere 3 in the chick embryo. J Neurosci 24 : 9383-9390

27.   Denavit-Saubié M, Champagnat J, Zieglgansbezrger W (1978) Effects of opiates and methionin-enkephalin on pontine and bulbar respiratory neurones of the cat. Brain Res 155: 55-67

28.   Di Pasquale E. Monteau R, Hilaire G (1992) In vitro study of central respiratory-like activity of the fetal rat. Exp Brain Res 89: 459-464.

29.   Domínguez del Toro E, Borday V, Davenne M, Neun R, Rijli F M, Champagnat J (2001) Generation of a novel functional neuronal circuit in Hoxa1 mutant mice. J Neurosci 21: 5637-5642

30.   Erickson JT, Conover J C, Borday V, Champagnat J, Katz D M (1996) Mice lacking BDNF exhibit visceral sensory neuron losses distinct from mice lacking NT4 and display a severe developmental deficit in control of breathing. J Neurosci 16: 5361-5371

31.   Feldman JL, Del Negro CA (2006) Looking for inspiration: new perspectives on respiratory rhythm. Nat Rev Neurosci 3: 232-242

32.   Ferland RJ, Eyaid W, Collura RV, Tully LD, Hill RS, Al-Nouri D, Al-Rumayyan A, Topcu M, Gascon G, Bodell A, Shugart YY, Ruvolo M, Walsh CA (2004) Abnormal cerebellar development and axonal decussation due to mutations in AHI1 in Joubert syndrome. Nat Genet 36:1008-1013

33.   Fortin G, Champagnat J, Lumsden A (1994) Onset and maturation of branchio-motor activities in the chick hindbrain. Neuroreport 5: 1149-1152

34.   Fortin G, Kato F, Lumsden A, Champagnat J (1995) Rhythm generation in the segmented hindbrain of chick embryos. J Physiol 486: 735-744

35.   Fortin G, Jungbluth S, Lumsden A, Champagnat J (1999) Segmental specification of GABAergic inhibition during development of hindbrain neural networks. Nat Neurosci 2 : 873-877

36.   Galli L, Maffei L (1988) Spontaneous impulse activity of rat retinal ganglion cells in prenatal life. Science 242 : 90-101

37.   Garaschuk O, Linn J, Eilers J, Konnerth A (2000) Large-scale oscillatory calcium waves in the immature cortex. Nat Neurosci 3: 452-459

38.   Gavalas A (2002) ArRAnging the hindbrain. Trends Neurosci 25: 61-64

39.   Giudicelli F, Taillebourg E, Charnay P, Gilardi-Hebenstreit P (2001) Krox-20 patterns the hindbrain through both cell-autonomous and non cell-autonomous mechanisms. Gene Dev 15: 567-580

40.   Glover JC (2001) Correlated patterns of neuron differentiation and Hox gene expression in the hindbrain: a comparative analysis. Brain Res Bull 55: 683-693

41.   Goridis C, Brunet JF (1999) Transcriptional control of neurotransmitter phenotype. Curr Opin Neurobiol 9: 47-53

42.   Goulding M, Pfaff SL (2005) Development of circuits that generate simple rhythmic behaviors in vertebrates. Curr Opin Neurobiol 15: 14-20

43.   Gourine AV, Llaudet E, Dale N, Spyer KM (2005) ATP is a mediator of chemosensory transduction in the central nervous system. Nature 436: 108-111

44.   Gray PA, Rekling JC, Bocchiaro CM, Feldman JL (1999) Modulation of respiratory frequency by peptidergic input to rhythmogenic neurons in the pre-Botzinger complex. Science 286: 1566-1568

45.   Gray PA, Janczewski WA, Mellen N, McCrimmon DR, Feldman JL (2001) Normal breathing requires pre-Bötzinger complex neurokinin-1 receptor-expressing neurons. Nat Neurosci 4: 927-930

46.   Greer JJ, Smith JC, Feldman J (1992) Respiratory and locomotor patterns generated in the fetal rat brainstem-spinal cord in vitro. J Neurophysiol 67: 996-999

47.   Gust J, Wright JJ, Pratt EB, Bosma MM (2003) Development of synchronized activity of cranial motor neurons in the segmented embryonic mouse hindbrain. J Physiol 550: 123-133

48.   Hamburger V, Hamilton HL (1951) A series of normal stages in the development of the chick embryo. J Morphol 88: 49-92

49.   Hanson MG, Landmesser LT (2003) Characterization of the circuits that generate spontaneous episodes of activity in the early embryonic mouse spinal cord. J Neurosci 23: 587-600

50.   Hunt PN, McCabe AK, Gust J, Bosma M (2006a) Spatial restriction of spontaneous activity towards the rostral primary initiating zone during development of the embryonic mouse hindbrain. J Neurobiol 66: 1225-1238

51.   Hunt PN, McCabe AK, Gust J, Bosma M (2006b) Primary role of the serotonergic midline system in synchronized spontaneous activity during development of the embryonic mouse hindbrain. J Neurobiol 66: 1239-1252

52.   Jacquin TD, Borday V, Schneider-Maunoury S, Topilko P, Ghilini G, Kato F, Charnay P, Champagnat J (1996) Reorganization of pontine rhythmogenic neuronal networks in Krox-20 knockout mice. Neuron 17: 747-758

53.   Janczewski WA, Onimaru H, Homma I, Feldman JL (2002) Opioid-resistant respiratory pathway from the preinspiratory neurons to abdominal muscles: in vivo and in vitro study in newborn rat. J Physiol 545: 1017-1026

54.   Jessell TM (2000) Neuronal specification in the spinal cord: inductive signals and transcriptional codes. Nat Rev Genet 1: 20-29

55.   Johnson SM, Smith JC, Funk GD, Feldman JL (1994) Pacemaker behavior of respiratory neurons in medullary slices from neonatal rat. J Neurophysiol 72 : 2598-2608.

56.   Jungbluth S, Koentges G, Lumsden A (1997) Coordination of early neural tube development by BDNF/trkB. Development 124: 1877-1885

57.   Katz LC, Shatz C.J (1996) Synaptic activity and the construction of cortical circuits. Science 274: 1133-1138

58.   Kobayashi K, Lemke RP, Greer JJ (2001) Ultrasound measurements of the fetal breathing movements in the rat. J Appl Physiol 91: 316-320

59.   Kobayashi S, Onimaru H, Inoue M, Inoue T, Sasa R (2005) Localization and properties of respiratory neurons in the rostral pons of the newborn rat. Neuroscience 134: 317-25.

60.   Koshiya N, Guyenet PG (1996) Tonic sympathetic chemoreflex after blockade of respiratory rhythmogenesis in the rat. J Physiol 491: 859-869

61.   Koshiya N, Smith JC (1999) Neuronal pacemaker for breathing visualized in vitro. Nature 400: 360-363

62.   Kovalchuk Y, Hanse E, Kafitz KW, Konnerth A (2002) Postsynaptic induction of BDNF-mediated long-term potentiation. Science 295: 1729-1734.

63.   Kullander K, Butt SJ, Lebret JM, Lundfald L, Restrepo CE, Rydstrom A, Klein R, Kiehn O (2003) Role of EphA4 and EphrinB3 in local neuronal circuits that control walking. Science 299: 1889-1892

64.   Lagier-Tourenne C, Boltshauser E, Breivik N, Gribaa M, Betard C, Barbot C, Koenig M (2004) Homozygosity mapping of a third Joubert syndrome locus to 6q23. J Med Genet 41: 273-277

65.   Li JY, Joyner AL (2001) *Otx2* and *Gbx2* are required for refinement and not induction of mid-hindbrain gene expression. Development 128: 4979-4991

66.   Lieske SP, Thoby-Brisson M, Telgkamp P, Ramirez JM (2000) Reconfiguration of the neural network controlling multiple breathing patterns: eupnea, sighs and gasps. Nat Neurosci 3: 600-607

67.   Liu X, Ernfors P, Wu H, Jaenisch R (1995) Sensory but not motor neuron deficits in mice lacking NT4 and BDNF. Nature 375: 238-241

68.   Lu B (2003) BDNF and activity-dependent synaptic modulation. Learn Memory 10: 86-98

69.   Lumsden A, Keynes R (1989) Segmental patterns of neuronal development in the chick hindbrain. Nature 337: 424-428

70.   Lumsden A, Krumlauf R (1996) Patterning the vertebrate neuraxis. Science 274 : 1109-1115

71.   Maden M (2002) Retinoid signalling in the development of the central nervous system. Nat Rev Neurosci 3: 843-853

72.   Manzke T, Guenther U, Ponimaskin EG, Haller M, Dutschmann M, Schwarzacher S, Richter DW (2003) 5-HT4 (a) receptors avert opioid-induced breathing depression without loss of analgesia. Science 301: 226-229

73.   McCrimmon DR, Milsom WK, Alheid GF (2004) The rhombencephalon and breathing: a view from the pons. Respir Physiol Neurobiol 143:103-337

74.   McKay LC, Janczewski WA, Feldman JL (2005) Sleep-disordered breathing after targeted ablation of preBötzinger complex neurons. Nat Neurosci 8: 1142-1144

75.   Mellen N M, Janczewski W A, Bocchiaro C M, Feldman J L (2003) Opioid-induced quantal slowing reveals dual networks for respiratory rhythm generation. Neuron 37: 821-826

76.   Morin-Surun M-P, Boudinot E, Gacel G, Champagnat J, Roques B P, Denavit-Saubie M (1984) Different effects of mu and delta opiate agonists on respiration. Eur J Pharmacol 98: 235-240

77.   Morin-Surun, MP, Boudinot E, Dubois C, Matthes HW, Kieffer BL, Denavit-Saubie M, Champagnat J, Foutz AS (2001). Respiratory function in adult mice lacking the mu-opioid receptor: role of delta-receptors. Eur J Neurosci 13: 1703-1710

78.   Mulkey DK, Stornetta RL, Weston MC, Simmons JR, Parker A, Bayliss DA, Guyenet PG (2004) Respiratory control by ventral surface chemoreceptor neurons in rats. Nat Neurosci 7: 1360-1369

79.   O'Donovan MJ, Landmesser L (1987) The development of hindlimb motor activity studied in the isolated spinal cord of the chick embryo. J Neurosci 7: 3256-3264

80.   O'Donovan MJ (1999) The origin of spontaneous activity in developing networks of the vertebrate nervous system. Curr Opin Neurobiol 9: 94-104

81.   Onimaru H, Homma I (2003) A novel functional neuron group for respiratory rhythm generation in the ventral medulla. J Neurosci 23: 1478-1486

82.   Oppenheim RW (1991) Cell death during development of the nervous system. Annu Rev Neurosci 14: 453-501

83.   Oppenheim RW, Yin QW, Prevette D, Yan Q (1992) Brain-derived neurotrophic factor rescues developing avian motoneurons from cell death. Nature 360: 755-757

84. Oyelese AA, Rizzo MA, Waxman SG, Kocsis JD (1997) Differential effects of NGF and BDNF on axotomy-induced changes in $GABA_A$-receptor mediated conductance and sodium currents in cutaneous afferent neurons. J Neurophysiol 78: 31-42

85. Pagliardini S, Ren J, Greer JJ (2003) Ontogeny of the pre-Botzinger complex in perinatal rats. J Neurosci 23: 9575-9584

86. Pascual O, Denavit-Saubie M, Dumas S, Kietzmann T, Ghilini G, Mallet J, Pequignot JM (2001) Selective cardiorespiratory and catecholaminergic areas express the hypoxia-inducible factor-1alpha (HIF-1alpha) under *in vivo* hypoxia in rat brainstem. Eur J Neurosci 14: 1981-1991

87. Pattyn A, Vallstedt A, Dias JM, Samad OA, Krumlauf R, Rijli FM, Brunet JF, Ericson J (2003) Coordinated temporal and spatial control of motor neuron and serotonergic neuron generation from a common pool of CNS progenitors. Gene Dev 17: 729-737

88. Pattyn A, Simplicio N, van Doorninck JH, Goridis C, Guillemot F, Brunet JF (2004) *Ascl1/Mash1* is required for the development of central serotonergic neurons. Nat Neurosci 7: 589-595

89. Peinado A (2000) Traveling slow waves of neural activity: a novel form of network activity in developing neocortex. J Neurosci 20: RC54

90. Pena F, Parkis MA, Tryba AK, Ramirez JM (2004) Differential contribution of pacemaker properties to the generation of respiratory rhythms during normoxia and hypoxia. Neuron 43: 105-117

91. Potts JT, Rybak IA, Paton JF (2005) Respiratory rhythm entrainment by somatic afferent stimulation. J Neurosci 25: 1965-1978

92. Powell FL, Scheid P (1986) Physiology of gas exchange in the avian respiratory system. In: Form and Function in Birds, Volume 4, AS King and J. McLellan eds, Academic Press, London, San Diego, Berkeley, Boston, Sydney, Tokyo, Toronto, pp. 393-437

93. Ramirez JM, Schwarzacher SW, Pierrefiche O, Olivera BM, Richter DW (1998) Selective lesioning of the cat pre-Bötzinger complex *in vivo* eliminates breathing but not gasping. J Physiol 507: 895-907

94. Rekling JC, Champagnat J, Denavit-Saubie M (1996) Electroresponsive properties and membrane potential trajectories of three types of inspiratory neurons in the newborn mouse brainstem in vitro. J Neurophysiol 75: 795-810

95. Ren J, Greer JJ (2003) Ontogeny of rhythmic motor patterns generated in the embryonic rat spinal cord. J Neurophysiol 89: 1187-1195

96. Rogalski SL, Appleyard SM, Pattillo A, Terman GW, Chavkin C (2000) TrkB activation by brain-derived neurotrophic factor inhibits the G-protein-gated inward rectifier Kir3 by tyrosine phosphorylation of the channel. J Biol Chem 18: 25082-25088

97. Schneider-Maunoury S, Topilko P, Seitandou T, Levi G, Cohen-Tannoudji M, Pournin S, Babinet C, Charnay P (1993) Disruption of *Krox-20* results in alteration of rhombomeres 3 and 5 in the developing hindbrain. Cell 17: 1199-1214

98. Sendtner M, Holtmann B, Kolbeck R, Thoenen H, Barde YA (1992) Brain-derived neurotrophic factor prevents the death of motoneurons in newborn rats after nerve section. Nature 360: 757-759

99. Severson CA, Wang W, Pieribone VA, Dohle CI, Richerson GB (2003) Midbrain serotonergic neurons are central pH chemoreceptors. Nat Neurosci 6: 1139-1140

100. Sherwood NT, Lesser SS, Lo DC (1997) Neurotrophin regulation of ionic currents and cell size depends on cell context. Proc Natl Acad Sci USA 94: 5917-5922

101.   Smith JC, Ellenberger HH, Ballanyi K, Richter DW, Feldman JL (1991) Pre-Bötzinger complex: a region that may generate respiratory rhythm in mammals. Science 254: 726-729

102.   Stornetta RL, Rosin DL, Wang H, Sevigny CP, Weston MC, Guyenet PG (2003) A group of glutamatergic interneurons expressing high levels of both neurokinin-1 receptors and somatostatin identifies the region of the pre-Botzinger complex. J Comp Neurol 455: 499-512

103.   Stornetta RL, Moreira TS, Takakura AC, Kanq BJ, Chang DA, West GH, Brunet JF, Mulkey DK, Bayliss DA, Guyenet PG (2006) Expression of *Phox2b* by brain-stem neurons involved in chemosensory integration in the adult rat. J Neurosci 26: 10305-10314

104.   Sturdy CB, Wild JM, Mooney R (2003) Respiratory and telencephalic modulation of vocal motor neurons in the zebra finch. J Neurosci 23: 1072-1086

105.   Suzue T (1984) Respiratory rhythm generation in the *in vitro* brain stem-spinal cord preparation of the neonatal rat. J Physiol 354: 173-183

106.   Tanabe Y, Jessell TM (1996) Diversity and pattern in the developing spinal cord. Science 274: 1115-1123.

107.   Thoby-Brisson M, Ramirez JM (2001) Identification of two types of inspiratory pacemaker neurons in the isolated respiratory neural network of mice. J Neurophysiol 86: 104-112

108.   Thoby-Brisson M, Telgkamp P, Ramirez JM (2000) The role of the hyperpolarization-activated current in modulating rhythmic activity in the isolated respiratory network of mice. J Neurosci 20: 2994-3005

109.   Thoby-Brisson M, Cauli B, Champagnat J, Fortin F, Katz DM (2003) Expression of functional TrkB receptors by rhythmically active respiratory neurons in the pre-Bötzinger complex of neonatal mice. J Neurosci 23: 7685-7689

110.   Thoby-Brisson M, Trinh JB, Champagnat J, Fortin G (2005) Emergence of the pre-Bötzinger respiratory rhythm generator in the mouse embryo. J Neurosci 25: 4307-4318.

111.   Viemari JC, Burnet H, Bevengut M, Hilaire G (2003) Perinatal maturation of the mouse respiratory rhythm-generator: in vivo and in vitro studies. Eur J Neurosci 17: 1233-1244.

112.   Voituron N, Frugiere A, Champagnat J, Bodineau L (2006) Hypoxia-sensing properties of the newborn rat ventral medullary surface in vitro. J Physiol 577: 55-68

113.   Wallén-Mackenzie A, Gezelius H, Thoby-Brisson M, Nygard A, Enjin A, Fujiyama F, Fortin G, Kullander K (2006) Vesicular glutamate transporter 2 is required for central respiratory rhythm generator but not for locomotor pattern generation. J Neurosci 26: 12294-12307

114.   Wang WG, Stornetta RL, Rosin DL, Guyenet PG (2001) Neurokinin-1 receptor-immunoreactive neurons of the ventral respiratory group in the rat. J Comp Neurol 434: 128-146

115.   Wilson RJ, Vasilakos K, Harris MB, Straus C, Remmers JE (2002) Evidence that ventilatory rhythmogenesis in the frog involves two distinct neuronal oscillators. J Physiol 540: 557-570

116.   Weese-Mayer DE, Bolk S, Silvestri JM, Chakravarti A (2002) Idiopathic congenital central hypoventilation syndrome: evaluation of brain-derived neurotrophic factor genomic DNA sequence variation. Am J Med Genet 107: 306-310

117.   Yvert B, Branchereau P, Meyrand P (2004) Multiple spontaneous rhythmic activ-
        ity patterns generated by the embryonic mouse spinal cord occur within a specific
        developmental time window. J Neurophysiol 91: 2101-2109

12. Transcription factor control of central respiratory neuron development

Bruno C. BLANCHI (1) and Michael H. SIEWEKE (2)

(1) Semel Institute for Neuroscience and Human Behavior, Mental Retardation Research Center, Neuroscience Research Building, University of California-Los Angeles, Los Angeles, CA, 90095-7332, U.S.A. (2) Centre d'Immunologie de Marseille Luminy, CNRS-INSERM-Univ. Med., Campus de Luminy, Case 906, 13288 Marseille Cedex 09, France

12.1 Introduction

Breathing is a vital and continuous behavior that is controlled by respiratory neurons in the brainstem that generate respiratory rhythm, modulate it in response to sensory input and coordinate it with other behaviors such as locomotion or phonation [7; 34]. Distinct neuronal populations of this network have been identified and extensively characterized by anatomical, physiological or pharmacological techniques. These different types of neurons with specialized functions in respiratory control have distinct developmental programs and gene expression profiles, which ultimately must be controlled by specific sets of transcription factors.

Targeted gene inactivation of several transcription factors in mice has been shown to result in developmental defects of specific groups of respiratory neurons and to produce corresponding respiratory phenotypes [10]. Most of these transcription factors affect the development of neurons involved in the modulation of breathing activity [24-26; 104; 114; 115], but recently the transcription factor *MafB* has also been found to influence the development of neurons directly contributing to respiratory rhythmogenesis itself [9].

The recent observation that the majority of patients with congenital central hypoventilation syndrome (CCHS) and Rett syndrome (RTT) carry mutations of the transcriptional regulators *PHOX2B* and *MECP2*, respectively [2; 4; 102; 118], is consistent with the developmental and respiratory phenotypes reported for mouse models with deletions of these factors (see Chapters 14, 16). The *PHOX2B* and *MECP2* studies thus indicate that findings from animal models can be highly relevant to human central respiratory diseases. In other disorders where a central respiratory control defect has also been suspected, such as sudden infant death syndrome (SIDS) or central sleep apnea (CSA), only few direct links to specific mutations have been described so far, and the genetic causes are likely to be more complex. However, dysfunction of specific brainstem neurons involved in breathing control has been recently reported in cases of SIDS, supporting the hypothesis that at least certain cases of sudden infant death might also be caused by developmental defects of the respiratory network [89; 123] (see Chapter 7).

The first part of this chapter will describe the major groups of neurons involved in central respiratory control. The second part will outline the transcription factor programs that specify the development of these neurons and the respiratory phenotypes associated with their inactivation in mutant mice. The chapter will concentrate on the specification of respiratory neurons during late development and will not cover the influence of early hindbrain development on breathing activity, which is reviewed in Chapter 11 and elsewhere [11; 15]. Finally, in the last part this chapter will discuss the implication of transcription factor deficiencies in human central respiratory disorders.

12.2 Brainstem populations of neurons participating in central breathing control

Several distinct populations of neurons are critically involved in the control of respiratory behavior. They have traditionally been characterized at the anatomic level and by experiments involving electrical or pharmacological stimulation and inhibition, lesion and resection, or activity recording of candidate groups of neurons. These physiological and anatomical studies have identified groups of pontine and brainstem neurons that participate in respiratory rhythmogenesis, modulate breathing activity and provide peripheral or central regulatory inputs [7; 34]. This section will discuss groups of neurons involved in 1) respiratory rhythm generation, 2) central chemoreception and 3) peripheral sensory pathways driving central modulation of breathing.

12.2.1 Groups of neurons involved in respiratory rhythm generation

The generation of the respiratory rhythm in mammals appears to involve the close collaboration of at least two groups of neurons located in the rostral ventral medulla, the pre-inspiratory (pre-I) neurons of the parafacial respiratory group (pFRG) and the inspiratory neurons of the pre-Bötzinger complex (pre-BötC) [35; 36; 55; 74; 84; 85] (Fig. 1).

The pre-BötC has been characterized in more detail and been shown to be essential for normal respiratory function *in vitro* and *in vivo*. Brainstem-spinal cord preparations from neonatal rodents generate a respiratory-like rhythmic activity. This respiratory-like activity persists in brainstem slices that contain the pre-

BötC as well as in pre-BötC island preparations [56; 107]. Accordingly, selective lesion of the pre-BötC eliminates the rhythmic respiratory activity *in vitro* as well as *in vivo* [99, 107, 125]. Furthermore, *in vivo* bilateral lesion of NK1R expressing pre-BötC neurons strongly impairs respiratory rhythmogenesis [42; 126].

Recent findings suggest that neurons of the pFRG also contribute to breathing rhythm generation. Some pFRG neurons have been found to discharge before the inspiratory burst and have been classified as pre-inspiratory neurons [84]. In addition, the pFRG is sufficient to maintain a respiratory-like rhythm in brainstem slice preparations, and pharmacological experiments suggest that pre-BötC and pFRG may act as two synchronized rhythm generators that together constitute a functional respiratory rhythmogenic network [35; 36; 55; 74; 84; 85].

12.2.2 Groups of neurons involved in central chemoreception

Several central chemoreceptive areas are distributed through the central nervous system [5; 79]. They can detect changes in the pH of cerebral interstitial fluid caused by variation of CO_2 levels. Thus focal acidification of central chemo-reception regions in animal models stimulates ventilation. Studies in anesthetised and conscious animals have shown that the retrotrapezoid nucleus (RTN), the me-dullary raphe (MR, rostral part), the nucleus of the solitary tract (nTS, caudal part) and area postrema (AP), the locus coeruleus (LC) and the rostral part of the ven-tral respiratory group (VRG) are involved in central chemoreception [79] (Fig. 1).

In addition, focal injection of antagonists of muscarinic, glutamatergic or purogenic receptors, or of agonist of GABAergic receptors into the RTN reduces the response to systemic CO_2 stimulation, suggesting a role for several RTN neu-ronal populations in central chemoreception [39, 78, 79]. Accordingly, systemic CO_2 stimulation induces increased firing rate of glutamatergic neurons in the RTN [76]. In the rostral MR, injection of 5HT1A receptor antagonists also reduces re-sponse to systemic CO_2 stimulation, suggesting a role for serotonergic neurons of the MR in chemoreception as well. However, the role of the MR could also be to modulate the effect of other chemoreceptive regions. Indeed MR injection of GABAergic agonist modulates the ventilatory response to hypoxia, presumably by modulating the processing of the information from the carotid bodies in the nTS. In addition injection of serotonergic antagonist in the caudal MR modulates the re-sponse to chemoreception by the RTN [79; 110; 111].

Finally, elimination of distinct neuronal sub-populations by using spe-cific toxin-ligand or toxin-antibody conjugates has allowed the study of the role of several potential central chemoreceptor neuronal populations. Elimination of neu-rokinin-1 receptor (NK1R) expressing neurons in the RTN or in the ventral sur-face of the medulla reduces the response to systemic CO_2 stimulation [79; 80].

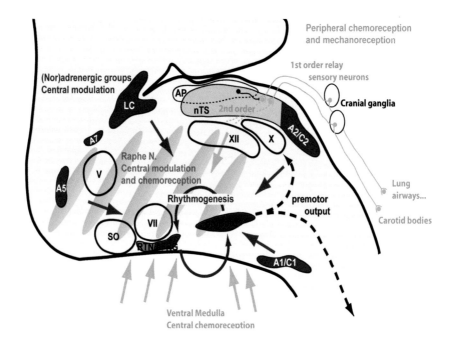

Fig. 1. Neuronal groups contributing to central control of breathing. Schematic representation of the brainstem in a sagittal view. Rhythmogenic groups of neurons in red (pre-BötC, pre-Bötzinger complex; pFRG, parafacial respiratory group). Groups of neurons involved in central chemoreception and modulation of the respiratory activity: Retrotrapezoid nucleus (RTN, which overlaps with the pFRG area); (Nor)adrenergic groups of neurons in blue (A1/C1; A2/C2; A5; LC, locus coeruleus; sLC, sub-locus coeruleus); Nucleus of the solitary tract in light orange (nTS); Raphe nuclei in brown hatching (Raphe N.). Central chemoreception in the ventral medulla is represented by orange arrows; first (1st) and second (2nd) order relay sensory neurons of the cranial ganglia and nTS respectively are represented in orange. First order relay sensory neurons carry chemoreception and baroreception information from the chemoreceptors of the carotid bodies and the baroreceptors of the lungs and airways. Area Postrema (AP); superior olive (SO); motor nuclei (V, VII, X, XII). (Modified from [10], Copyright (2005), with permission from Elsevier.).

Similarly, reduced response to systemic CO_2 is observed after elimination of serotonin transporter protein (SERT) expressing neurons in the MR or after elimination of tyrosine hydroxylase (TH) expressing catecholaminergic neurons of the LC (A6), sub-LC (A7) and A5 via fourth ventricle injection of anti-dopamine β-hydroxylase (DBH) antibody coupled to saporin (SAP) [66; 81]. Together these observations demonstrate the role of glutamatergic, GABAergic, cholinergic and purinergic neurons of the RTN, of 5HT neurons of the MR, catecholaminergic neurons of the pons (A5, LC and A7), as well as of the VRG and nTS in central chemoreception or its modulation.

12.2.3 Groups of neurons involved in peripheral sensory pathways driving modulation of breathing

Peripheral chemoreceptors, notably located in the carotid bodies, detect blood hypoxemia and, in response, stimulate breathing. Under normal conditions they also provide a tonic excitatory input to the brainstem breathing network [37]. Chemosensitive cells in the carotid bodies form synapses with primary visceral sensory neurons whose soma are located in the petrosal ganglion. These chemoafferent neurons project onto second order visceral sensory neurons of the nTS, which in turn modulate the activity of the central respiratory network. In addition, mechanoreceptors of the lungs and airways permanently adapt the tone of the airways and can modulate central command of breathing through regulatory reflex loops that also project to the nTS [57] (Fig. 1).

Catecholaminergic neurons located in the brainstem and the pons produce the neurotransmitters adrenaline and noradrenaline and display a predominantly depressive effect on respiratory frequency [1; 33]. In rodent neonates, (nor)adrenergic neurons of the A5 and LC (A6) groups in the pons and the A1/C1 group in the ventral medulla have been shown to modulate the respiratory rhythm generation and the maturation of the respiratory network. They also integrate peripheral and central sensory information as well as higher order mood related or voluntary input. Specifically, noradrenergic A5 neurons repress the activity of the central respiratory network. In contrast, LC and A1/C1 neurons appear to have a facilitatory effect on the activity of the respiratory rhythm command. In addition, both A5 and LC activities appear to be crucial for the proper perinatal development of the brainstem respiratory network [49; 114; 128] (Fig. 1).

Ultimately, the central respiratory network generates central impulses that, via spinal cord motoneurons and brainstem motor nuclei, drive the rhythmic contractions of respiratory pump and intercostal muscles, and controls the tone of airway muscles [7].

12.3 Mouse mutants of transcription factors governing development of respiratory neurons and breathing control

12.3.1 Mouse mutants of transcription factors with defects in the development of modulatory respiratory neurons

Several groups of brainstem neurons function as modulators of the central rhythm generator activity in response to peripheral and central sensory or voluntary input. Some insights into the transcriptional code which control their differentiation have been gained using loss and gain of function experiments in animal models. Mutations of several of these transcription factors were found cause respiratory defects in knockout mice and in some cases in humans [10] (Fig. 4 and Table 1).

Development of (nor)adrenergic brainstem neurons

The locus coeruleus (LC or A6) and the A1/C1, A2/C2, A5 and A7 groups in the brainstem harbor (nor)adrenergic neurons, which have both inhibitory and facilitatory effects on respiratory frequency and are involved in perinatal development of the respiratory network [1; 33; 49; 114]. Several transcription factors required for the specification of all or certain sub-populations of these (nor)adrenergic brainstem neurons have been identified (Fig. 2).

The basic helix-loop-helix (bHLH) transcription factor *Mash-1* (mammalian achaete-scute homologue-1) is a mammalian homologue of the drosophila achaete-scute proneural complex (*Asc*). *Asc* controls the generation of neural precursors in both the central (CNS) and peripheral (PNS) nervous system in drosophila. In mice, *Mash1* is expressed in subsets of CNS neuronal progenitors along the neural tube and in progenitors of the PNS. *Mash1*$^{-/-}$ mice fail to develop LC, A1/C1, A2/C2, A5 and A7 as assessed by Nissl, DBH, TH or c-RET staining [25; 43; 50].

Phox2a and Phox2b (paired-like homeobox 2a and 2b) proteins are transcription factors of the Q50 paired-like homeodomain containing family. In the brainstem, *Phox2a* and *Phox2b* are expressed in all groups of (nor)adrenergic neurons, in the nTS and area postrema (AP) as well as in subsets of interneurons [13]. They are also expressed in most cranial motor nuclei. *Phox2a* deletion in mice causes a failure of LC development, whereas the other brainstem (nor)adrenergic neurons appear to develop normally [75]. In contrast, *Phox2b* deficient mice fail to develop all brainstem (nor)adrenergic centers (A1/C1, A2/C2, A5, A7 and LC) [13; 91; 92] (see Chapter 3).

Rnx (*Hox11l2/tlx3*) is, together with *Tlx1* (*Hox11*) and *Enx* (*Tlx2/Hox11l1*), a member of the *Tlx* family of homeodomain transcription factors. *Rnx* is transiently expressed in the developing (nor)adrenergic centers in the brainstem and *Rnx*$^{-/-}$ mice fail to develop all of these groups of (nor)adrenergic neurons except the LC [97].

Finally, experiments performed in zebrafish have shown that the transcription factor *Tfap2a* is required for the proper development of brainstem (nor)adrenergic neurons. *Tfap2a* mutant zebrafish show an almost complete absence of TH and DBH expression in the LC, medulla oblongata and area postrema regions [51] (Fig. 2 and Fig. 4).

Analysis of the timing of their expression, precise examination of knockout animals as well as gain-of-function experiments start to reveal the genetic interaction between these transcription factors in the specification of the different groups of brainstem (nor)adrenergic neurons. *Mash1* precedes *Phox2a* and *Phox2b* expression throughout the nervous system, while, in the LC, *Phox2a* precedes *Phox2b* expression. *Mash1*$^{-/-}$ mice lack expression of *Phox2a*, *Phox2b* and DBH in the LC area, whereas *Phox2a*$^{-/-}$ and *Phox2b*$^{-/-}$ mice lack expression of *Phox2b* and DBH, or DBH only, respectively. In addition, knock-in of *Phox2b* coding sequence in *PhoxPa* locus rescues the development of the LC, while knock-in of *Phox2a* in *Phox2b* locus is unable to sustain (nor)adrenergic specification in the LC. Together these observations suggest that *Mash1*, *Phox2a* and *Phox2b* act in a linear cascade and that the main function of *Phox2a* is to switch on *Phox2b* expression during the development of the LC [19; 38; 50; 90; 92] (Fig. 2). In contrast, correct development of the other brainstem (nor)adrenergic groups of neurons (A1/C1, A2/C2, A5 and A7) do not require *Phox2a* expression. In these regions, *Mash1* activity is required to induce the expression of *Phox2b* which control *Phox2a* and DBH expression [13; 19; 38; 50; 75; 92] (Fig. 2).

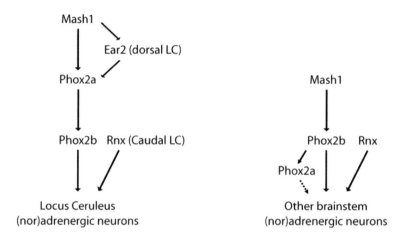

Fig. 2: Transcription factors governing brainstem (nor)adrenergic neuron development. Transcription factors *Mash1* and *Phox2b* are critically required for the development of all brainstem (nor)adrenergic groups of neurons. *Phox2a* is required for locus coeruleus (LC) (nor)adrenergic differentiation after *Mash1* expression but before *Phox2b* expression. In contrast, the other brainstem (nor)adrenergic groups of neurons (A1/C1, A2/C2 and A5) do not require *Phox2a* expression. *Rnx* is required for the development of the caudal LC and A1/C1, A2/C2 and A5. *Rnx* acts in a transcription factor cascade independent from the *Mash1-Phox2a/2b* cascade. Finally, *Ear2* is required together with *Mash1* for the development of rostral LC (nor)adrenergic neurons. Transcription factors important for the development of (nor)adrenergic neurons in other species are not shown but mentioned in the text.

Mash-1, *Phox2a* and *Phox2b* genes appear to be necessary but not sufficient for (nor)adrenergic fate specification. They are also expressed in non (nor)adrenergic neurons and over-expression of *Phox2a* can only specify (nor)adrenergic neurons during a certain time window and in a specific region of the developing brain in zebrafish [44]. These observations suggest the need for additional signaling events or cofactors.

Mice deficient for *Rnx* still express *Mash1*, *Phox2a* and *Phox2b* but lack (nor)adrenergic groups of neurons except the rostral LC. Thus *Rnx* also appears to act at the end point of a genetic cascade, but not downstream of *Phox2b*, since it precedes *Phox2b* expression in (nor)adrenergic progenitors. *Phox2b* and *Rnx* rather appear to be independently required for (nor)adrenergic neuronal fate specification, since *Phox2b* and *Rnx* expression in (nor)adrenergic progenitors is not affected in *Rnx* and *Phox2b* mutants respectively [97] (Fig. 2). In zebrafish, both *Tfap2a* and *Phox2a* are required for (nor)adrenergic neuron development in the LC. In addition, P*hox2a* and *Tfap2a* expression are not affected in *Tfap2a* and *Phox2a* mutant respectively. These observations suggest that *Tfap2a* may represent a parallel transcriptional program required for (nor)adrenergic specification in the LC. In the medulla, *Tfap2a* and *Phox2a* are expressed by partially overlapping sub-population of (nor)adrenergic neurons and only *Tfap2a* is required for their development [51]. Finally, mice deficient for the orphan nuclear receptor *Ear2* (*Nr2f6*) present a dramatic reduction in the number of DBH and TH expressing neurons in the LC, primarily in its dorsal division [117]. *Phox2a* and *Phox2b* expression are also strongly impaired in *Ear2*$^{-/-}$ animals. This observation positions *Ear2* between *Mash1* and *Phox2a/2b* in the transcription factor cascade that controls the differentiation of at least a sub-population of LC (nor)adrenergic neurons (Fig. 2 and Fig. 4).

Development of the nTS, the AP, visceral sensory neurons and carotid bodies

Peripheral sensory input notably from chemoreceptors of the carotid bodies and mechanoreceptors of the lungs and airways is carried via first order relay visceral sensory neurons located in distal cranial ganglia. First order relay sensory neurons project onto second order sensory neurons that form the nTS. The AP also receives innervations from a subset of visceral sensory neurons. NTS and AP constitute the central relay for visceral information but are also involved in central chemoreception. They have a stimulatory influence on the central rhythm generating network of the ventral medulla in response to peripheral input and provide, under normal conditions, tonic excitatory stimulation to the rhythm generators. Some of the same players involved in (nor)adrenergic fate specification are also required for the development of the nTS, the AP, and visceral sensory neurons that project onto these structures as well as chemoreceptor cells of the carotid bodies [26; 75; 91; 95; 97].

Mice lacking the proneural transcription factor *Mash1* present largely atrophic nTS and AP, whereas nTS and AP do not form at all in *Rnx* and *Phox2b* deficient mice (Fig. 4). *Mash1*$^{-/-}$ mice are characterized by an almost complete absence of *Phox2b*, *Rnx* and *Lmx1b* expressing cells that are usually present in the developing nTS and AP [95]. Moreover the generation of the few remaining postmitotic nTS precursor cells is delayed compared to wild-type mice. Thus *Mash1* appears to be responsible for the generation of nTS sensory neuron precursors. The fact that some post-mitotic nTS precursor cells are still generated in the absence of *Mash1* suggests a role for another factor in this process. *Neurogenin-2* (*Ngn2*) expression is able to compensate for the absence of *Mash1* proneural activity in other systems [88]. However, expression of *Ngn2* from the *Mash1* locus does not restore nTS formation. This observation suggests that *Mash1* may have a role in the specification of nTS sensory neurons in addition to its proneural activity [95]. Whether this function is to induce expression of *Phox2b*, like in (nor)adrenergic specification, or *Rnx* is not known but ectopic expression of

Mash1 can induce *Rnx* expression and development of sensory neurons in the nTS area in the chicken embryo [52].

The nTS and AP express *Mx* and *Phox2b*, and both *Rnx* and *Phox2b* deficient mice show a complete loss of nTS and AP formation [26; 97] (Fig. 4). Expression of these two transcription factors appears to be controlled independently from each other since, at early stages of nTS development, *Rnx* and *Phox2b* expressions are maintained or partially maintained in *Phox2b* and *Rnx* mutants, respectively [26; 97]. Together these observations indicate that nTS and AP development originate from *Mash1* dependant postmitotic precursor cells that subsequently express *Rnx*, *Phox2b* and *Lmx1b*. *Rnx* and *Phox2b* are independently required for the further specification of nTS and AP sensory neurons. Whether *Lmx1b* is necessary for nTS formation has not been reported yet. These similarities in the transcription factor requirements for the development of the nTS and (nor)adrenergic groups may reflect the existence of a putative common progenitor that might express both *Rnx* and *Phox2b* [97]. These similarities also indicate that additional factors restricted to either (nor)adrenergic or nTS fate programs must exist.

Phox2b, *Rnx*, as well as *Phox2a*, are expressed in other neurons of the afferent visceral system which carry the peripheral sensory input that modulates breathing. They are expressed in the geniculate, petrosal and nodose cranial ganglia where the first-order relay visceral sensory neurons are located. *Phox2b* and *Phox2a* deficient mice present a profound atrophy of these cranial ganglia, the *Phox2b* phenotype being more severe with an almost complete loss of these structures [26; 75; 90] (Fig. 4). *Rnx* deficiency, in contrast, does not appear to affect the generation of the first order relay visceral sensory neurons. This might be due to a potential compensatory activity of another *Tlx* factor, *Enx*, which is expressed in these neurons whereas it is not expressed in the nTS. However, in *Rnx* deficient mice, visceral sensory neurons aberrantly project to the dorsal medulla, probably due to the absence of nTS [97]. Finally, *Phox2b* is also critical for the development of the carotid bodies. The glomus cells express both *Phox2b* and *Phox2a*, and *Phox2b*$^{-/-}$ mice present a degeneration of the carotid bodies that never express *Phox2a* and *TH* [26] (see Chapter 3).

Development of serotonergic brainstem neurons

Serotonergic neurons are localized in several anatomically defined regions along the midline of the brainstem called medullary raphe nuclei B1 (caudally) to B9 (rostrally). They are characterized by the production of the neurotransmitter 5-hydroxytryptamin (serotonin, 5HT). Raphe nuclei have been described as central chemosensitive areas and have an important modulatory effect on respiratory rhythmogenesis. Several transcription factors controlling the specification of brainstem serotonergic neurons have been recently identified. Most of these factors are required for the correct development of all raphe nuclei serotonergic groups. However, some selectivity has been observed and could help to discriminate between different sub-classes of medullary serotonergic neurons with potentially different functions.

Mice lacking expression of the zinc finger transcription factor *Gli-2* only generate fifty percent of the 5HT brainstem neurons [71]. Similarly, zebrafish lacking *Fox2a* (*Monorail*, *Mol*) transcription factor present an almost complete loss of serotonergic neurons in the brainstem [82]. These transcription factors are involved in the induction of the floor plate and 5HT neuron precursors originate close to this region. Thus failure in floor plate specification might be responsible for the serotonergic differentiation defect observed in these mice rather then a cell autonomous requirement for these transcription factors. Accordingly, *Gli-2*$^{-/-}$ mice also present a defect in the generation of midbrain dopaminergic neurons. In addition, exogenous expression of Shh (Sonic Hedgehog) in *Mol*$^{-/-}$ zebrafish restores the specification of 5HT neurons [21, 71, 82].

Several other transcription factor deletions result in a more specific absence or severe reduction of brainstem 5HT neurons: the homeodomain factor *Nkx2.2* [12], the bHLH factor *Mash1* [94], the zinc finger proteins *Gata-2* [22] and *Gata-3* [22, 112], the lim domain factor *Lmx1b* [29] and the *Ets* factor *Pet-1* [48] (Fig. 3 and Fig. 4).

Nkx2.2 is expressed in several ventral hindbrain cell populations during early development [103]. It is required for the early specification of non-committed progenitors whose direction towards 5HT fate requires additional factors and/or down-regulation of other lineage specific factors, one of which may be *Phox2b* [38]. *Nkx2.2* appears to be essential for serotonergic specification in the medullary raphe except for 5HT neurons of the dorsal raphe nuclei (B6 and B7). In these nuclei the homeodomain transcription factor *Nkx2.9* is expressed and has been proposed to compensate for the absence of *Nkx2.2* [12]. In addition *Nkx6.1* appears to collaborate with *Nkx2.2* to induce serotonergic specification. Thus *Nkx6.1* knock-down in the rostral hindbrain of chick embryos disturbs serotonergic neuron development. However *Nkx6* deficient mice, that lack expression of both *Nkx6.1* and *Nkx6.2*, normally generate 5HT brainstem neurons, suggesting differences in this pathway between species [22, 93].

Observation of *Mash1* deficient mice revealed that this factor is also essential for the generation of brainstem serotonergic neurons. *Mash1* appears to be required for both the generation of postmitotic precursors through its proneural activity and for the specification of the serotonergic fate (Fig. 3 and Fig. 4). In *Mash1*$^{-/-}$ embryos, *Nkx2.2*, *Nkx2.9*, and *Phox2b* expression is maintained, whereas subsequent expression of *Pet-1*, *Lmx1b*, *Gata-2* and *Gata-3* is lost in serotonergic domains. In these mice, postmitotic 5HT precursors that would have repress *Phox2b* expression and start expressing β-III-tubulin fail to develop. Expression of *Ngn2* from the endogenous *Mash1* locus, which as been shown to be able to replace *Mash1* proneural activity in other systems [88], is able to restore the generation of postmitotic 5HT precursors. In contrast *Ngn2* only very partially restores serotonergic neuron differentiation, indicating that *Mash1* has an additional role in the specification of the serotonergic precursors. Finally *Mash1* ectopic expression in the rostral hindbrain, even together with *Nkx2.2*, fails to induce serotonergic differentiation, suggesting that another factor must collaborate with *Nkx2.2* and *Mash1* to induce rostral 5HT neuron specification [94].

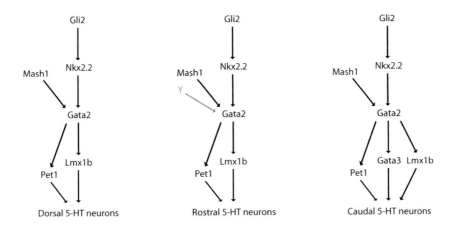

Fig. 3. Transcription factors governing raphe nuclei serotonergic neuron development. Transcription factor *Gli-2* is required for the correct development of all populations of raphe development nuclei serotonergic neurons. *Nkx2.2* and *Mash-1*, *Gata-2*, *Lmx1b* and *Pet-1* are sequentially expressed along the differentiation of the brainstem serotonergic neurons. These factors are all required for the differentiation of raphe nuclei 5HT neurons. In addition *Gata-3* is required for 5HT neuron differentiation in the caudal raphe nuclei. Finally, in rostral raphe nuclei, an unknown factor (Y) appears to collaborate with *Mash1* and *Nkx2.2* for the induction of 5HT neuron differentiation. Transcription factors important for the development of serotonergic neurons in other species are not shown but mentioned in the text.

Gata-2 and *Gata-3* appear to be required for the development of 5HT neurons downstream of *Nkx2.2* and *Mash1* with a more selective requirement of *Gata-3* for caudal serotonergic brainstem neurons [22; 112] (Fig. 3 and Fig. 4). Analysis of serotonergic specification in *Gata-2$^{-/-}$* mice has been performed in hindbrain explants since these mice die between embryonic day E9.5 and E10.5 due to a general hematopoietic system defect. No 5HT neuron differentiation could be detected in cultured *Gata-2$^{-/-}$* explants [22]. *Gata-3$^{-/-}$* mice die during embryonic development but can be rescued by treatment of the pregnant mouse with noradrenalin precursor [68]. Another strategy consisted in studying *Gata-3$^{-/-}$* chimeric mice. Both studies demonstrated the importance of *Gata-3* for the generation of 5HT brainstem neurons, especially in the caudal raphe nuclei [94; 112]. *Gata-3$^{-/-}$* mice present a 30% reduction of the number of 5HT neurons in the rostral raphe nuclei and an 80% reduction of the number of caudal raphe 5HT neurons. Importantly even in the caudal raphe nuclei the expression pattern of *Pet-1* and *Lmx1b* is preserved. *Lmx1b* is expressed in 5HT neuron precursors after *Gata-2* and *Gata-3* but before *Pet-1*, and *Lmx1b$^{-/-}$* mice fail to develop any brainstem serotonergic neurons. In these mice, *Pet-1* and *Gata-3* expression are lost completely or only in caudal areas, respectively [29]. Finally *Pet-1* is specifically expressed in serotonergic neurons, precedes the detection of 5HT and directly controls the promoters of several key genes of 5HT synthesis and metabolism, suggesting that it is critical for the terminal specification of these neurons [47; 96]. *Pet-1$^{-/-}$* mice present a 70% reduction of brainstem serotonergic neurons [48]. In addition, *Lmx1b* and *Gata-3* expression are unaltered in *Pet-1*[-/-] developing embryo.

In summary, *Nkx2.2*, together with *Mash1*, induces the specification of 5HT neuron precursors, and the expression of *Gata-2* and *Gata-3*. *Gata-2* expression is critical at this stage, whereas *Gata-3* is only required for the specification of caudal 5HT neurons. Then *Lmx1b* and *Pet-1* expression are induced and are critical for further serotonergic specification. These last transcription factors appear to act in parallel pathways, may be together with *Gata-3*, in caudal 5HT neurons, since their expression is not lost in *Gata-3* and *Pet-1* deficient mice (Fig. 3). Finally they induce the expression of 5HT-specific genes and appear to be involved in the later maintenance of the serotonergic identity [18; 21].

Respiratory phenotypes of mouse mutants with defects in modulatory neurons

Mice deficient for *Mash1*, *Phox2a* or *Rnx* expression all die within the first day of life with different respiratory abnormalities and several observations argue for respiratory problems as a primary cause of death [24; 25; 43; 75; 92; 104; 114] (Table 1).

Mash1

Mash1[-/-] mice present an increased mortality rate within the first minutes after birth (42% of knock-out versus 14% of wild-type littermates) whereas all *Mash1*[-/-] eventually die within the first day. The proportion of *Mash1*[-/-] newborns developing gasps is increased compared to control mice, suggesting that *Mash1*[-/-] newborns have an impaired oxygenation. Moreover *Mash1*[-/-] and *Mash1*[+/-] newborns have an elevated breathing frequency, consistent with the loss of (nor)adrenergic neurons observed for example in the A5 pontine group, which has a depressive effect on the medullary respiratory rhythm generator [25]. In addition, *Mash1*[+/-] newborn mice present a decreased response to hypercapnia compared to their wild type littermates [24]. This phenotype is only observed in male mutants. In contrast response to hypercapnia in adults and response to hypoxia are comparable to controls.

These data indicate that *Mash1* is involved in the development of the respiratory control neuronal network. However, the reported respiratory defects appear to be insufficient to explain the increased mortality of *Mash1*[-/-] mice. The critical role of *Mash1* in the development of nTS and AP [95] might result in defective central or peripheral chemoreception and an impaired ability to adapt respiratory rhythm to hypoxia or hypercapnia. Thus the ability to respond to these stimulations needs to be evaluated in *Mash1*[-/-] newborn mice. Alternatively, impaired tonic drive from the peripheral chemoreceptors or lack of facilitatory effect from the (nor)adrenergic groups such as LC could also result in the impaired breathing activity observed in *Mash1*[-/-] newborns [24, 49].

Phox2a

Surgically delivered E18.5 $Phox2a^{-/-}$ mice present a hypoventilation *in vivo*, which is accompanied by a severe decrease of the minute volume, and a decreased *in vitro* respiratory-like rhythm in medulla-spinal cord preparations [114]. Lack of *Phox2a* expression selectively affects LC development but not the other brainstem (nor)adrenergic neuron groups. Accordingly, blockade of α1-adrenoceptors in wild-type mouse embryos during late gestation results in *in vivo* and *in vitro* neonatal respiratory deficits similar to those observed in $Phox2a^{-/-}$ animals. This observation suggest that the hypoventilation observed in *Phox2a* animals is caused by an abnormal maturation of the respiratory network during the late gestational phase which depends on (nor)adrenergic facilitation by the LC [114].

In vitro $Phox2a^{-/-}$ preparations present an altered response to hypoxic aCSF application, which, in wild-type (WT) preparations, decreases the respiratory rhythm-like activity [113]. This abnormal *in vitro* response to hypoxia may reflect an impaired development of the medullary respiratory network caused by defects in LC development. In contrast, *in vivo* response to hypoxia in $Phox2a^{-/-}$ newborn mice seems to be comparable to control animals ($122 \pm 7\%$ versus $129 \pm 7\%$ of the control values which are 65 ± 11 and 104 ± 6 cycles per minute in $Phox2a^{-/-}$ versus WT animals respectively) [114]. However, since the central depression of the respiratory rhythm in response to hypoxia is lost in *Phox2a* animals and since *Phox2a* deficient mice present a severe atrophy of the petrosal ganglia [75], which contain sensory neurons carrying peripheral chemoreception from the carotid bodies, it is questionable whether the *in vivo* response to hypoxia is complete in *Phox2a* animals and whether a partially impaired peripheral chemoreception could participate in the lethal phenotype of these animals.

Rnx

$Rnx^{-/-}$ newborn mice present cyanosis and die within 24h even when respiration was initiated by transient mechanical stimulation [104]. Respiration of $Rnx^{-/-}$ newborns is characterized by a hyperventilation with short inspiration duration and by a high incidence of long episodes of apnea. Recording of respiratory-like output from medulla-spinal cord preparations demonstrated the central origin of their breathing defect. C4 phrenic root recording shows an enhanced frequency of respiratory bursts with shorter inspiratory duration than controls. Accordingly to the apneas observed *in vivo*, the central respiratory rhythm recorded in $Rnx^{-/-}$ preparations is interrupted by frequent pauses of the respiratory-like activity. In addition, membrane potential of ventral medulla inspiratory neurons correlates with the respiratory-like activity recorded at the phrenic root, and failures of inspiratory neuron firing in $Rnx^{-/-}$ preparation correlates with the respiratory-like activity arrests. However, the absence of *Rnx* expression in the rostral ventrolateral medulla, reported by Qian et al. [97], suggests that the development of rhythmic respiratory neurons themselves may not be affected in $Rnx^{-/-}$ mice. Alternatively, defects in the development of the nTS, AP and (nor)adrenergic groups of neurons are likely to participate in the breathing activity perturbations observed in absence of *Rnx*. *Rnx*$^{-/-}$ mice show an increased respiratory frequency which is consistent with the lack of brainstem (nor)adrenergic neurons and their depressive function on respiratory frequency. In addition, periods of elevated frequency are interrupted by recurrent apneas, which may be related to the simultaneous deletion of the nTS and to the loss of peripheral sensory excitatory input, which is known to cause irregular respiratory activity.

Phox2b

Phox2b$^{+/-}$ neonates present abnormal ventilatory responses to hypoxia, hyperoxia and hypercapnia as well as sleep related breathing anomalies [26; 30; 98]. In addition, Dauger et al. have reported a partial loss of TH expressing chemoafferent neurons in the petrosal ganglion of these mice [26]. Alternatively *Phox2b*$^{+/-}$ respiratory phenotypes might be related to gene dosage dependent defects in other structures which are affected in *Phox2b*$^{-/-}$ mice, such as carotid bodies, second order visceral sensory neurons of the nTS and AP or (nor)adrenergic groups of neurons, resulting in impaired peripheral or central chemoreception or central excitatory drive [26; 91]. The phenotype of *Phox2b*$^{+/-}$ mice is discussed in more detail in Chapter 14, especially their potential as a model for congenital central hypoventilation syndrome (CCHS), which is associated with heterozygous *PHOX2B* mutations.

Serotonergic program

So far, among the mouse strains deficient for transcription factors that affect serotonergic fate only *Mash1*$^{-/-}$ deficient mice have been analyzed for respiratory phenotypes. *Mash1*$^{-/-}$ mice present defective specification of several other neuronal sub-populations involved in breathing control, precluding an interpretation of a possible contribution of the lack of 5HT neurons to their respiratory phenotype. The absence of studies of the other 5HT neuron deficient mouse models is partly due to early embryonic lethality, as in the case of *Gata-2* and *Gata-3*. *Nkx2.2* and *Lmxb1* deficient neonates die within a week or the first day of life, respectively, but also have potentially life threatening defects in cell types other than 5HT neurons that could contribute to the phenotype [16; 29; 109]. Interestingly about 30% to 40% of *Pet-1*$^{-/-}$ mice die within the first week after birth. After this period of vulnerability, no abnormal death of *Pet-1* deficient mice is observed [48]. Since serotonergic neuron abnormalities have been reported in SIDS, a detailed analysis of respiratory function in *Pet-1*$^{-/-}$ mice would be especially warranted [89; 123].

12.3.2 Mouse mutants of transcription factors with respiratory phenotypes and defects in the development of the rhythm generating respiratory neurons

Despite the central function of rhythmogenic brainstem neurons of the pre-BötC and pFRG in the neuronal control of the respiratory function [35], knowledge about the development and specification of these groups of rhythmogenic neurons is extremely limited compared to the known developmental pathways of modulatory neurons (see above). This may be partially due to the originally functional rather than biochemical definition of the pre-BötC and pFRG groups, and the resulting difficulty in defining a clearly characterized population of rhythmogenic neurons. Only in the past few years, molecular markers of the

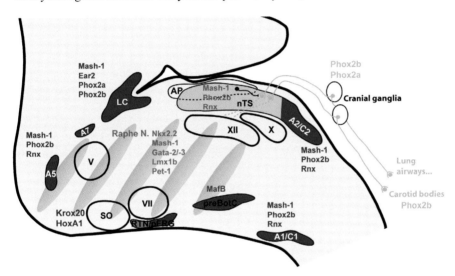

Fig. 4. Transcription factors controlling the development of brainstem respiratory neurons. Schematic representation of the respiratory groups of neurons in a sagittal view of the brainstem with transcription factors required for their development. Transcription factors required for rhythmogenic neuron development in red (*MafB*), for (nor)adrenergic neuron differentiation in blue (*Mash1*, *Ear2*, *Phox2b*, *Phox2a*, *Rnx*), for differentiation of the serotonergic neurons of the Raphe nuclei in light brown (*Nkx2.2*, *Mash1*, *Gata-2*, *Gata-3*, *Lmx1b*, *Pet-1*), for nTS and area postrema development in brown (*Mash1*, *Phox2b*, *Rnx*), and for relay sensory neurons and carotid bodies development in orange (*Phox2b*, *Phox2a*). *Krox20* and *Hoxa1*, which are required for the correct development of the rostral ventral medulla are represented in black. Transcription factors important for the development of these neurons in other species are not shown but mentioned in the text. (Modified from [10], Copyright (2005), with permission from Elsevier.).

anatomic location of the pre-BötC [41; 46; 108] and, in the case of the NK1 and 5HT4a receptors and of the transcription factor *MafB*, of functionally important subpopulations of neurons in this region [9; 41; 42; 69] have been identified. They provide important markers for future developmental studies and cell lineage tracing experiments.

MafB / Kreisler

Until recently nothing was known, about the transcriptional control of developmental programs that guide cell identity and organization of pre-BötC neurons. The first inroad towards the dissection of such transcriptional programs was made with the discovery that the basic leucin zipper transcription factor of the *Maf* family *MafB* is critically important for the development of a functional rhythmogenic pre-BötC [9]. *MafB* deficient neonates present extensive cyanosis and prolonged apneas, only interrupted by occasional gasps, and die within minutes to 2 hours after birth, representing one of the most severe possible phenotype for mice that otherwise reach birth following a mendelian ratio. These observations suggest a critical role for *MafB* in the development of vital structures of the breathing system. No defects are found in the peripheral respiratory apparatus of $MafB^{-/-}$ mice. They do not present any anatomical of histological defect of the heart, upper airways, ribcage or lungs and their diaphragm appears to be correctly innervated by the phrenic nerve. On the other hand severe defects are found in central breathing control. Thus, respiratory-like rhythmic motoneuron output from $MafB^{-/-}$ newborn brainstem preparations *in vitro* show a severely reduced frequency. Furthermore, in $MafB^{-/-}$ preparations, the central respiratory-like activity does not respond to electric stimulation of the pre-BötC or to substance P or hypoxia modulation. In addition, electrolytic lesions of the pre-BötC in $MafB^{+/+}$ brainstem preparations severely reduce phrenic bursts frequency to levels found in $MafB^{-/-}$ samples but have no additional effect on the activity of $MafB^{-/-}$ preparations. Together these observations indicate that *MafB* is required for the development of a functional rhythmogenic pre-BötC. Interestingly, the loss of preBötC function in $MafB^{-/-}$ mice is accompanied by a severe anatomical disorganization of this group of neurons. *MafB* is expressed by a subpopulation of pre-BötC neurons, some of them co-expressing the marker NK1R. In addition *MafB* deficient mice present a significant loss of NK1R positive neurons in the pre-BötC and a disruption of the normally compact anatomy of this group of neurons (Fig. 4). In contrast, *MafB* deficiency does not cause any apparent defect in the development of neuronal groups involved in respiratory modulation.

The underlying mechanisms of the defective pre-BötC development in absence of *MafB* are not completely resolved. However, it was observed that *MafB* deficiency results in a marked reduction of NK1R expressing pre-BötC neurons without causing increased cell death in the pre-BötC region during late embryonic development. Together with the previously reported function of *Maf* factors in cell fate decisions and control of progenitor proliferation in various systems [60; 101; 105], these observations suggest a possible role of *MafB* in the proliferation or specification of precursors of pre-BötC neurons critically involved in breathing rhythmogenesis. Alternatively or in addition, *MafB* might be required for the formation of correct cellular contacts in the pre-BötC, since it has been shown to control expression of diverse adhesion molecules in various systems [67]. Whatever the detailed mechanism, with molecular markers now available and the identification of a critically required transcription factor for functional pre-BötC development, the dissection of developmental pathways of pre-BötC neurons is within reach.

The *MafB* gene is also affected by the x-ray generated classical *Kreisler* mutation, which affects a regulatory region but not the coding sequence of the gene [20]. As a consequence homozygosity for the *Kreisler* mutation (*Kr/Kr*) does not result in complete *MafB* deficiency. It rather causes a specific loss of expression in early hindbrain development but does not abolish expression in other tissues [31], including the pre-BötC neurons of the hindbrain at late developmental stages up to birth (B. Blanchi and M. Sieweke unpublished observations). Despite severe defects in early hindbrain development, interestingly *Kr/Kr* mice live to adulthood [72] and have only a minor breathing phenotype under specific challenge conditions [14]. By contrast, *MafB$^{-/-}$* mice lacking *MafB* coding sequence and therefore *MafB* activity in all cells where it is normally expressed, including preBötC neurons, die from central breathing failure at birth, as discussed above [9]. This clearly shows that the early function of *MafB* in hindbrain patterning is dispensable, whereas expression at later developmental stages during the specification of respiratory neurons is critical for breathing function.

Other factors

Deletion of the transcription factors *Hoxa1* and *Krox20* in mouse results in hypoplasia of a rostral ventral medulla region that might contain the rhythmogenic neurons of the pFRG and seventy percent of *Krox20$^{-/-}$* and all *Hoxa1$^{-/-}$* mice die within the first day after birth presenting severe respiratory defects [28; 54]. This has been attributed to defective early hindbrain development in absence of *Krox20* and *Hoxa1* [11; 15] (see Chapter 11). Given the example of *MafB* (see above), it may be possible that these transcription factors also play additional roles in later development of specific respiratory neurons. Further analysis of the expression of these factors at later embryonic development stages and their inducible deletion at specific developmental stages would therefore be of particular interest.

12.3.3 Mouse mutants of transcription factors with respiratory phenotype of unresolved underlying developmental defect

Nurr1

Mice deficient for the transcription factor *Nurr1*, an orphan nuclear receptor, have a severely impaired breathing activity characterized by hypoventilation, increased occurrence of apneas and defective response to hypoxia, and die within 24 hours after birth. The underlying developmental defect for this phenotype is unknown. Although these mice fail to generate any midbrain dopaminergic neurons [129], it is not likely to be the cause of the early lethality observed in *Nurr1$^{-/-}$* mice, since mice deficient for dopamine expression survive several weeks after birth [130]. *Nurr1* is also expressed in other nervous system regions involved in breathing control. It is expressed in the nTS from E13.5, in the dorsal motor nuclei X (dmnX) from E10.5, in the carotid bodies as well as in petrosal ganglia explants [83]. So far the only cellular or anatomical abnormality reported in these regions in *Nurr1*⁻ᐟ⁻ mice is a decreased expression of *Ret* in the dmnX [116]. In contrast, anatomy of the carotid bodies and TH immunostaining, as well as formation of the dmnX and nTS, appear normal in these mice.

Nurr1⁻ᐟ⁻ mice present an impaired breathing rhythm *in vivo*, whereas respiratory rhythm generator activity, recorded on brainstem-spinal cord preparations, is mostly normal. In addition, *Nurr1*⁻ᐟ⁻ mice have a defective *in vivo* response to hypoxia characterized by the absence of any frequency response. Brainstem-spinal cord preparations from *Nurr1*⁻ᐟ⁻ mice also present a partially altered response to hypoxia [83]. Together these observations suggest that the breathing defect observed in *Nurr1*⁻ᐟ⁻ mice might be caused by an impaired tonic drive from the sensory system in normoxic condition and a defective chemosensitivity, especially in the periphery.

Pbx3

Mice deficient for *Pbx3*, a TALE class homeodomain transcription factor, present an impaired ventilation characterized by largely reduced respiratory frequency and minute volume and by irregular amplitude of breathing movements. Recording of respiratory-like output at the level of the phrenic root shows an increased frequency of the respiratory bursts generated by *Pbx3*⁻ᐟ⁻ medulla-spinal cord preparations. However, the amplitude of these phrenic bursts is severely reduced in *Pbx3*⁻ᐟ⁻ preparations. This could explain the irregular amplitude of breathing movements and the reduction of the minute volume observed *in vivo* in *Pbx3*⁻ᐟ⁻ mice. *Pbx3* is highly expressed in the developing central nervous system including many areas of the brainstem. Among them *Pbx3* is expressed in an area of the caudal ventral medulla, that could overlap with the ventral respiratory column, where it co-localizes with the transcription factor *Rnx*, whose mutation leads to fatal central hypoventilation in mice [100, 104]. In addition, *Pbx3* has been shown to interact with *Rnx* in a transcriptionally active complex. On the other hand Pbx proteins are involved in early midbrain patterning and in maintenance of the midbrain-hindbrain boundary [32]. *Pbx3* deficient newborns develop apneas and marked cyanosis, and die within a few hours after birth [100]. The respiratory defect in *Rnx*⁻ᐟ⁻ mice has been proposed to result from defective development of the nTS rather than from a previously proposed defect affecting neurons that generate the respiratory rhythm [97, 104]. Despite the co-expression of *Pbx3* and *Rnx* in the caudal ventral medulla, further investigation of potential functional defects in other groups of neurons that control breathing has to be done to fully understand the respiratory defect of *Pbx3*⁻ᐟ⁻ mice.

Mecp2

Methyl-CpG binding protein 2 (*MECP2*) gene mutations in human lead to an X-linked neurological disorder called Rett syndrome (RTT) [4]. RTT patients present respiratory irregularities characterized by hyperventilation and apneas [58; 124]. In mice, *Mecp2* mutation causes a progressive neurological disorder that recapitulates features of RTT [17; 45; 131]. *Mecp2$^{-/y}$* male mutant mice present a slow and irregular breathing rhythm with frequent apneas and die around 8 weeks [115]. Respiratory disturbances observed in these mice appear to be at least partially caused by deficient noradrenergic and serotonergic modulation accompanied by a loss of brainstem TH expressing neurons [115]. In addition, an augmented response to hypoxia has been observed in *Mecp2$^{+/-}$* heterozygous female mice suggesting an increased peripheral chemoreception in these mice [8]. Respiratory anomalies of *Mecp2* deficient mice are discussed in more detail in Chapter 16.

Table 1. : Transcription factor mutations affecting specific central respiratory neurons in mice and identified corresponding mutations in human respiratory syndromes. (Modified from [10], Copyright (2005), with permission from Elsevier.)

Gene	Overall phenotype of knockout mice	Associated cellular disorder in the nervous system	Mutations in human respiratory syndromes
	*PreBötzinger complex defect*		
MafB	death within 2 hours after birth[9] no breathing rhythm except gasps *in vivo* *in vitro* preparations: severely reduced rhythm, no response to electric and peptidergic stimuli or hypoxia	Loss of cellularity and organisation in the preBötC[9]	Not determined
	*Defects in NA neurons and NTS*		
Phox2b	death during embryonic development[92] +/-: altered response to hypercapnia, hypoxia and hyperoxia during wakefulness[26,98] hypoventilation and apnea during sleep[30]	Loss of NTS, NA groups of neurons and visceral sensory neurons[26,92]	polyAla expansion or mutation/deletion found in most CCHS patients[2,6,102,119] no mutation found in 128 SIDS cases[61,121]
Phox2a	death within the first day after birth[75] reduced respiratory frequency and *in vitro* response to hypoxia[114]	Loss of LC and visceral sensory neurons[75]	one mutation in 77 CCHS patients[102,119] one mutation in 92 SIDS cases[121]
Rnx	death within the first day after birth[104] alternation of tachypnea and apnea periods[104]	Loss of NTS and NA groups of neurons except rostral LC[97]	no mutation in 108 CCHS patients[3,59,70,119] one mutation in 92 SIDS cases [121]
Mash-1	death within the first day after birth increased breathing frequency[25]	Loss of NTS, NA groups of neurons and 5-HT neurons[50,94,95]	mutation or polyAla deletion found in six CCHS patients, of whom three also had polyAla expansion in Phox2b[27,102,119] no mutation found in 92 SIDS cases[121]
	*Defects in 5-HT neurons*		
Pet-1	30% death during the first week after birth[48]	Loss of most 5-HT neurons[48]	Not determined
Lmx1b	death within the first week after birth[16]	Loss of 5-HT neurons[29]	Not determined
Gata-2	death during embryonic development	Loss of 5-HT neurons[22]	Not determined
Gata-3	death during embryonic development[68]	Loss of most 5-HT neurons in caudal raphe nuclei[94]	Not determined
Nkx2.2	death within the first week after birth[109]	Loss of most 5-HT neurons[12]	Not determined

12.3.4 Transcription factor mutations in human central respiratory disorders

Several human respiratory disorders are known to be or suspected to be caused by defects in the neuronal network that controls breathing. Thus, transcription factors that govern the specification of groups of neurons responsible for respiratory control are potential candidates for a role in Congenital Central Hypoventilation Syndrome (CCHS), Central Sleep Apnea Syndrome (CSAS), Sudden Infant Death Syndrome (SIDS) or Rett Syndrome (RTT) [10]. *PHOX2B* and *MECP2* mutations have been recently shown to be responsible for CCHS and RTT, whereas an increasing amount of data suggests that defects in serotonergic neurons may contribute to the pathogenesis of SIDS.

CCHS

Congenital central hypoventilation syndrome (CCHS) is a rare respiratory disorder characterized by persistent hypoventilation in absence of primary peripheral cardiorespiratory defects [40; 122]. It appears to be caused by a defective autonomic control of breathing activity which results in impaired response to hypercapnia and hypoxia. The majority of CCHS patients show heterozygous mutation of the *PHOX2B* transcription factor which mainly involve polyalanine expansions [2; 6; 102; 120] (Table 1). This is in accordance with the impaired development of several populations of neurons, involved in peripheral chemoreception and central modulation of breathing, observed in *Phox2b* deficient mice [26; 91] and with the respiratory phenotype of *Phox2b$^{+/-}$* mice [26; 30; 98] (see above and Chapter 14). A small number of CCHS patients do not present mutations in the coding sequence of *PHOX2B* and may be associated with other gene mutations, possibly in the same differentiation pathways. Indeed, 6 mutations in *HASH1*, the human homologue of *Mash1*, have been found in CCHS, three of which were not accompanied by a simultaneous polyalanine expansion in *PHOX2B* [27; 102]. Furthermore, one mutation in *PHOX2A* [102] but no mutations in *RNX* have been reported in CCHS patients [3; 59; 70; 118] (Table 1). CCHS and the role of *PHOX2B* in respiratory control are described in more detail in Chapters 3, 4 and 14.

SIDS

Sudden Infant Death Syndrome (SIDS) is the leading cause of death during the first year of life in the developed world [23; 53]. Despite a significant research effort, the pathophysiology underlying SIDS is poorly understood. Although this disorder may be complex, genetic susceptibilities are likely to contribute to the disease, which is further elaborated in Chapter 7. A screen for mutations of *PHOX2B*, *PHOX2A*, *HASH1* and *RNX* transcription factors in 92 SIDS cases revealed only one mutation in both *PHOX2A* and *RNX* [61; 121]. However several observations indicate that brainstem serotonergic neurons may be more important in SIDS than the principal modulatory neurons specified by these transcription factors. Neuropathological analyses have reported increased numbers of serotonergic neurons in the raphe nuclei and the ventral surface of the medulla, and decreased serotonin receptor binding in several brainstem regions involved in respiratory control [62; 64; 86; 87; 89]. Consistent with a role of the serotonergic system in SIDS, polymorphisms in regulatory regions of the serotonin transporter (5HTT) gene have been reported [77; 119; 120]. Together, these observations suggest a role for the impaired function of medullary 5HT neurons in SIDS [63; 64; 89; 123]. The elucidation of the transcription factor programs that drive medullary serotonergic neuron specification in mice, and the characterization of the physiological and respiratory consequences of the mutation of these factors may provide a useful basis for an improved understanding of the molecular and cellular mechanisms underlying SIDS. Several transcription factor deficient mice, which fail to properly develop medullary 5HT neurons, die at birth or during the early postnatal period, e.g. *Mash1*, *Lmx1b* and *Pet-1* deficient mice. Transcription factors *Pet-1* and *Lmx1b* are of particular interest since all *Lmx1b$^{-/-}$* and 30% to 40% of *Pet-1$^{-/-}$* mice die during the first week of life. These mice show a partial (*Pet-1*, 70%) or complete (*Lmx1b*) loss of raphe nuclei 5HT neurons [29; 48]. The fact that only 30% of *Pet-1* mice die during the first week, whereas the other 70% lives into adulthood, is consistent with a proposed multi-risk model of SIDS etiology. In this model, sudden death occurs when several risk factors accumulate; a genetic succeptibility, an exogenous stressor (position to sleep, asphyxia, hypoxia) and a life period of high vulnerability (early postnatal life). It would therefore be of interest to determine whether mutations of transcription factors involved in serotonergic neuron specification are found in SIDS cases.

Rett syndrome

Rett syndrome (RTT) is an X-linked neurological disorder caused by mutations in the Methyl CpG binding Protein 2 (*MECP2*) gene [4] and characterized by loss of language and motor skills, mental retardation, seizures, and respiratory defects [131]. Breathing activity of RTT patients is characterized by irregular ventilation with episodes of normal breathing interrupted by periods of hyperventilation, forced breathing or apneas[58; 124]. Surprisingly, breathing irregularities in RTT were shown to be limited to periods of wakefulness, while breathing stabilizes during sleep. In *Mecp2* deficient mice irregular breathing appears to be at least partially caused by impaired noradrenergic and serotonergic modulation of medullary respiratory network activity [115]. Interestingly, application of noradrenalin restores normal in vitro respiratory-like activity in brainstem preparations from *Mecp2* deficient animals [115]. This finding opens therapeutic perspectives for the treatment of respiratory defects in RTT patients. Respiratory disorders in RTT and function of the *MECP2* transcriptional regulator are reviewed in more detail in Chapter 16.

212 Bruno C. BLANCHI and Michael H. SIEWEKE

CSAS

Central Sleep Apnea Syndrome (CSAS) is characterized by periodic depression of the respiratory drive which results in irregular ventilation and apneas, followed by compensatory periods of hyperventilation [65; 127]. The molecular basis or potential genetic susceptibilities for this disorder are unknown, however, impaired development or activity of central respiratory control is likely to be involved in the occurrence of central sleep apneas (CSA). In particular, CSA may be caused by defective breathing rhythmogenesis. McKay et al. have recently demonstrated that ablation of NK1R neurons in the pre-BötC, which leads to ataxic breathing in the awake animal, results in respiratory disturbances characterized by apneas during sleep [73]. Thus, animals with defective development of the rhythmogenic neurons in the pre-BötC and pFRG may be useful models for CSAS. Given that the transcription factor *MafB* is required for the development of a functional pre-BötC and that *MafB* deficiency in mice leads to a lethal central respiratory defect associated with a loss of NK1R neurons in the pre-BötC, deficiency for *MafB* or down stream targets could play a role in central sleep apnea.

12.4. Conclusion

During recent years, transcription factor mutant mouse models and developmental studies, in combination with physiological methods, electrophysiological measurements, and pharmacological or imaging techniques, have proven to be a powerful approach to provide new insight into the organization and development of the network of respiratory neurons. Furthermore, genetic analysis of human respiratory disorders has identified mutations of nuclear factors, whose deletion in mice caused phenotypes largely consistent with the symptoms observed in human disease, indicating that mouse models provide valuable information that is applicable to human disorders.

Like in many other tissues, the same transcription factor can have functions at different stages of respiratory neuron development from pattern formation in the early hindbrain to the specification of functionally different, terminally differentiated neurons at birth. The influence of early hindbrain patterning on breathing function at birth has been discussed elsewhere [11; 15] (see Chapter 11), whereas this chapter has focused on the later specification of functionally different neuron populations. The case of *MafB* may serve as a good example for the notion that the timing of transcription factor deficiency can be critical for the resulting respiratory phenotype. As outlined earlier the *MafB* gene is also affected by the x-ray generated *Kreisler* (*Kr*) mutation, which causes a specific loss of expression in the early hindbrain but not in the pre-BötC neurons at late developmental stages. Despite severe defects in early hindbrain development, *Kr/Kr* mice live to adulthood [72] and have only a minor breathing phenotype [14], whereas *MafB$^{-/-}$* mice, which lack *MafB* expression also in later development, including in pre-BötC neurons, die from central breathing failure at birth [9]. This indicates that normal breathing function requires *MafB* activity in late respiratory neuron specification but not in early hindbrain patterning. Expression at different stages of development is also documented for most of the transcription factors discussed in this chapter and it is not always resolved whether early, late or both functions contribute to respiratory phenotypes at birth. It will therefore be important in the future to create new mouse models for stage and lineage specific deletion of these factors to precisely dissect at what stage of development a transcription factor is critically required to assure a particular breathing function.

Important new information may also come from lineage tracing studies that would define the contribution of specific transcription factor expressing populations to the network of respiratory neurons. A powerful tool to address such questions may become the expression of toxin receptors from the promoters of transcription factors, which would enable toxin-mediated ablation of cell populations with a specific transcription factor signature. This approach may be further elaborated by using Cre recombinase or toxin receptor sub-units expressed from different gene loci to include timing or combinatorial aspects in the analysis. In general the future will likely see an increasingly sophisticated analysis of the described transcription factor pathways, likely taking into account more detailed promoter analysis, the potential importance of expression levels as well as inhibitory or synergistic cross-talk between the participating factors that may be due to direct physical contact or dependent on indirect mediators, as has been shown in other developmental systems [106].

This increased resolution should both improve our molecular understanding of the developmental pathways and may provide new approaches for the genetic and anatomical analysis of human pathologies. With increasing insight into the network of transcription factors specifying respiratory neurons, more complex and potentially multi-factorial respiratory disorders, including cases of SIDS and CSAS, may also become amendable to genetic analysis. Finally, the fact that some of the identified factors can affect neurons of developmentally distinct but functionally cooperative cell lineages may make these factors clinically interesting diagnostic markers in the short term and attractive targets of therapeutic intervention in the long term.

References

1.   Al-Zubaidy ZA, Erickson RL, Greer JJ (1996) Serotonergic and noradrenergic effects on respiratory neural discharge in the medullary slice preparation of neonatal rats. Pflugers Arch 431: 942-949
2.   Amiel J, Laudier B, Attie-Bitach T, Trang H, de Pontual L, Gener B, Trochet D, Etchevers H, Ray P, Simonneau M, Vekemans M, Munnich A, Gaultier C, Lyonnet S (2003) Polyalanine expansion and frameshift mutations of the paired-like homeobox gene PHOX2B in congenital central hypoventilation syndrome. Nat Genet 33: 459-461

3.   Amiel J, Pelet A, Trang H, de Pontual L, Simonneau M, Munnich A, Gaultier C, Lyonnet S (2003) Exclusion of RNX as a major gene in congenital central hypoventilation syndrome (CCHS, Ondine's curse). Am J Med Genet A 117: 18-20

4.   Amir RE, Van den Veyver IB, Wan M, Tran CQ, Francke U, Zoghbi HY (1999) Rett syndrome is caused by mutations in X-linked MECP2, encoding methyl-CpG-binding protein 2. Nat Genet 23: 185-188

5.   Ballantyne D, Scheid P (2000) Mammalian brainstem chemosensitive neurones: linking them to respiration *in vitro*. J Physiol 525: 567-577

6.   Berry-Kravis EM, Zhou L, Rand CM, Weese-Mayer DE (2006) Congenital central hypoventilation syndrome: PHOX2B mutations and phenotype. Am J Respir Crit Care Med 174: 1139-1144

7.   Bianchi AL, Denavit-Saubie M, Champagnat J (1995) Central control of breathing in mammals: neuronal circuitry, membrane properties, and neurotransmitters. Physiol Rev 75: 1-45

8.   Bissonnette JM, Knopp SJ (2006) Separate respiratory phenotypes in methyl-CpG-binding protein 2 (Mecp2) deficient mice. Pediatr Res 59: 513-518

9.   Blanchi B, Kelly LM, Viemari JC, Lafon I, Burnet H, Bevengut M, Tillmanns S, Daniel L, Graf T, Hilaire G, Sieweke MH (2003) MafB deficiency causes defective respiratory rhythmogenesis and fatal central apnea at birth. Nat Neurosci 6: 1091-1100

10.   Blanchi B, Sieweke MH (2005) Mutations of brainstem transcription factors and central respiratory disorders. Trends Mol Med 11: 23-30

11.   Borday C, Chatonnet F, Thoby-Brisson M, Champagnat J, Fortin G (2005) Neural tube patterning by Krox20 and emergence of a respiratory control. Respir Physiol Neurobiol 149: 63-72

12.   Briscoe J, Sussel L, Serup P, Hartigan-O'Connor D, Jessell TM, Rubenstein JL, Ericson J (1999) Homeobox gene Nkx2.2 and specification of neuronal identity by graded Sonic hedgehog signalling. Nature 398: 622-627

13.   Brunet JF, Pattyn A (2002) Phox2 genes - from patterning to connectivity. Curr Opin Genet Dev 12: 435-440

14.   Chatonnet F, del Toro ED, Voiculescu O, Charnay P, Champagnat J (2002) Different respiratory control systems are affected in homozygous and heterozygous kreisler mutant mice. Eur J Neurosci 15: 684-692

15.   Chatonnet F, Dominguez del Toro E, Thoby-Brisson M, Champagnat J, Fortin G, Rijli FM, Thaeron-Antono C (2003) From hindbrain segmentation to breathing after birth: developmental patterning in rhombomeres 3 and 4. Mol Neurobiol 28: 277-294

16.   Chen H, Lun Y, Ovchinnikov D, Kokubo H, Oberg KC, Pepicelli CV, Gan L, Lee B, Johnson RL (1998) Limb and kidney defects in Lmx1b mutant mice suggest an involvement of LMX1B in human nail patella syndrome. Nat Genet 19: 51-55

17.   Chen RZ, Akbarian S, Tudor M, Jaenisch R (2001) Deficiency of methyl-CpG binding protein-2 in CNS neurons results in a Rett-like phenotype in mice. Nat Genet 27: 327-331

18.   Cheng L, Chen CL, Luo P, Tan M, Qiu M, Johnson R, Ma Q (2003) Lmx1b, Pet-1, and Nkx2.2 coordinately specify serotonergic neurotransmitter phenotype. J Neurosci 23: 9961-9967

19.   Coppola E, Pattyn A, Guthrie SC, Goridis C, Studer M (2005) Reciprocal gene replacements reveal unique functions for Phox2 genes during neural differentiation. Embo J 24: 4392-4403

20.   Cordes SP, Barsh GS (1994) The mouse segmentation gene kr encodes a novel basic domain-leucine zipper transcription factor. Cell 79: 1025-1034

21.   Cordes SP (2005) Molecular genetics of the early development of hindbrain serotonergic neurons. Clin Genet 68: 487-494

22.   Craven SE, Lim KC, Ye W, Engel JD, de Sauvage F, Rosenthal A (2004) Gata2 specifies serotonergic neurons downstream of sonic hedgehog. Development 131: 1165-1173

23.   Daley KC (2004) Update on sudden infant death syndrome. Curr Opin Pediatr 16: 227-232

24.   Dauger S, Renolleau S, Vardon G, Nepote V, Mas C, Simonneau M, Gaultier C, Gallego J (1999) Ventilatory responses to hypercapnia and hypoxia in Mash-1 heterozygous newborn and adult mice. Pediatr Res 46: 535-542

25.   Dauger S, Guimiot F, Renolleau S, Levacher B, Boda B, Mas C, Nepote V, Simonneau M, Gaultier C, Gallego J (2001) MASH-1/RET pathway involvement in development of brainstem control of respiratory frequency in newborn mice. Physiol Genomics 7: 149-157

26.   Dauger S, Pattyn A, Lofaso F, Gaultier C, Goridis C, Gallego J, Brunet JF (2003) Phox2b controls the development of peripheral chemoreceptors and afferent visceral pathways. Development 130: 6635-6642

27.   De Pontual L, Nepote V, Attie-Bitach T, Al Halabiah H, Trang H, Elghouzzi V, Levacher B, Benihoud K, Auge J, Faure C, Laudier B, Vekemans M, Munnich A, Perricaudet M, Guillemot F, Gaultier C, Lyonnet S, Simonneau M, Amiel J (2003) Noradrenergic neuronal development is impaired by mutation of the proneural HASH-1 gene in congenital central hypoventilation syndrome (Ondine's curse). Hum Mol Genet 12: 3173-3180

28.   Del Toro ED, Borday V, Davenne M, Neun R, Rijli FM and Champagnat J (2001) Generation of a novel functional neuronal circuit in Hoxa1 mutant mice. J Neurosci 21: 5637-5642

29.   Ding YQ, Marklund U, Yuan W, Yin J, Wegman L, Ericson J, Deneris E, Johnson RL, Chen ZF (2003) Lmx1b is essential for the development of serotonergic neurons. Nat Neurosci 6: 933-938

30.   Durand E, Dauger S, Pattyn A, Gaultier C, Goridis C, Gallego J (2005) Sleep-disordered breathing in newborn mice heterozygous for the transcription factor Phox2b. Am J Respir Crit Care Med 172: 238-243

31.   Eichmann A, Grapin-Botton A, Kelly L, Graf T, Le Douarin NM, Sieweke M (1997) The expression pattern of the mafB/kr gene in birds and mice reveals that the kreisler phenotype does not represent a null mutant. Mech Dev 65: 111-122

32.   Erickson T, Scholpp S, Brand M, Moens CB, Jan Waskiewicz A (2006) Pbx proteins cooperate with Engrailed to pattern the midbrain-hindbrain and diencephalic-mesencephalic boundaries. Dev Biol 301: 504-517

33.   Errchidi S, Monteau R, Hilaire G (1991) Noradrenergic modulation of the medullary respiratory rhythm generator in the newborn rat: an in vitro study. J Physiol 443: 477-498

34.   Feldman JL, Mitchell GS, Nattie EE (2003) Breathing: rhythmicity, plasticity, chemosensitivity. Annu Rev Neurosci 26: 239-266

35.   Feldman JL, Del Negro CA (2006) Looking for inspiration: new perspectives on respiratory rhythm. Nat Rev Neurosci 7: 232-242

36.   Feldman JL, Janczewski WA (2006) The Last Word: Point: Counterpoint authors respond to commentaries on "the parafacial respiratory group (pFRG)/pre-Botzinger complex (preBotC) is the primary site of respiratory rhythm generation in the mammal". J Appl Physiol 101: 689

37.   Forster HV, Pan LG, Lowry TF, Serra A, Wenninger J, Martino P (2000) Important role of carotid chemoreceptor afferents in control of breathing of adult and neonatal mammals. Respir Physiol 119: 199-208

38.   Goridis C, Rohrer H (2002) Specification of catecholaminergic and serotonergic neurons. Nat Rev Neurosci 3: 531-541

39.   Gourine AV, Llaudet E, Dale N, Spyer KM (2005) ATP is a mediator of chemosensory transduction in the central nervous system. Nature 436: 108-111

40.   Gozal D (1998) Congenital central hypoventilation syndrome: an update. Pediatr Pulmonol 26: 273-282

41.   Gray PA, Rekling JC, Bocchiaro CM, Feldman JL (1999) Modulation of respiratory frequency by peptidergic input to rhythmogenic neurons in the pre-Botzinger complex. Science 286: 1566-1568

42.   Gray PA, Janczewski WA, Mellen N, McCrimmon DR, Feldman JL (2001) Normal breathing requires preBotzinger complex neurokinin-1 receptor-expressing neurons. Nat Neurosci 4: 927-930

43.   Guillemot F, Lo LC, Johnson JE, Auerbach A, Anderson DJ, Joyner AL (1993) Mammalian achaete-scute homolog 1 is required for the early development of olfactory and autonomic neurons. Cell 75: 463-476

44.   Guo S, Brush J, Teraoka H, Goddard A, Wilson SW, Mullins MC, Rosenthal A (1999) Development of noradrenergic neurons in the zebrafish hindbrain requires BMP, FGF8, and the homeodomain protein soulless/Phox2a. Neuron 24: 555-566

45.   Guy J, Hendrich B, Holmes M, Martin JE, Bird A (2001) A mouse Mecp2-null mutation causes neurological symptoms that mimic Rett syndrome. Nat Genet 27: 322-326

46.   Guyenet PG, Sevigny CP, Weston MC, Stornetta RL (2002) Neurokinin-1 receptor-expressing cells of the ventral respiratory group are functionally heterogeneous and predominantly glutamatergic. J Neurosci 22: 3806-3816

47.   Hendricks T, Francis N, Fyodorov D, Deneris ES (1999) The ETS domain factor Pet-1 is an early and precise marker of central serotonin neurons and interacts with a conserved element in serotonergic genes. J Neurosci 19: 10348-10356

48.   Hendricks TJ, Fyodorov DV, Wegman LJ, Lelutiu NB, Pehek EA, Yamamoto B, Silver J, Weeber EJ, Sweatt JD, Deneris ES (2003) Pet-1 ETS gene plays a critical role in 5HT neuron development and is required for normal anxiety-like and aggressive behavior. Neuron 37: 233-247

49.   Hilaire G, Viemari JC, Coulon P, Simonneau M, Bevengut M (2004) Modulation of the respiratory rhythm generator by the pontine noradrenergic A5 and A6 groups in rodents. Respir Physiol Neurobiol 143: 187-197

50.   Hirsch MR, Tiveron MC, Guillemot F, Brunet JF, Goridis C (1998) Control of noradrenergic differentiation and Phox2a expression by MASH1 in the central and peripheral nervous system. Development 125: 599-608

51.   Holzschuh J, Barrallo-Gimeno A, Ettl AK, Durr K, Knapik EW, Driever W (2003) Noradrenergic neurons in the zebrafish hindbrain are induced by retinoic acid and require tfap2a for expression of the neurotransmitter phenotype. Development 130: 5741-5754

52.   Hornbruch A, Ma G, Ballermann MA, Tumova K, Liu D, Cairine Logan C (2005) A BMP-mediated transcriptional cascade involving Cash1 and Tlx-3 specifies first-order relay sensory neurons in the developing hindbrain. Mech Dev 122: 900-913

53.   Hunt CE (2001) Sudden infant death syndrome and other causes of infant mortality: diagnosis, mechanisms, and risk for recurrence in siblings. Am J Respir Crit Care Med 164: 346-357

54. Jacquin TD, Borday V, Schneider-Maunoury S, Topilko P, Ghilini G, Kato F, Charnay P, Champagnat J (1996) Reorganization of pontine rhythmogenic neuronal networks in Krox-20 knockout mice. Neuron 17: 747-758

55. Janczewski WA, Feldman JL (2006) Distinct rhythm generators for inspiration and expiration in the juvenile rat. J Physiol 570: 407-420

56. Johnson SM, Koshiya N, Smith JC (2001) Isolation of the kernel for respiratory rhythm generation in a novel preparation: the pre-Botzinger complex "island". J Neurophysiol 85: 1772-1776

57. Jordan D (2001) Central nervous pathways and control of the airways. Respir Physiol 125: 67-81

58. Julu PO, Kerr AM, Apartopoulos F, Al-Rawas S, Engerstrom IW, Engerstrom L, Jamal GA, Hansen S (2001) Characterisation of breathing and associated central autonomic dysfunction in the Rett disorder. Arch Dis Child 85: 29-37

59. Kanai M, Numakura C, Sasaki A, Shirahata E, Akaba K, Hashimoto M, Hasegawa H, Shirasawa S, Hayasaka K (2002) Congenital central hypoventilation syndrome: a novel mutation of the RET gene in an isolated case. Tohoku J Exp Med 196: 241-246

60. Kelly LM, Englmeier U, Lafon I, Sieweke MH, Graf T (2000) MafB is an inducer of monocytic differentiation. Embo J 19: 1987-1997

61. Kijima K, Sasaki A, Niki T, Umetsu K, Osawa M, Matoba R, Hayasaka K (2004) Sudden infant death syndrome is not associated with the mutation of PHOX2B gene, a major causative gene of congenital central hypoventilation syndrome. Tohoku J Exp Med 203: 65-68

62. Kinney HC, Filiano JJ, White WF (2001) Medullary serotonergic network deficiency in the sudden infant death syndrome: review of a 15-year study of a single dataset. J Neuropathol Exp Neurol 60: 228-247

63. Kinney HC (2005) Abnormalities of the brainstem serotonergic system in the sudden infant death syndrome: a review. Pediatr Dev Pathol 8: 507-524

64. Kinney HC, Myers MM, Belliveau RA, Randall LL, Trachtenberg FL, Fingers ST, Youngman M, Habbe D, Fifer WP (2005) Subtle autonomic and respiratory dysfunction in sudden infant death syndrome associated with serotonergic brainstem abnormalities: a case report. J Neuropathol Exp Neurol 64: 689-694

65. Kohnlein T, Welte T, Tan LB, Elliott MW (2002) Central sleep apnoea syndrome in patients with chronic heart disease: a critical review of the current literature. Thorax 57: 547-554

66. Li A, Nattie E (2006) Catecholamine neurones in rats modulate sleep, breathing, central chemoreception and breathing variability. J Physiol 570: 385-396

67. Li MA, Alls JD, Avancini RM, Koo K, Godt D (2003) The large Maf factor Traffic Jam controls gonad morphogenesis in Drosophila. Nat Cell Biol 5: 994-1000

68. Lim KC, Lakshmanan G, Crawford SE, Gu Y, Grosveld F, Engel JD (2000) Gata3 loss leads to embryonic lethality due to noradrenaline deficiency of the sympathetic nervous system. Nat Genet 25: 209-212

69. Manzke T, Guenther U, Ponimaskin EG, Haller M, Dutschmann M, Schwarzacher S, Richter DW (2003) 5HT4(a) receptors avert opioid-induced breathing depression without loss of analgesia. Science 301: 226-229

70. Matera I, Bachetti T, Cinti R, Lerone M, Gagliardi L, Morandi F, Motta M, Mosca F, Ottonello G, Piumelli R, Schober JG, Ravazzolo R, Ceccherini I (2002) Mutational analysis of the RNX gene in congenital central hypoventilation syndrome. Am J Med Genet 113: 178-182

71.   Matise MP, Epstein DJ, Park HL, Platt KA, Joyner AL (1998) Gli2 is required for induction of floor plate and adjacent cells, but not most ventral neurons in the mouse central nervous system. Development 125: 2759-2770

72.   McKay IJ, Muchamore I, Krumlauf R, Maden M, Lumsden A, Lewis J (1994) The kreisler mouse: a hindbrain segmentation mutant that lacks two rhombomeres. Development 120: 2199-2211

73.   McKay LC, Janczewski WA, Feldman JL (2005) Sleep-disordered breathing after targeted ablation of preBotzinger complex neurons. Nat Neurosci 8: 1142-1144

74.   Mellen NM, Janczewski WA, Bocchiaro CM, Feldman JL (2003) Opioid-induced quantal slowing reveals dual networks for respiratory rhythm generation. Neuron 37: 821-826

75.   Morin X, Cremer H, Hirsch MR, Kapur RP, Goridis C, Brunet JF (1997) Defects in sensory and autonomic ganglia and absence of locus coeruleus in mice deficient for the homeobox gene Phox2a. Neuron 18: 411-423

76.   Mulkey DK, Stornetta RL, Weston MC, Simmons JR, Parker A, Bayliss DA, Guyenet PG (2004) Respiratory control by ventral surface chemoreceptor neurons in rats. Nat Neurosci 7: 1360-1369

77.   Narita N, Narita M, Takashima S, Nakayama M, Nagai T, Okado N (2001) Serotonin transporter gene variation is a risk factor for sudden infant death syndrome in the Japanese population. Pediatrics 107: 690-692

78.   Nattie E, Li A (2000) Muscimol dialysis in the retrotrapezoid nucleus region inhibits breathing in the awake rat. J Appl Physiol 89: 153-162

79.   Nattie E, Li A (2006) Central chemoreception 2005: a brief review. Auton Neurosci 126-127: 332-338

80.   Nattie EE, Li A (2002) Substance P-saporin lesion of neurons with NK1 receptors in one chemoreceptor site in rats decreases ventilation and chemosensitivity. J Physiol 544: 603-616

81.   Nattie EE, Li A, Richerson G, Lappi DA (2004) Medullary serotonergic neurones and adjacent neurones that express neurokinin-1 receptors are both involved in chemoreception in vivo. J Physiol 556: 235-253

82.   Norton WH, Mangoli M, Lele Z, Pogoda HM, Diamond B, Mercurio S, Russell C, Teraoka H, Stickney HL, Rauch GJ, Heisenberg CP, Houart C, Schilling TF, Frohnhoefer HG, Rastegar S, Neumann CJ, Gardiner RM, Strahle U, Geisler R, Rees M, Talbot WS, Wilson SW (2005) Monorail/Foxa2 regulates floorplate differentiation and specification of oligodendrocytes, serotonergic raphe neurones and cranial motoneurones. Development 132: 645-658

83.   Nsegbe E, Wallen-Mackenzie A, Dauger S, Roux JC, Shvarev Y, Lagercrantz H, Perlmann T, Herlenius E (2004) Congenital hypoventilation and impaired hypoxic response in Nurr1 mutant mice. J Physiol 556: 43-59

84.   Onimaru H, Homma I (2003) A novel functional neuron group for respiratory rhythm generation in the ventral medulla. J Neurosci 23: 1478-1486

85.   Onimaru H, Homma I (2006) Point: Counterpoint: The parafacial respiratory group (pFRG)/pre-Botzinger complex (preBotC) is the primary site of respiratory rhythm generation in the mammal. Point: the PFRG is the primary site of respiratory rhythm generation in the mammal. J Appl Physiol 100: 2094-2095

86.   Ozawa Y, Okado N (2002) Alteration of serotonergic receptors in the brain stems of human patients with respiratory disorders. Neuropediatrics 33: 142-149

87.   Panigrahy A, Filiano J, Sleeper LA, Mandell F, Valdes-Dapena M, Krous HF, Rava LA, Foley E, White WF, Kinney HC (2000) Decreased serotonergic receptor binding in rhombic lip-derived regions of the medulla oblongata in the sudden infant death syndrome. J Neuropathol Exp Neurol 59: 377-384

88.   Parras CM, Schuurmans C, Scardigli R, Kim J, Anderson DJ, Guillemot F (2002) Divergent functions of the proneural genes Mash1 and Ngn2 in the specification of neuronal subtype identity. Genes Dev 16: 324-338

89.   Paterson DS, Trachtenberg FL, Thompson EG, Belliveau RA, Beggs AH, Darnall R, Chadwick AE, Krous HF, Kinney HC (2006) Multiple serotonergic brainstem abnormalities in sudden infant death syndrome. JAMA 296: 2124-2132

90.   Pattyn A, Morin X, Cremer H, Goridis C, Brunet JF (1997) Expression and inter-actions of the two closely related homeobox genes Phox2a and Phox2b during neurogenesis. Development 124: 4065-4075

91.   Pattyn A, Morin X, Cremer H, Goridis C, Brunet JF (1999) The homeobox gene Phox2b is essential for the development of autonomic neural crest derivatives. Nature 399: 366-370

92.   Pattyn A, Goridis C, Brunet JF (2000) Specification of the central noradrenergic phenotype by the homeobox gene Phox2b. Mol Cell Neurosci 15: 235-243

93.   Pattyn A, Vallstedt A, Dias JM, Samad OA, Krumlauf R, Rijli FM, Brunet JF, Ericson J (2003) Coordinated temporal and spatial control of motor neuron and serotonergic neuron generation from a common pool of CNS progenitors. Genes Dev 17: 729-737

94.   Pattyn A, Simplicio N, van Doorninck JH, Goridis C, Guillemot F, Brunet JF (2004) Ascl1/Mash1 is required for the development of central serotonergic neu-rons. Nat Neurosci 7: 589-595

95.   Pattyn A, Guillemot F, Brunet JF (2006) Delays in neuronal differentiation in Mash1/Ascl1 mutants. Dev Biol 295: 67-75

96.   Pfaar H, von Holst A, Vogt Weisenhorn DM, Brodski C, Guimera J, Wurst W (2002) mPet-1, a mouse ETS-domain transcription factor, is expressed in central serotonergic neurons. Dev Genes Evol 212: 43-46

97.   Qian Y, Fritzsch B, Shirasawa S, Chen CL, Choi Y, Ma Q (2001) Formation of brainstem (nor)adrenergic centers and first-order relay visceral sensory neurons is dependent on homeodomain protein Rnx/Tlx3. Genes Dev 15: 2533-2545

98.   Ramanantsoa N, Vaubourg V, Dauger S, Matrot B, Vardon G, Chettouh Z, Gault-ier C, Goridis C, Gallego J (2006) Ventilatory response to hyperoxia in newborn mice heterozygous for the transcription factor Phox2b. Am J Physiol 290: R1691-1696

99.   Ramirez JM, Schwarzacher SW, Pierrefiche O, Olivera BM, Richter DW (1998) Selective lesioning of the cat pre-Botzinger complex in vivo eliminates breathing but not gasping. J Physiol 507: 895-907

100.   Rhee JW, Arata A, Selleri L, Jacobs Y, Arata S, Onimaru H, Cleary ML (2004) Pbx3 deficiency results in central hypoventilation. Am J Pathol 165: 1343-1350

101.   Sadl V, Jin F, Yu J, Cui S, Holmyard D, Quaggin S, Barsh G, Cordes S (2002) The mouse Kreisler (Krml1/MafB) segmentation gene is required for differentia-tion of glomerular visceral epithelial cells. Dev Biol 249: 16-29

102.   Sasaki A, Kanai M, Kijima K, Akaba K, Hashimoto M, Hasegawa H, Otaki S, Koizumi T, Kusuda S, Ogawa Y, Tuchiya K, Yamamoto W, Nakamura T, Ha-yasaka K (2003) Molecular analysis of congenital central hypoventilation syn-drome. Hum Genet 114: 22-26

103.   Shimamura K, Hartigan DJ, Martinez S, Puelles L, Rubenstein JL (1995) Longi-tudinal organization of the anterior neural plate and neural tube. Development 121: 3923-3933

104.   Shirasawa S, Arata A, Onimaru H, Roth KA, Brown GA, Horning S, Arata S, Okumura K, Sasazuki T, Korsmeyer SJ (2000) Rnx deficiency results in congeni-tal central hypoventilation. Nat Genet 24: 287-290

105.   Sieweke MH, Tekotte H, Frampton J, Graf T (1996) MafB is an interaction part-
        ner and repressor of Ets-1 that inhibits erythroid differentiation. Cell 85: 49-60
106.   Sieweke MH, Graf T (1998) A transcription factor party during blood cell differ-
        entiation. Curr Opin Genet Dev 8: 545-551
107.   Smith JC, Ellenberger HH, Ballanyi K, Richter DW, Feldman JL (1991) Pre-
        Botzinger complex: a brainstem region that may generate respiratory rhythm in
        mammals. Science 254: 726-729
108.   Stornetta RL, Rosin DL, Wang H, Sevigny CP, Weston MC, Guyenet PG (2003)
        A group of glutamatergic interneurons expressing high levels of both neurokinin-
        1 receptors and somatostatin identifies the region of the pre-Botzinger complex. J
        Comp Neurol 455: 499-512
109.   Sussel L, Kalamaras J, Hartigan-O'Connor DJ, Meneses JJ, Pedersen RA, Ruben-
        stein JL, German MS (1998) Mice lacking the homeodomain transcription factor
        Nkx2.2 have diabetes due to arrested differentiation of pancreatic beta cells. De-
        velopment 125: 2213-2221
110.   Taylor NC, Li A, Nattie EE (2005) Medullary serotonergic neurones modulate
        the ventilatory response to hypercapnia, but not hypoxia in conscious rats. J
        Physiol 566: 543-557
111.   Taylor NC, Li A, Nattie EE (2006) Ventilatory effects of muscimol microdialysis
        into the rostral medullary raphe region of conscious rats. Respir Physiol Neuro-
        biol 153: 203-216
112.   Van Doorninck JH, van Der Wees J, Karis A, Goedknegt E, Engel JD, Coesmans
        M, Rutteman M, Grosveld F, De Zeeuw CI (1999) GATA-3 is involved in the
        development of serotonergic neurons in the caudal raphe nuclei. J Neurosci 19:
        RC12
113.   Viemari JC, Burnet H, Bevengut M, Hilaire G (2003) Perinatal maturation of the
        mouse respiratory rhythm-generator: in vivo and in vitro studies. Eur J Neurosci
        17: 1233-1244
114.   Viemari JC, Bevengut M, Burnet H, Coulon P, Pequignot JM, Tiveron MC,
        Hilaire G (2004) Phox2a gene, A6 neurons, and noradrenaline are essential for
        development of normal respiratory rhythm in mice. J Neurosci 24: 928-937
115.   Viemari JC, Roux JC, Tryba AK, Saywell V, Burnet H, Pena F, Zanella S,
        Bevengut M, Barthelemy-Requin M, Herzing LB, Moncla A, Mancini J, Ramirez
        JM, Villard L, Hilaire G (2005) Mecp2 deficiency disrupts norepinephrine and
        respiratory systems in mice. J Neurosci 25: 11521-11530
116.   Wallen AA, Castro DS, Zetterstrom RH, Karlen M, Olson L, Ericson J, Perlmann
        T (2001) Orphan nuclear receptor Nurr1 is essential for Ret expression in mid-
        brain dopamine neurons and in the brain stem. Mol Cell Neurosci 18: 649-663
        Warnecke M, Oster H, Revelli JP, Alvarez-Bolado G, Eichele G (2005) Abnor-
        mal development of the locus coeruleus in Ear2(Nr2f6)-deficient mice impairs
        the functionality of the forebrain clock and affects nociception. Gene Dev 19:
        614-625
117.   Weese-Mayer DE, Berry-Kravis EM, Zhou L, Maher BS, Silvestri JM, Curran
        ME, Marazita ML (2003) Idiopathic congenital central hypoventilation syn-
        drome: analysis of genes pertinent to early autonomic nervous system embryolo-
        gic development and identification of mutations in PHOX2B. Am J Med Genet
        123: 267-278
118.   Weese-Mayer DE, Zhou L, Berry-Kravis EM, Maher BS, Silvestri JM, Marazita
        ML (2003) Association of the serotonin transporter gene with sudden infant death
        syndrome: a haplotype analysis. Am J Med Genet 122: 238-245

119. Weese-Mayer DE, Berry-Kravis EM, Maher BS, Silvestri JM, Curran ME, Marazita ML (2003) Sudden infant death syndrome: association with a promoter polymorphism of the serotonin transporter gene. Am J Med Genet 117: 268-274

120. Weese-Mayer DE, Berry-Kravis EM, Zhou L, Maher BS, Curran ME, Silvestri JM, Marazita ML (2004) Sudden infant death syndrome: case-control frequency differences at genes pertinent to early autonomic nervous system embryologic development. Pediatr Res 56: 391-395

121. Weese-Mayer DE, Berry-Kravis EM, Marazita ML (2005) In pursuit (and discovery) of a genetic basis for congenital central hypoventilation syndrome. Respir Physiol Neurobiol 149: 73-82

122. Weese-Mayer DE (2006) Sudden infant death syndrome: is serotonin the key factor? JAMA 296: 2143-2144

123. Weese-Mayer DE, Lieske SP, Boothby CM, Kenny AS, Bennett HL, Silvestri JM, Ramirez JM (2006) Autonomic nervous system dysregulation: breathing and heart rate perturbation during wakefulness in young girls with Rett syndrome. Pediatr Res 60: 443-449

124. Wenninger JM, Pan LG, Klum L, Leekley T, Bastastic J, Hodges MR, Feroah TR, Davis S, Forster HV (2004) Large lesions in the pre-Botzinger complex area eliminate eupneic respiratory rhythm in awake goats. J Appl Physiol 97: 1629-1636

125. Wenninger JM, Pan LG, Klum L, Leekley T, Bastastic J, Hodges MR, Feroah T, Davis S, Forster HV (2004) Small reduction of neurokinin-1 receptor-expressing neurons in the pre-Bötzinger complex area induces abnormal breathing periods in awake goats. J Appl Physiol 97: 1620-1628

126. White DP (2005) Pathogenesis of obstructive and central sleep apnea. Am J Respir Crit Care Med 172: 1363-1370

127. Zanella S, Roux JC, Viemari JC, Hilaire G (2006) Possible modulation of the mouse respiratory rhythm generator by A1/C1 neurones. Respir Physiol Neurobiol 153: 126-138

128. Zetterstrom RH, Solomin L, Jansson L, Hoffer BJ, Olson L, Perlmann T (1997) Dopamine neuron agenesis in Nurr1-deficient mice. Science 276: 248-250

129. Zhou QY, Palmiter RD (1995) Dopamine-deficient mice are severely hypoactive, adipsic, and aphagic. Cell 83: 1197-1209

130. Zoghbi HY (2003) Postnatal neurodevelopmental disorders: meeting at the synapse? Science 302: 826-830

13. Lessons from mutant newborn mice with respiratory control deficits

Claude GAULTIER

Service de Physiologie, Hôpital Robert Debré, 48 Bd Sérurier 75019 Paris France

13.1 Introduction

New concepts about the development of respiratory control are evolving from the results of innovative research, much of which involves genetics. The contribution of genetic factors to developmental respiratory control disorders is being actively studied, despite major obstacles such as the difficulty in characterizing these disorders, large number of genes involved in respiratory control, and interactions between genetic and environmental factors.

Gene targeting techniques have been used successfully in studies of ontogenic processes. An increasing number of studies in newborn mice with targeted gene deletions have helped to understand how gene disruptions affect the development of respiratory control. Furthermore, combined studies in mutant newborn mice and humans have provided pathogenic information on genetically determined disorders of respiratory control development in humans (see Chapters 4, 14-16).

This chapter starts with a brief review of the respiratory abnormalities seen in patients with developmental respiratory control disorders. Next, the main developmental characteristics of the newborn mouse are described. Finally, the respiratory phenotypes of mutant newborn mice with targeted deletions of genes involved in the development of rhythmogenesis and/or chemosensitivity to oxygen and carbon dioxide are discussed. The objective is to show how studies of these phenotypes conducted *in vivo* and *in vitro* have brought new insights into the mechanisms underlying the development of respiratory control in health and disease.

13.2 Developmental respiratory control disorders

Normal breathing is essential in mammals to ensure that cells receive sufficient oxygen during development. Therefore, all developmental disorders of respiratory control that manifest during early development may compromise brain oxygenation, thereby leading to irreversible cognitive impairment [5].

Several clinical phenotypes of developmental respiratory control disorders have been identified in humans [38].

Abnormal respiratory control is the main clinical symptom at birth, for instance in congenital central hypoventilation syndrome (CCHS), whose clinical phenotype is described in Chapter 4. CCHS was initially described as a respiratory control disorder characterized by central hypoventilation during sleep and blunted ventilatory responses to sustained hypercapnia and hypoxia [60]. CCHS, subsequently, was shown to encompass a wide range of autonomic nervous system abnormalities.

Respiratory abnormalities may occur as part of a broader disease phenotype, for instance in Prader-Willi syndrome, Rett syndrome, or Riley-Day syndrome. Patients with Prader-Willi syndrome exhibit sleep-disordered breathing with apnoeas and episodes of hypoventilation [67]. Furthermore, patients with Prader-Willi syndrome have reduced ventilatory responses to hypoxia with absent peripheral chemoreceptor responses [41]. Severe blunting of carbon dioxide chemosensitivity is only an occasional finding that may be related to obesity, which is a feature of Prader-Willi syndrome [3]. Rett syndrome is characterized by an abnormal breathing pattern with episodes of hyperventilation followed by central apnoea and oxygen desaturation during wakefulness, contrasting with the absence of patent respiratory abnormalities during sleep [56]. The respiratory abnormalities emerge during early childhood [97]. Abnormalities in heart rate and baroreflex control also occur in Rett syndrome [48; 99]. Therefore, autonomic dysfunction may account for the high risk of sudden death in girls with Rett syndrome. Finally, abnormal ventilatory and cerebrovascular responses to hypoxia have been reported in patients with Riley-Day syndrome or familial dysautonomia [7].

Abnormal respiratory control plays a role in sudden infant death syndrome (SIDS) [88]. Cardiorespiratory abnormalities have been reported in SIDS victims, including obstructive apnoeas during sleep [50], complete airway obstruction preceding death [88], deficient ventilatory responses to hypoxia [46], and abnormal autonomic heart-rate control [35]. Furthermore, multiple brainstem serotonergic abnormalities have been found in SIDS victims [51; 72]. Finally, interactions between genetic predisposition and environmental factors may contribute to the pathogenesis of SIDS. Chapter 7 reviews new insights into the genetic basis for SIDS.

Obstructive sleep apnoea syndrome is mainly associated with upper airway abnormalities in infants and children. No significant impairments of the ventilatory responses to hypercapnia or hypoxia were found in children with obstructive apnoeas [57]. However, respiratory control abnormalities have been reported in familial obstructive apnoea syndrome [80], suggesting that genetic variations in respiratory control may predispose to obstructive apnoea syndrome (see Chapter 8). Sleep-disordered breathing, including obstructive apnoeas, occurs in children with attention-deficit hyperactivity disorder (ADHD) [20]. Genetic studies showed a strong association between ADHD and the dopamine transporter gene [29].

Respiratory control is immature at birth. As a result, apnoeas and periodic breathing are common in newborns, most notably those born prematurely. Most premature infants have brief central apnoeas, especially during active sleep [37]. However, severe apnoeas occur in a subset of premature infants. The variability in the severity of apnoeas in premature infants may reflect genetically determined variants of respiratory control development. An early study reported decreased ventilatory responses to hypercapnia in premature infants with apnoeas [40].

Finally, sleep may be associated with deteriorations in gas exchange in patients who have respiratory disorders with abnormal lung mechanics. In these patients, genetically determined variants of respiratory control may contribute to the pathogenesis of sleep-disordered breathing (see Chapter 2).

13.3 The newborn mouse

Mice are the preferred mammalian species for manipulating genes. In genetically engineered mice, analysis of the respiratory phenotype at birth is crucial for two main reasons. First, null mutants for most genes of interest die within a few hours after birth, possibly from respiratory failure, which leaves very little time to investigate the respiratory control phenotype. Second, respiratory control is the outcome of numerous adaptation processes, including plasticity and learning [39]. These processes may correct the respiratory impairment shortly after birth, masking the effects of gene mutations [25; 27].

Multiple respiratory control abnormalities may occur in mutant newborn mice, complicating the interpretation of the role for individual components of respiratory control. Selecting genetically appropriate controls for comparison with genetically engineered newborn mice is an important issue. The phenotypic effects of a mutation probably depend on the genetic background, as demonstrated in other settings [61].

Survival differs across mutant newborn mice. Death may occur *in utero*, for instance with mutations in genes for paired-like homeobox 2b (*Phox2b*) or endothelin-converting enzyme 1 (*Ece1*) [73; 100]. In general, heterozygous mutant mice develop normally, which allows for longitudinal testing.

The maturation of rhythmogenesis during the perinatal period was investigated recently in mice [92]. Studies were conducted *in vivo* and *in vitro* (pontomedullary and medullary preparations) from embryonic day 16 (E16) to the postnatal period. E16 medullary networks can produce rhythmic activity *in vitro*. Between E16 and 18, maturation affects the rhythm generator responses to stimulation by transmitters such as substance P, serotonin, and noradrenalin. Foetuses delivered on E16 are unable to breathe. However, this inability to breathe may reflect immaturity not only of respiratory control, but also of respiratory effectors.

Newborn mice show a very immature pattern of breathing reminiscent of the respiratory instability seen in human preterm infants [59]. As with other newborn mammals [83], ventilatory responses to hypercapnia are present at birth in mice. Peripheral chemoreceptor resetting occurs about 12 h after birth in newborn mice [83]. Respiratory control is influenced by the state of alertness. Sleep states cannot be identified by electroencephalography or electro-oculography in newborn rodents [49]. However, behavioural sleep states can be determined in newborn mice [30] (see Chapter 14).

13.4 Mutant newborn mice as models of abnormal respiratory rhythm

Respiratory rhythm is thought to be generated by a network of neurons in the rostral ventral lateral medulla, which includes two generators: the pre-Bötzinger complex generator (pre-BötC) and the pre-I rhythm generator (pFRG) [68]. Studies of breathing rhythmicity *in vivo* and *in vitro* in mutant mouse embryos or newborn mice have produced new information about the genes involved in the development of rhythm generation. Table 1 shows the breathing pattern abnormalities seen in mutant newborn mice investigated *in vivo*, such as an increase in apnoeas, abnormal respiratory frequency, and respiratory frequency variability.

Chapter 11 provides a detailed discussion of new insights into the genes involved in hindbrain segmentation during early embryonic stages, such as *Krox20* [47]. *Krox20* deletion leads to severe breathing instability at birth with apnoeas that are more numerous and 10 times longer in null mutant newborns than in wild type newborns. These respiratory control abnormalities usually result in death during the second day of life.

Studies in mutant mice have shed light on the development of neurons in the pre-BötC. Little was known about the genes that specify the identity of rhythm-generating neurons in the pre-BötC until a recent study in null mutant newborn mice lacking the transcription factor *MafB* [10]. This study shows that *MafB* is a marker for a subpopulation of pre-BötC neurons. Null mutant *MafB* newborn mice die from central apnoeas soon after birth [10] (see Chapter 12).

Other studies have shed light on the role of neurotrophic factors in the development of rhythmogenesis. Null mutant newborn mice lacking the brain-derived neurotrophic factor (Bdnf) have a low respiratory rate and numerous apnoeas [31]. Studies of brainstem spinal cord preparations from *Bdnf* mutant newborn mice have shown discharge frequency attenuation, which is more marked in homozygous than in heterozygous animals [6]. *Bdnf* appears critical to respiratory rhythmogenesis in neonatal mice, its effect being mediated by the Bdnf receptor tyrosine kinase B in pre-BötC neurons. Exposure of neurons in the neonatal pre-BötC to exogenous Bdnf specifically modifies the membrane properties of rhythmically active neurons [89].

The role for *N*-methyl-D-aspartate (NMDA) receptors in functional respiratory rhythm-generating neurons has been examined in newborn mice lacking the NMDA1 gene [36]. *In vitro* preparations from mutant newborn mice generate a respiratory rhythm identical to the control preparations, indicating that NMDA receptors may be not indispensable to respiratory rhythm development at birth. *In vivo*, however, mutant newborn mice lacking the NMDA1 receptor gene have respiratory depression at birth [76]. Furthermore, NMDA1 receptor deficiency during prenatal development leads to exaggerated synaptic long-term depression in the nucleus tractus solitarius, suggesting that respiratory depression may be mediated by abnormal chemosensitivity.

Table 1. Breathing pattern abnormalities under normoxic conditions in mutant newborn mice investigated *in vivo*.

+, Abnormal number and/or duration of apnoeas compared to wild-type littermates; *, significantly decreased, increased compared to wild-type littermatess ; ** in hypothermic conditions ; for other abbreviations, see text.

[1]Numbers in brackets correspond to references listed at the end of the chapter.

Newborn Mice		Age At Study	Respiratory Frequency	Respiratory Frequency Variability	Apneas
Krox20$^{-/-}$	[41][1]	Day 1	Decreased*		+
MafB$^{-/-}$	[10]	Birth			+
Bdnf$^{-/-}$	[31]	Day 2	Decreased*	Increased*	+
GAD67$^{-/-}$	[55]	A few hours	Decreased*	Increased*	+
Mash1$^{+/-}$	[26]	12 hours	Increased*		
Rnx$^{-/-}$	[86]	Day 1	Increased*		+
Nurr1$^{-/-}$	[65]	First day	Decreased*		+
PACAP$^{-/-}$	[23]	Day 4			+**
Phox2b$^{+/-}$	[30]	Day 5			+

A role for GABA in the development of respiratory rhythm generation was looked for in null mutant newborn mice lacking the GABA-synthesizing enzyme 67-kDa glutamic acid decarboxylase (GAD67) [55]. *In vivo* and *in vitro* studies were performed in homozygous mutant GAD67 newborn mice. Irregular breathing and periodic gasp-like respirations were noted in the null mutant animals. *In vitro* studies of GAD67$^{-/-}$ brainstem-spinal cord preparations showed shorter inspiratory C4 burst durations, compared to wild-type preparations. Whole cell recordings demonstrated decreased firing of inspiratory neurons in the ventral medulla of GAD67$^{-/-}$ mice. These data suggest that GABAergic transmission may be nonessential for respiratory rhythm generation but may contribute to maintain a regular respiratory rhythm and normal inspiratory pattern in neonatal mice.

Several studies in mutant newborn mice focused on the role for respiratory drive modulation by neurotransmitters. Loss of the adrenergic A5 nucleus due to gene inactivation leads to abnormally fast breathing, as observed in null newborn mice deficient for *Mash1* [26] or *Rnx* [86], which die soon after birth, and in null mutant mouse embryos lacking the glial cell line-derived neurotrophic factor

(*Gdnf*) [45]. *Phox2a* gene deficiency in mice results in loss of A6 nucleus neurons, causing respiratory rhythm abnormalities *in vivo* and *in vitro* and death of null mutant newborn mice [93]. These data suggest that insults causing abnormal A6 development during gestation may lead to severe respiratory deficits in human newborns. Finally, a study in null mutant foetuses lacking the *Ret* gene has established that *Ret* contributes to the prenatal maturation of the adrenergic A5 and A6 nuclei [94]. *Ret* inactivation has no obvious effects on the *in vivo* foetal breathing pattern. However, *Ret* inactivation causes a decrease in breathing frequency of *in vitro* foetal preparations. Null mutant newborn mice lacking *Ret* die soon after birth, whereas heterozygous mice have a normal breathing pattern at birth and develop normally. Therefore, respiratory phenotypes of those mutant newborn mice show that maturation of neonatal rhythm generation appears to require an appropriate balance between excitatory input from the A6 nucleus and inhibitory input from the A5 nucleus [44].

Serotonin has long time been reported as a main determinant of rhythmogenesis and apneas in newborn mammals [43]. Excessive exposure to serotonin can be achieved by inactivating the gene for monoamine oxidase-A, the main serotonin-degrading enzyme. Monoamine-oxidase gene deficiency results in respiratory abnormalities at birth with unstable respiratory rhythm, defective serotonin modulation, and abnormal phrenic motoneuron activity [14]. These breathing pattern abnormalities persist during adulthood [16]. Activation of serotonin-2A receptors is required for respiratory rhythm generation *in vitro* [74]. However, null mutant newborn mice lacking the gene for serotonin-2A receptors had no breathing pattern abnormalities, suggesting *in vivo* compensatory mechanisms [77].

The role of substance P, which modulates respiratory rhythm by acting through neurokinin-1 receptors (NK1) has been examined in newborn mutant mice. Mutant newborn mice lacking the NK1 receptor gene showed that these receptors were not necessary for producing a resting respiratory rhythm at birth [78]. Moreover, in a study of medullary slice preparations from 4- to 12-day-old mice lacking the preprotachykinin gene, which codes for the substance P precursor, the pre-BötC generated eupneic activity under normoxic conditions, suggesting adaptation or compensation for long-term substance P deficiency.

On other hand, one study has shown that excess acetylcholine leads to increased ventilation and increased tidal volumes in the first days of postnatal life, and that these abnormalities persist to adulthood in mutant mice lacking the acetylcholinesterase gene [12].

Finally, abnormally unstable breathing occurs in a number of other mutant newborn mice. Two mutant mice lacking genes involved in dopamine neurotransmission have been studied. Null-mutant newborn mice lacking the transcription factor *Nurr1,* which governs the development of dopaminergic neurons, have increased breathing instability with apnoeas, and they die soon after birth [65]. On the other hand, null mutant adult mice lacking the dopamine transporter protein gene breathed significantly more slowly than the wild type mice [34].

The role of pituitary adenylate cyclase-activating intestinal peptide (PACAP) superfamily, a potent stimulator of cAMP production and subsequent protein kinase-A activation, has been studied. Null mutant mice lacking PACAP exhibit hypoventilation during the neonatal period and are prone to sudden death preceded by prolonged apnoeas between 1 and 3 weeks after birth [23]. Furthermore, the mutant animals are vulnerable to changes in body temperature. Hypothermia aggravates the respiratory depression. This mutant newborn model shares similarities with the phenotype of SIDS victims.

Recent studies in mutant newborn mice have supported the existence of a causal link between genetic abnormalities and respiratory rhythm abnormalities found in human disorders such as CCHS, Prader-Willi syndrome, Rett syndrome (see Chapters 4, 14-16). Patients with Prader-Willi syndrome, who experience abnormal respiratory events during sleep [67], carry mutations in the encoding NECDIN gene [62]. Necdin null mutant newborn mice have abnormal *in vivo* and *in vitro* rhythmic respiratory activity [81]. Necdin deficiency in newborn mice has been shown to alter the modulating effects on pre-BötC function of transmitters such as serotonin and norepinephrine [71]. Such impairments in the modulation of respiratory rhythm may underlie the pathogenesis of respiratory control abnormalities in Prader-Willi patients who carry NECDIN mutations. Mutations in the gene for methyl-CpG binding protein 2 (MECP2) are associated with Rett syndrome [82]. Interestingly, $Mecp2^{-/y}$ mutant newborn mice breathe normally at birth and develop abnormal respiratory control later on, due in part to deficient noradrenergic and serotonergic modulation of the medullary respiratory network [95]. The recent discovery that genetic overexpression of Bdnf can ameliorate some of the functional deficits observed in *Mecp2* null mice is of major interest [19]. *PHOX2B* is the major disease-causing gene of CCHS, and most CCHS patients carry a heterozygous *PHOX2B* polyalanine expansion mutation [2]. The respiratory phenotype of heterozygous *Phox2b* mutant newborn mice shares features with CCHS, such as sleep-disordered breathing [30] (see Chapter 14).

13.5 Mutant newborn mice with abnormal chemosensitivity

13.5.1 Chemosensitivity to oxygen

The arterial chemoreflex is supported by a sensory pathway that includes three relays: the carotid bodies, the petrosal ganglia, and the nucleus of the solitary tract. Furthermore, oxygen-sensing neurons in the central nervous system play a role in chemosensitivity to oxygen [66]. Studies in mutant newborn mice have helped to understand the genes involved in the ventilatory response to sustained hypoxia or to hyperoxia in the neonatal period. Newborn mammals exhibit a biphasic ventilatory response during sustained hypoxia, with an initial hyperpneic phase followed by a decline in ventilation below the normoxic level (hypoxic ventilatory decline). Table 2 shows the abnormal ventilatory responses to sustained hypoxia and to hyperoxia reported in mutant newborn mice.

The neurotrophic factors Bdnf and Gdnf exert coordinated effects that ensure the survival of neurons involved in the arterial chemoreflex [32]. Mutant newborn mice lacking *Bdnf* have arterial chemoreflex deficiencies, as shown by the absence of ventilatory responses to hyperoxia [31].

Table 2. Ventilatory responses to hypercapnia, hypoxia and hyperoxia in mutant newborn mice investigated *in vivo*

VR : ventilatory response ; HVR: hyperpneic ventilatory response ; HVD : hypoxic ventilatory decline; beta-2nACh, beta-2 subunit of the nicotinic acetylcholine receptors ; NS, not significantly different from wild-type littermates ; *, significantly decreased compared to wild-type littermates; ** significantly decreased in males ; for others abbreviations, see text.

[1]Numbers in brackets correspond to references listed at the end of the chapter.

Newborn Mice		Age at study	Hypercapnic VR	Hypoxic		Hyperoxic VR
				HVR	HVD	
Bdnf$^{-/-}$	[31][1]	First 4 days	Present			Absent
Edn1$^{-/-}$	[53]	First day	Blunted	Blunted		
Ednra$^{-/-}$	[21]	First day	Blunted	Blunted		
Ece$^{+/-}$	[84]	A few hours	NS	Decreased*		
Nurr1$^{-/-}$	[65]	First day	NS	Blunted		
PACAP$^{-/-}$	[23]	Day 4	Blunted	Blunted		
Mash1$^{+/-}$	[25]	12h	Decreased**	NS		
Ret$^{-/-}$	[17]	A few hours	Decreased*	NS		
Ret$^{+/-}$	[1]	12h	NS	NS	Augmented	
Phox2b$^{+/-}$	[27]	Day 2	Decreased	NS	Augmented	
	[79]	Day 5				Augmented
	[27]	Day 10	NS			
Beta-2n AchR$^{-/-}$	[28]	48h	NS		Attenuated	
5-HTT$^{-/-}$	[34]	Day 1-3		Augmented	Attenuated	
Kir2-2$^{-/-}$	[69]	Day 14-15	Decreased*			

Ventilatory responses to hypoxia have been investigated in newborn null mutant mice lacking the genes for the endothelin 1 (*Edn1*) and 3 pathways. Null mutant mice lacking *Edn1* and endothelin receptor a (*Ednra*) have blunted ventilatory response to hypoxia [21; 53]. Furthermore, heterozygous *Edn1* mutant mice have decreased ventilatory responses to hypoxia in adulthood. In contrast, inactivation of endothelin 3 and endothelin-receptor b does not modify the ventilatory responses to hypoxia [54; 63]. Finally, heterozygous mutant mice that lack the endothelin-converting-enzyme 1 gene (*Ece1*$^{-/-}$) exhibit abnormal hyperpneic ventilatory responses to hypoxia that persist during adulthood [84].

Numerous neurotransmitters are expressed in the carotid bodies, such as substance P, acetylcholine, dopamine, and enkephalins. Studies in mutant newborn mice have provided new information about the biological significance of these neurotransmitters. In mutant newborn mice lacking the neurokinin-1 receptor (NK1) gene, the response to short-lasting hypoxia is normal at birth but deficient in adulthood [78]. This finding indicates that long-term substance P deficiency due to absence of NK1 leads to regulatory mechanisms that produce deficient responses during hypoxia contrasting with apparently normal activity under normoxic conditions during adulthood. Long-term substance P deprivation in mutant mice lacking the gene for preprotachykinin-A (PPT-A) results in a deficient response to anoxia of the isolated respiratory network from adult mutant PPT-A mice [87]. The inability of these mutant mice to compensate for stressful hypoxic conditions after the neonatal period is reminiscent of SIDS, in which respiratory control seems adequate under normal conditions but breaks down when an exogenous stressor occurs [51; 72]. Finally, preliminary data show that null mutant newborn mice with inactivation of the gene for tachykinin-1 (TAC1), which results in deficiencies in neurokin A and substance P, exhibit an abnormal response to intermittent hypoxia on postnatal days 8-10 [8]. On the opposite, an excess of substance P due to inactivation of the gene for neural endopeptidase, which degrades substance P within the carotid bodies, results in exaggerated ventilatory responses to hypoxia in mutant adult mice [42].

Two studies have looked for ventilatory response abnormalities in null mutant mice lacking genes involved in dopamine transmission. Responses to hypoxia were blunted in null mutant newborn mice lacking *Nurr1*, a gene involved in dopamine transmission and expressed in the carotid bodies and nucleus tractus solitarius [65], and null mutant adult mice lacking the dopamine transporter protein [96]. These findings support a role for dopamine as an inhibitory modulator of the carotid-body ventilatory response to hypoxia.

Ventilatory responses to hypoxia have been studied in null mutant adult mice lacking the genes for muscarinic receptors M1 or M3 through which acetylcholine is acting [13]. Mutant mice lacking M3, but not M1, have deficient ventilatory responses to hypoxia, but M3 mutants had normal responses.

Finally, null mutant newborn mice lacking the gene for PACAP have not only an abnormal respiratory rhythm, but also a blunted ventilatory response to hypoxia, suggesting a role for this gene in modulating chemosensitivity to oxygen [23]. A recent *in vitro* study sheds light on the mechanisms of action of PACAP, showing that PACAP stimulates the carotid bodies [24].

The role for potassium (K^+) channels in chemosensitivity to oxygen was studied recently using gene inactivation. The findings obtained after inactivation of the K^+ channel Kv1.1 established a role for this channel in afferent chemosensitivity to oxygen [52]. Null mutant mice have exaggerated ventilatory responses to hypoxia, compared to wild type mice. *In vitro*, the sensory discharge during hypoxia is greater with carotid bodies from null mutant mice than from wild type mice. Missense mutations in the gene for the Kv1.1 channel are associated with episodic ataxia type 1 syndrome in humans. Whether these patients exhibit exaggerated arterial chemoreflexes remains to be determined.

Studies in mutant newborn mice have shown that the two phases of biphasic ventilatory responses involve different genes in the neonatal period. Hypoxic hyperpneic ventilatory responses are not affected in newborn mice with heterozygous disruption of *Mash1* [25], *Ret* [1], or *Phox2b* [27], whereas the hypoxic ventilatory decline is augmented in newborn mice that are heterozygous for *Ret* or

Phox2b disruption [1; 27]. Preliminary data suggest that inactivation of the gene for the Kir6.2 potassium channel may result in an exaggerated hypoxic ventilatory decline in mutant mice [70]. Conversely, the hypoxic ventilatory decline is decreased by disruptions in other genes. The hypoxic ventilatory decline is small in null mutant newborn mice lacking the beta-2 subunit of the nicotinic acetylcholine receptors [28], which have recently been shown to reproduce many of the abnormalities caused by perinatal nicotine exposure [22]. Finally, preliminary data suggest that loss of the serotonin transporter protein in 5HTT null mutant newborn mice results in attenuation of the hypoxic ventilatory decline [34].

Ventilatory responses to hypoxia were studied in heterozygous *Mecp2* mutant mice [9]. The increase in ventilation in response to hypoxia was greater in these mice than in the wild type controls and was followed by marked ventilatory depression (see Chapter 16).

Recent study shows that the response of the isolated respiratory networks to hypoxia was decreased in Necdin null mutant newborn mice as compared to wild type preparations [101].

Finally, chemosensitivity to oxygen has been investigated in mice with cystic fibrosis due to absence of the CF transmembrane conductance regulator gene (*Cftr*) [11]. Ventilatory responses to hypoxia were blunted. Patients with cystic fibrosis have decreased ventilatory responses to hypoxia, which are traditionally attributed to decreased performance of the respiratory effector subjected to abnormal respiratory loads caused by chronic airway obstruction [15]. However, the abnormal chemosensitivity to oxygen in *Cftr* mutant mice suggests that patients with cystic fibrosis may have also deficient chemosensitivity to oxygen.

13.5.2 Chemosensitivity to carbon dioxide

Central chemoreceptors are believed to be widely distributed [33]. There is little agreement on the time pattern of involvement of each site in sensitivity to CO_2, and neither are the site-specific mechanisms known [85]. Table 2 shows the abnormal ventilatory responses to sustained hypercapnia reported in mutant newborn mice.

Studies looking at respiratory phenotypes in newborn mice with targeted gene deletions have started to establish links between the expression of specific genes and the development of CO_2 sensitivity. Inactivation of genes such as *Edn1*, *Ednra*, *Ret*, *Mash1*, and *Phox2b* results in impaired chemosensitivity to CO_2. Null mutant newborn mice lacking the *Edn1* and *Ednra* genes have a blunted ventilatory response to hypercapnia [53; 54]. The ventilatory response to hypercapnia remains deficient in adulthood in heterozygous *Edn1* mice [53]. Blunted ventilatory responses to hypercapnia occur in null mutant newborn mice lacking the *Ret* gene, but not in the heterozygous mutants [1; 17]. Deficient ventilatory responses to hypercapnia have been observed in heterozygous *Mash1* mutant newborn mice [24]. Heterozygous mutant mice lacking the *Phox2b* gene (whose human orthologue is mutated in CCHS patients) exhibit a reduced ventilatory response to hypercapnia on postnatal day 2 [25]. However, in contrast to the persistence of the deficient ventilatory response to hypercapnia in CCHS patients, this response recovered on P10 in the mutant mice, suggesting the existence of compensatory mechanisms of unknown origin (see Chapter 14) [26].

Mutant newborn mice deficient in pituitary adenylate cyclase-activating polypeptide (PACAP) have a blunted ventilatory response to hypercapnia, suggesting that the PACAP signalling pathway may also contribute to the mechanisms underlying chemosensitivity to CO_2 [23].

Interestingly, a recent study examined the influence of states of alertness in mutant mice lacking the prepro-orexin gene [64]. The null adult mutant ORX mice lack both orexin-A and-B. Their ventilatory responses to hypercapnia are comparable to those of wild type mice during slow-wave sleep and rapid-eye movement sleep but decrease during quiet wakefulness, suggesting a role for orexin in chemosensitivity to CO_2 during wakefulness.

The role for K^+ channels in the mechanisms underlying chemosensitivity to CO_2 dioxide has been examined in mutant mice. Mutant mice lacking the gene for the K^+ channel Kir2-2 were studied between postnatal days 9 and 18 [69]. They exhibited smaller increases in the ventilatory responses to hypercapnia on days 14-15 than same-age wild type pups, suggesting a transient role for Kir2.2 in central chemosensitivity during postnatal development.

Finally, a study in mutant mice lacking the gene for acetylcholinesterase found enhanced chemosensitivity to CO_2 associated with an excess of acetylcholine, one of the neurotransmitters thought to be involved in central chemosensitivity [12].

13.6 Clinical relevance of mutant newborn mice models

The discovery that *Phox2b* governs the development of the autonomic nervous system [73] led clinicians to look for *PHOX2B* mutations in patients with CCHS [2]. However, the fact that heterozygous mutant *Phox2b* newborn mice exhibit only part of the features of the human disease suggests a need for developing mice that carry a *Phox2b* polyalanine expansion mutation similar to that found in CCHS patients. Hopefully, these knock-in mutant newborn mice will prove to be a good model for testing new therapeutic strategies *in vivo* and *in vitro*. Whether drugs that enhance heat shock protein expression, thereby improving the function of the mutant PHOX2B protein as shown by *in vitro* experiments [4; 91], are effective *in vivo* in knock-in newborn mice is an important issue (see Chapter 6).

Studies of the respiratory phenotype of *Mecp2* mutant newborn mice have shed light on the mechanisms underlying the autonomic abnormalities present in Rett syndrome. Furthermore, the beneficial effects of BDNF overexpression in *Mecp2* null mutant mice may open up avenues towards new treatments for Rett syndrome (see Chapter 16). The development of null mutant *necdin* newborn mice has provided an effective tool for unravelling the pathogenesis of the respiratory control abnormalities seen in patients with Prader-Willi syndrome (see Chapter 15). The mutant model for hyperdopaminergic state may provide new insight into the respiratory control abnormalities associated with ADHD in children.

PACAP-deficient neonatal mice have been suggested as a SIDS-like model, chiefly because they are prone to sudden death after the neonatal period. These mice have a severe deficiency in chemosensitivity, which has usually not been reported in human SIDS victims. Nevertheless, an interesting feature of these mutants is their vulnerability to changes in environmental temperature. Hypothermia is among the risk factors thought to be involved in the pathogenesis of SIDS. Future studies in mutant mice lacking genes that have been found to be mutated in some SIDS victims (see Chapter 7) should investigate the effects of environmental factors on the respiratory phenotype. Interactions between a genetic predisposition to respiratory control abnormalities and environmental factors such as nicotine exposure during foetal life should be studied in mutant newborn mice lacking selected genes.

A major health issue is the increasing rate of prematurity in human newborns. Newborn mice constitute a good model of respiratory control immaturity. The severity of apnoea of prematurity varies considerably across individuals, suggesting that preterm infants with severe apnoeas may carry mutations or polymorphisms of genes involved in the maturation of rhythm generation. Genetic studies have not yet been performed in infants with apnoea of prematurity.

Finally, genetics may govern not only the normal programming of respiratory control development, but also the processing of adaptation and plasticity that occurs when the infant's homeostasis is disturbed by prenatal and postnatal insults. Respiratory plasticity is probably a key factor in the development of respiratory control (see Chapter 17). A model of respiratory plasticity is intermittent hypoxia, which has been studied in neonatal rats [75]. Future studies in mutant newborn mice should examine whether respiratory plasticity is affected by inactivation of selected genes.

13.7 Conclusion

Neonatal phenotype determination in mutant newborn mice is mandatory to explore the molecular mechanisms involved in developmental abnormalities of respiratory control, which are not fully reflected by adult phenotypes, and the impact of these abnormalities on neurological development. So far, studies of respiratory phenotypes have helped to understand how gene disruption may disturb one or several components of respiratory control during postnatal development. Furthermore, studies in mutant newborn mice combined with studies in humans have provided valuable pathogenic information on genetically determined disorders of respiratory control (e.g., congenital central hypoventilation syndrome, Prader-Willi syndrome, Rett syndrome, and Attention-Deficit Hyperactivity Disorders). Nevertheless, none of the genetically engineered newborn mice developed to date fully replicate the complex phenotype of human respiratory control disorders. It is clearly important to extend these studies to the early disturbances of respiratory control related to immaturity, in order to improve our understanding of apnoeas of prematurity. Furthermore, studies in mutant newborn mice should help to elucidate the interactions between genetic and environmental factors, which play a major role in the pathogenesis of sudden infant death. Finally, mutant newborn mice should prove useful for testing new therapeutic strategies, which is a major challenge for future research studies.

References

1.   Aizenfisz S, Dauger S, Durand E, Vardon G, Levacher B, Simonneau M, Pachnis V, Gaultier C, Gallego J (2002) Ventilatory responses to hypercapnia and hypoxia in heterozygous c-ret newborn mice. Respir Physiol Neurobiol 131: 213-222

2.   Amiel J, Laudier B, Attie-Bitach T, Trang H, De Pontual L, Gener B, Trochet D, Etchevers H, Rau P, Simonneau M, Vekemans M, Munnich A, Gaultier C, Lyonnet S (2003) Polyalanine expansion and frameshift mutations of the paired-like homeobox gene *PHOX2B* in congenital central hypoventilation syndrome. Nat Genet 33: 459-461

3.   Arens R, Gozal D, Omlin KJ, Livingston FR, Liu J, Keens TG, Ward SL (1994) Hypoxic and hypercapnic ventilatory response in Prader-Willi syndrome. J Appl Physiol 77: 2231-2236

4.   Bachetti T, Bocca P, Borghini S, Matera S, Prigione I, Ravazzolo R, Ceccherini I (2007) Geldamycine promotes nuclear localisation and clearance of PHOX2B misfolded proteins containing polyalanine expansions. Int J Biochem Cell Biol 39: 327-339

5.   Back SA, Rivkees SA (2004) Emerging concepts in periventricular white matter injury. Semin Perinatol 28: 405-414

6.   Balkowiec A, Katz DM (1998) Brain-derived neurotrophic factor is required for normal development of the central respiratory rhythm in mice. J Physiol 510: 527-533

7.   Bernardi L, Hilz M, Stemper B, Passino C, Welsch G, Axelrod FB (2003) Respiratory and cerebrovascular responses to hypoxia and hypercapnia in familial dysautonomia. Am J Respir Crit Care Med 167: 141-149

8.   Berner J, Sharev Y, Lagercrantz H, Bilkei-Gorzo A, Hökfelt T, Wickström R (2006) Altered hypoxic response in newborn mice with targeted deletion of the tachykinin 1 gene (TAC1$^{-/-}$) (Abstract). Xth Oxford Conference Lake Louise Canada 17- 24 September: p38

9.   Bissonnette JM, Knopp SJ (2006) Separate respiratory phenotype in methyl-CpG-binding protein 2 (Mecp2) deficient mice. Pediatr Res 59: 513-518

10.   Blanchi B, Kelly LM, Viemari JC, Lafon I, Burnet H, Bevengut M, Tillmans S, Daniel L, Graf T, Hilaire G, Sieweke MH (2003) MafB deficiency causes defective respiratory rhythmogenesis and fatal apnea at birth. Nat Neurosci 6: 1091-1099

11.   Bonora M, Bernaudin JF, Guernier C, Brahimi-Horn MC (2004) Ventilatory responses to hypercapnia and hypoxia in conscious cystic fibrosis knockout mice Cftr$^{-/-}$. Pediatr Res 55: 738-746

12.   Boudinot E, Emery MJ, Mouisel E, Chatonnet A, Champagnat J, Escourrou P, Foutz AS (2004) Increased ventilation and CO_2 chemosensitivity in acetylcholinesterase knockout mice. Respir Physiol Neurobiol 140: 231-241

13.   Boudinot E, Yamada M, Wess J, Champagnat J, Foutz AS (2004) Ventilatory pattern and chemosensitivity in M1 and M3 muscarinic receptor knockout mice. Respir Physiol Neurobiol 139: 237-245

14.   Bou-Flores C, Lajard AM, Monteau R, De Mayer E, Seif I, Lanoir J, Hilaire G (2000) Abnormal phrenic motoneuron activity and morphology in neonatal monoamine oxydase A-deficient transgenic mice: possible role of a serotonin excess. J Neurosci. 20: 4646-4656

15.   Bureau M, Liepen L, Begin R (1981) Neural drive and ventilatory strategy of breathing in normal children, and in patients with cystic fibrosis and asthma. Pediatrics 68: 187-194

16.   Burnet H, Bevengut M, Chakri F, Bou-Flores C, Coulon P, Gaytan S, Pasaro R, Hilaire G (2001) Altered respiratory activity and respiratory regulations in adult monoamine oxidase A-deficient mice. J Neurosci 21: 5212-5221

17.   Burton MD, Kawashima A, Brayer JA, Kazemi H, Hannon DC, Schuchardt A, Costantini F, Pachnis V, Kinane TB (1997) RET proto-ongogene is important for the development of respiratory CO2 sensitivity. J Auton Nerv Syst 63: 137-143

18.   Chatonnet F, Boudinot E, ChatonnetA, Champagnat J, Foutz AS (2004) Breathing without acethylcholinesterase. Adv Exp Med Biol 551: 165-170

19.   Chang Q, Khare G, Dani V, Nelson S, Jaenisch R (2006) The disease progression of Mecp2 mutant mice is affected by the level of BDNF expression. Neuron 49: 341-348

20.   Chervin RD, Archbold KH, Dillon JE, Panahi P, Pituch KJ, Dahl RE, Guilleminault C (2002) Inattention, hyperactivity, and symptoms of sleep-disordered breathing. Pediatrics 109: 449-456

21.   Clouthier DE, Hosoda K, Richarson JA, Williams SC, Yanagisawa H, Kuwaki T, Kumada M, Hammer RE, Yanagisawa M (1998) Cranial and cardiac neural crest defects in endothelin-A receptor deficient mice. Development 125: 813-824

22.   Cohen G, Roux JC, Graihle R, Malcolm G, Changeux JP, Lagercrantz H (2005) Perinatal exposure to nicotine causes deficits associated with a loss of nicotinic receptor function. Proc Natl Acad Sci USA 102: 3817-3821

23.   Cummings KJ, Pendlebury JD, Sherwood NM, Wilson RJA (2003) Sudden neonatal death in PACAP-deficient mice is associated with reduced respiratoy chemoresponse and susceptibility to apnoea. J Physiol 555: 15-26

24.   Cummings KJ, Day T, Wilson RJA (2006) PACAP stimulates the carotid body leading to an increase in neuronal ventilation (Abstract). X[th] Oxford Conference Lake Louise Canada 17- 24 September: p53

25.   Dauger S, Renolleau S, Vardon G, Nepote V, Mas C, Simonneau M., Gaultier C, Gallego J (1999) Ventilatory responses to hypercapnia and hypoxia in Mash-1 heterozygous newborn and adult mice. Pediatr Res 46: 535-542

26.   Dauger S, Guimiot F, Renolleau S, Levacher B, Boda B, Mas C, Nepote V, Simonneau M, Gaultier C, Gallego J (2001) MASH-1/RET pathway involvement in development of brain stem control of respiratory frequency in newborn mice. Physiol Genomics 7: 149-157

27.   Dauger S, Pattyn A, Lofaso F, Gaultier C, Goridis C, Gallego J, Brunet JF (2003) Phox2b controls the development of peripheral chemoreceptors and afferent visceral pathways. Development 130: 6635-6642

28.   Dauger S, Durand E, Cohen G, Lagercrantz H, Changeux J.P, Gaultier, C, Gallego J (2004) Control of breathing in newborn mice lacking the beta-2nAChR subunit. Acta Physiol Scand 181: 1-8

29.   DiMaio S, Grizenko N, Joober R (2003) Dopamine genes and attention-deficit hyperactivity disorder: a review. J Psychiatry Neurosci 28: 27-38

30.   Durand E, Dauger S, Pattyn A, Gaultier C, Goridis C, Gallego J (2005) Sleep-disordered breathing in newborn mice heterozygous for the transcription factor Phox2b. Am J Respir Crit Care Med 172: 238-243

31.   Erickson JT, Conover J., Borday V, Champagnat J, Barbacid M, Yancopoulos G, Katz DM (1996) Mice lacking brain-derived neurotrophic factor exhibit visceral sensory neuron losses distinct from mice lacking NT4 and display a severe developmental deficit in control of breathing. J Neurosci 16: 5361-5371

32.   Erickson JT, Brosenitsch TA, Katz DM (2001) Brain-derived neurotrophic factor and glial cell line-derived neurotrophic factor are required simultaneously for survival of dopaminergic primary sensory neurons in vivo. J Neurosci 15: 581-589

33.   Feldman JL, Mitchell GS, Nattie EE (2003) Breathing: rhythmicity, plasticity, chemosensitivity. Annu Rev Neurosci 26: 239-266

34.   Fisher JT, Escudero CA, Simms TM, Vincent SG (2005) Impact of loss of the serotonin transporter protein on the ventilatory response to hypoxia in the newborn mouse. FASEB J 19: A650

35.   Franco P, Szliwowski H, Dramaix M, Kahn A (1999) Decreased autonomic responses to obstructive sleep events in future victims of sudden infant death syndrome. Pediatr Res 46: 33-39

36.   Funk GD, Johnson SM, Smith JC, Dong XW, Lai J, Feldman JL (1997) Functional respiratory rhythm generating networks in neonatal mice lacking NMDAR1 gene. J Neurophysiol 78: 1414-1420

37.   Gaultier C (1999) Sleep apnoea in infants. Sleep Med Rev 3: 303-312

38.   Gaultier C (2004) Genes and genetics in respiratory control. Paediatr Respir Rev 5: 166-172

39.   Gaultier C, Gallego J (2005) Development of respiratory control: evolving concepts and perspectives. Respir Physiol Neurobiol 149: 3-15

40.   Gerhardt T, Bancalari E (1979) Ventilatory responses to CO_2 in premature infants with apnea. Pediatr Res 13: 534-539

41.   Gozal D, Arens R, Omlin KJ, Davidson-Ward SL, Keens TG (1994) Absent peripheral chemosensitivity in Prader-Willi syndrome. J Appl Physiol 77: 2231-2236

42.   Grasemann H, Lu B, Jiao A, Boudreau J, Gerard NP, De Sanctis GT (1999) Targeted deletion of the neural endopeptidase gene alters ventilatory responses to acute hypoxia in mice. J Appl Physiol 87: 1266-1271

43.   Hilaire G, Morin D, Lajard AM, Monteau R (1993) Changes in serotonin metabolism may elicit obstructive apnoea in the newborn rat. J Physiol 466: 367-381

44.   Hilaire G, Viemari JC, Coulon P, Simonneau M, Bévengut M (2004) Modulation of the respiratory rhythm generator by the pontine noradrenergic A5 and A6 groups in rodents. Respir Physiol Neurobiol. 143: 187-197

45.   Huang L, Guo H, Hellard DT, Katz DM (2005) Glial cell line-derived neurotrophic factor (GDNF) is required for differentiation of pontine noradrenergic neurons and patterning of central respiratory output. Neuroscience 130: 95-105

46.   Hunt CE, McCulloch K, Brouillette RT (1981) Diminished hypoxic ventilatory responses in near-miss sudden infant death syndrome. J Appl Physiol 50: 1313-1317

47.   Jacquin TD, Borday V, Schneider-Maunoury S, Topiilko P, Ghilini G, Kato F, Charnay P, Champagnat J (1996) Reorganisation of pontine rhythmogenic neuronal networks in Krox-20 knockout mice. Neuron 17: 747-758

48.   Julu PO, Kerr AM, Apartopoulos F, Al-Rawas S, Witt Engerström I, Enger-strömI, Jamal GA, Hansen S (2001) Characterisation of breathing and associated central autonomic dysfunction in the Rett syndrome. Arch Dis Child 85: 29-37

49.   Karlsson KA, Blumberg MS (2002) The union of the state: myoclonic twitching is coupled with nucleal atonia in infant rats. Behav Neurosci 116: 912-917

50.   Kato I, Groswasser J, Franco P, Scaillet S, Kelmanson I, Togari H, Kahn A (2001) Developmental characteristics of infants who succumb to sudden infant death syndrome. Am J Respir Crit Care Med 15: 1464- 1469

51.   Kinney HC, Filiano JJ, White WF (2001) Medullary serotonergic network defi-ciency in the sudden infant death syndrome: review of a 15-year study of a single dataset. J Neuropathol Exp Neurol. 60: 228-247

52.   Kline DD, Buniel MC, Glazebrook P, Peng YJ, Ramirez-Navarro A, Prabhakar NR, Kunze DL (2005) Kv1.1 deletion augments the afferent hypoxic chemosen-sory pathway and respiration. J Neurosci 25: 2289-3399

53.   Kuwaki T, Cao WH, Kurihara Y, Kirihara H, Ling GY, Onodera M, Ju KH, Ya-zaki Y, Kumlada M (1996) Impaired ventilatory responses to hypoxia and hyper-capnia in mutant mice deficient in endothelin-1. Am J Physiol 270: R1279-R1286

54.   Kuwaki T, Ling GY, Onodera M, Nakamura, Ju KH, Cao WH, Kumada M, Kuri-hara H, Kurihara Y, Yazaki Y, Ohuchi T, Yanagisawa M, Fukuda Y (1999) En-dothelin in the central control of cardiovascular and respiratory functions. Clin Exp Pharmacol Physiol 26: 989-994

55.   Kuwana S, Okada Y, Sugawara Y, Tsunekawa N, Obata K (2003) Disturbances of neural respiratory control in neonatal mice lacking GABA synthetizing en-zyme 67-kDa isoform of glutamic acid decarboxylase. Neurosci 120: 861-870

56.   Marcus CL, Carroll JL, McColley SA, Loughlin GM, Curtis S, Pyzik P, Naidu S (1994) Polysomnographic characteristics of patients with Rett syndrome. J Pedi-atr 125: 218-224

57.   Marcus CL, Lutz J, Carroll JL, Bamford O (1998) Arousal and ventilatory re-sponses during sleep in children with obstructive sleep apnea. J Appl Physiol 84: 1926-1936

58.   Matera I, Bachetti T, Puppo F, Di Duca M, Morandi F, Casiraghi GM, Cilio MR, Hennekam R, Hofstra R, Schober JG, Ravazzolo R, Ottonello G, Ceccherini I (2004) PHOX2B mutations and polyalanine expansions correlate with the sever-ity of the respiratory phenotype and associated symptoms in both congenital and late-onset central hypoventilation syndrome. J Med Genet 41: 373-380

59.   Matrot B, Durand E, Dauger S, Vardon G, Gaultier C, Gallego J (2005) Auto-matic classification of activity and apneas using whole-body plethysmography in newborn mice. J Appl Physiol 98: 365-370

60.   Mellins RB, Balfour HH, Turino GM, Winters RW (1970) Failure of autonomic control of ventilation (Ondine's curse) of an infant born with this syndrome and review of the literature 49: 497-504

61.   Nadeau JH (2001) Modifier genes in mice and humans. Nat Rec 2: 165-174

62.   Nakada Y, Taniura H, Uetsuki T, InazawaJ, Yoshikawa K (1998) The human chromosomal gene for *Necdin*, a neuronal growth suppressor, in the Prader-Willi syndrome deletion region. Gene 213: 65-72

63.   Namakura A, Kuwaki T, Kuriyama T, Yanagisawa M, Fukuda Y (2001) Normal ventilation and ventilatory responses to chemical stimuli in juvenile mutant mice deficient in endothelin-3. Respir Physiol 124: 1-4

64. Namakura A, Zhang W, Yanagisawa M, Fukuda Y, Kuwaki T (2006) Vigilance state-dependent attenuation of hypercapnic chemoreflex and exaggerated sleep apnea in orexin knockout mice. J Appl Physiol 102: 241-248

65. Nsegbe E, Wallen-Mackensie AS, Roux JC, Shvarev Y, Lagercrantz H, Perlman T, Herlenius E (2004) Congenital hypoventilation hypoxic response in Nurr1 mutant mice. J Physiol 556: 43-59

66. Neubauer JA, Sunderram J (2004) Oxygen-sensing neurons in the central nervous system. J Appl Physiol 96: 367-374

67. Nixon GM, Brouillette RT (2002) Sleep and breathing in Prader-Willi syndrome. Pediatr Pulmonol 34: 209-217

68. Onimaru H, Kumagawa Y, Homma I (2006) Respiration-related rhythmic activity in the rostral medulla of newborn rats. J Neurophysiol 96: 55-61

69. Oyamada Y, Yamaguuchi K, Murai M, Hakuno H, Ishizaka A (2005) Role of Kir2.2 in hypercapnic ventilatory response during postnatal development of mouse. Respir Physiol Neurobiol 145: 143-151

70. Oyamada Y, Yamaguchi K, Murai M, Ishizaka A, Okada Y (2006) Potassium channels in the central control of breathing. Adv Exp Med Biol 580: 339-344

71. Pagliardini S, Ren J, Wevrick R, Greer JJ (2005) Developmental abnormalities of neuronal structure and function in prenatal mice lacking the Prader-Willi syndrome gene *necdin*. Am J Pathol 167: 175-191

72. Paterson DS, Trachtenberg FL, Thompson EG, Beggs AH, Darnall R, Chadwick AE, Krous HF, Kinney HC (2006) Multiple serotonergic brainstem abnormalities in sudden infant death syndrome. JAMA 296: 2124-2132

73. Pattyn A, Morin X, Cremer H, Goridis C, Brunet JF (1999) The homeobox gene Phox2b is essential for the development of autonomic neural crest derivatives. Nature 399: 366-370

74. Pena F, Ramirez JM (2002) Endogenous activation of serotonin-2A receptors is required for respiratory rhythm generation in vitro. J Neurosci 22: 11055-11064

75. Peng YJ, Rennison J, Prabhakar NR (2004) Intermittent hypoxia augments carotid body and ventilatory response to hypoxia in neonatal rat pups. J Appl Physiol 97: 2020-2025

76. Poon CS, Zhou Z, Champagnat J (2000) NMDA receptor activity in utero averts respiratory depression and anomalous long-term depression in newborn mice. J Neurosci 20: RC73

77. Popa D, Lena C, Favre V, Prenat C, Gingrich J, Escourrou P, Hamon M, Adien J (2005) Contribution of 5-HT2 receptor subtypes to sleep-wakefulness and respiratory control, and functional adaptations in knock-out mice lacking 5-HT2A receptors. J Neurosci 25: 11231-11238

78. Ptack K, Burnet H, Blanchi B, Sieweke M, De Felipe C, Hunt SP., Monteau R, Hilaire G (2002) The murine neurokinin NK1 receptor gene contributes to the adult facilitation of ventilation. J Neurosci 16: 2245-2252

79. Ramanantsoa N, Vaubourg V, Dauger S, Matrot B, Vardon G, Chettouh Z, Gaultier C, Goridis C, Gallego J (2006) Ventilatory response to hyperoxia in newborn mice heterozygous for the transcription factor Phox2b. Am J Physiol Regul Integr Comp Physiol 290: R1691-R1696

80. Redline S, Leitner J, Arnold J, Tishler PV, Altose MD (1997) Ventilatory-control abnormalities in familial sleep apnea. Am J Respir Crit Care Med 156: 155-160

81. Ren J, Lee S, Pagliardini S, Gerard M., Stewart CL, Greer JJ, Wevrick R (2003) Absence of Ndn, encoding the Prader-Willi syndrome-deleted gene *necdin*, results in congenital deficiency of central respiratory drive in neonatal mice. J Neurosci 23: 1569-1573

82. Renieri A, Meloni I, Longo I, Ariani F, Mari F, pescucci C, Cambi F (2003) Rett
     syndrome: the complex nature of a monogenic disease. J Mol Genet 81: 346-354
83. Renolleau S, Dauger S, Autret F, Vardon G, Gaultier C, Gallego J (2001a) Matu-
     ration of baseline breathing and of hypercapnic and hypoxic ventilatory responses
     in newborn mice. Am J Physiol Integr Comp Physiol 281: R1746- R1753
84. Renolleau S, Dauger S, Vardon G, Levacher B, Simonneau M, Yanagisawa M,
     Gaultier C, Gallego J (2001b) Impaired ventilatory responses to hypoxia in mice
     deficient in endothelin-converting-enzyme-1. Pediatr Res 49: 705-712
85. Richerson GB, Wang W, Hodges MR, Dohle CI, Diez-Sampedro A (2005) Hom-
     ing in on the specific phenotype(s) of central respiratory chemoreceptors. Exp
     Physiol 90: 259- 269
86. Shirasawa S, Arata A, Onimaru H, Roth KA, Brown GA, Horning, S, Arata S.,
     Okumura K, Sasazuki T, Korsmeyer SJ (2000) Rnx deficiency results in congeni-
     tal central hypoventilation. Nat Genet 24: 287-290
87. Telgkamp P, Cao YQ, Basbaum AI, Ramirez JM (2002) Long-term deprivation
     of substance P in PPT-A mutant mice alters the anoxic response of the isolated
     respiratory network J Neurophysiol 88: 206-213
88. Thach B (2005) The role of respiratory control disorders in SIDS. Respir Physiol
     Neurobiol 149: 343-353
89. Thoby-Brisson M, Cauli B, Champagnat J, Fortin G, Katz DM (2003) Expression
     of functional tyrosine kinase B receptors by rhythmically active respiratory neu-
     rons in the pre-Bötzinger complex of neonatal mice. J Neurosci 23: 7685-7689
90. Trochet D, O'Brien LM, Gozal D, Nordenskjold A, Laudier B, Svensson PJ,
     Uhrig S, Cole T, Munnich A, Gaultier C, Lyonnet S, Amiel J (2005a) PHOX2B
     genotype allows for prediction of tumour risk in congenital central hypoventila-
     tion syndrome. Am J Hum Genet 76: 421-426
91. Trochet D, Hong SJ, Lim JK, Brunet JF, Munnich A, Kim KS, Lyonnet S,
     Goridis C, Amiel J (2005b) Molecular consequences of PHOX2B missense,
     frameshift and alanine expansion mutations leading to autonomic dysfunction.
     Hum Mol Genet 14: 3697-3708
92. Viemari JC, Burnet H, Bévengut M, Hilaire G (2003) Perinatal maturation of the
     mouse respiratory rhythm-generator: in vivo and in vitro studies. Eur J Neurosc
     17: 1233-1244
93. Viemari JC, Bévengut M, Burnet H, Coulon P, Pequignot JM, Tiveron MC,
     Hilaire G (2004) Phox2a gene, A6 neurons, and noradrenaline are essential for
     development of normal respiratory rhythm in mice. J Neurosci 24: 928-937
94. Viemari JC, Maussion G, Bévengut M, Burnet H, Pequignot JM, Népote V,
     Pachnis V, Simonneau M, Hilaire G (2005a) Ret deficiency in mice impairs the
     development of A5 and A6 neurons and the functional maturation of the respira-
     tory rhythm. J Neurosci 22: 2403-2412
95. Viemari JC, Roux JC, Tryba AK, Saywell V, Burnet H, Pena F, Zanella S,
     Bevengut M, Barthelemy-Requin M, Herzing LB, Moncia A, Mancini J, Ramirez
     JM, Villard L, Hilaire G (2005b) Mecp2 deficiency disrupts norepinephrine and
     respiratory systems in mice. J Neurosci 25: 11521-11530
96. Vincent SG, Waddell AE, Caron MG, Walker JK, Fisher JT (2007) A murine
     model of hyperdopaminergic state displays altered respiratory control. FASEB J
     21: 1463-1471
97. Weaving LS, Ellaway CJ, Gécz J, Christodoulou (2005) Rett syndrome: clinical
     review and genetic update J Med Genet 42: 1-7

98.   Weese-Mayer DE, Berry-Kravis EM, Zhou L, Maher BS, Silvestri JM, Curran ME, Mazarita ML (2003) Idiopathic congenital central hypoventilation syndrome: analysis of genes pertinent of early autonomic nervous system embryologic development and identification of mutations PHOX2B. Am J Med Genet 123: 267-278

99.   Weese-Mayer DE, Lieske SP, Boothby CM, Kenny AS, Bennett HL, Silvestri JM, Ramirez JM (2006) Autonomic nervous system dysregulation: breathing and heart rate perturbation during wakefulness in young girls with Rett syndrome. Pediatr Res 60: 443-449

100. Yanagisawa H, Yanagisawa M, Kapur RP, Richardson JA, Williams SC, Clouthier DE, De Wit D, Emoto N, Hammer RE (1998) Dual genetic pathways of endotheline-mediated intercellular signalling revealed by targeted disruption of endothelin converting enzyme-1 gene. Development 125: 825-836

101. Zanella S, Roux JC, Muscatelli F, Hilaire G (2006) Necdin gene, respiratory disturbances and Prader-Willi syndrome (Abstract). Xth Oxford Conference Lake Louise Canada 17-24 September: p 62

14. Tentative mouse model for the congenital central hypoventilation syndrome: heterozygous *phox2b* mutant newborn mice

Jorge GALLEGO, Nélina RAMANANTSOA and Vanessa VAUBOURG

Inserm U676, Hôpital Robert Debré, Université Paris 7
48 Bd Sérurier, 75019, Paris, France

14.1 Introduction

Central congenital hypoventilation syndrome (CCHS or Ondine's curse, see Chapter 4) is a rare disease that is generally present from birth and manifests as hypoventilation during sleep with apneas and cyanotic episodes, in the absence of primary neuromuscular or lung disease [1; 7]. Hypoventilation is also present during wakefulness in the most severe cases. Throughout life, patients with CCHS have absent or markedly reduced ventilatory responses to sustained hypercapnia [42] and, to a lesser extent, to sustained hypoxia [42]. These respiratory impairments are generally ascribed to impaired central integration of chemosensory inputs to the brainstem, rather than to failure of chemoreceptor activity, which is at least partially present [26; 33; 52]. About 16% of patients have Hirschsprung disease (HSCR), and patients with CCHS may exhibit a variety of autonomic disorders [56], indicating that CCHS is a consequence of abnormal neural crest development. Despite major advances in unraveling the pathophysiology of CCHS, no preventive or curative treatment is available. The mortality rate remains very high, especially in patients who also have HSCR. Survivors have a lifelong dependency on mechanical ventilation. Finally, developmental abnormalities, motor and speech delays, and learning disabilities have been reported in more than 25% of patients with CCHS [56].

The association of CCHS with HSCR suggested a role for genes involved in the development and migration of neural crest cells. Studies done to investigate this hypothesis identified *PHOX2B* as the disease-causing gene for CCHS [3; 34; 57; 59]. Over 90% of patients have a heterozygous *PHOX2B* mutation consisting in a polyalanine-repeat expansion. Mutations in other genes account for the remaining 10% of cases. *PHOX2B* is a master regulator of the noradrenergic phenotype and of neuronal relays in autonomic medullary reflex pathways [44], including peripheral chemosensitive pathways [18; 53] (see Chapter 3). To investigate the role for *PHOX2B* in the development of respiratory control, the respiratory phenotype of neonatal mice with one invalidated *Phox2b* allele (*Phox2b$^{+/-}$*) was investigated. In all the studies discussed below, *Phox2b$^{+/-}$* mutant pups were compared to their wild-type *Phox2b$^{+/+}$* littermates, and phenotypic traits were determined without previous knowledge of genotypes.

The *Phox2b$^{+/-}$* mice were examined in the neonatal period. As a rule, physiological and behavioral screening of adult mutant mice is regarded as a powerful tool for uncovering the function of genes and their role in human diseases. However, the role for neonatal screening of mice as the tool of choice for studying developmental disorders and pediatric treatments is less well recognized. This chapter describes new methods for neonatal screening in mice and discusses their application to the newborn mouse model of CCHS.

14.2 Neonatal phenotype determination

Phox2b$^{+/-}$ pups survive and are fertile, whereas homozygous *Phox2b* knock-out mice (*Phox2b$^{-/-}$*) die *in utero* around embryonic day 14 [43]. Body weight and mouth temperature measured were normal in 2-day-old *Phox2b$^{+/-}$* mice [18]. At 5 days of age, however, the mutant pups had slightly lower body weights and lower temperatures than wild-type pups, suggesting neonatal adaptation disorders [21]. In many mutant mouse models (generally null mutants), respiratory disorders are associated with extremely severe phenotypes and poor survival. In contrast, the *Phox2b$^{+/-}$* model shows limited phenotypic abnormalities that consists solely in disturbances of postnatal respiratory control, as detailed below.

14.3 Non-invasive ventilatory phenotyping in newborn mice

Breathing is usually quantified based on breath duration, inspiratory and expiratory duration, tidal volume, and ventilation. Tidal volume and ventilation are divided by body weight to adjust for inter- and intra-individual differences in growth, which are particularly marked during early development. Breathing variables are measured both under baseline conditions, i.e., while the animals are breathing air; and during exposure to chemical stimuli such as hypercapnia, hypoxia, or hyperoxia, to test for chemosensitivity. The number and duration of apneas and periodic breathing episodes are indicators of respiratory instability, a general characteristic of the neonatal period in newborn mammals (see Chapter 9).

The small birth weight (about 1-2 g) and tidal volume (3-4) in newborn mice the use of measurement devices that are widely employed for larger animals including adult mice (e.g., pneumotachometers, thermistors, respiratory inductance spirometers, and magnetometers). Two methods provide valid measurements in newborn mice: head-out plethysmography (e.g., [13]) and whole-body plethysmography (e.g., [35]).

In head-out plethysmography, the newborn mouse is placed in a chamber and its head is slipped outside the chamber through an opening with an airtight seal around the neck. The amount of air that moves in and out of the chamber as a result of breathing is roughly proportional to the changes in chest volume. This method provides a relatively direct measure of breathing. Its main drawback is that the animal must be tightly restrained to ensure that no air leakage occurs around the neck. The effects of the neck collar on upper airway resistance in such tiny animals are difficult to control. Furthermore, restraint is generally regarded as a potent stressor that has marked effects on the baseline breathing pattern, at least in adult mice [17]. The effect of restraint has not been studied in newborn mice but likely involves profound changes in arousal states and breathing pattern.

Restraint can be avoided by using whole-body flow barometric plethysmography, which consists in placing the animal in a chamber and measuring the pressure changes in the chamber. Pressure in the chamber increases during inspiration and decreases during expiration. According to theory, the pressure increase during inspiration is caused by addition of water vapor to the inspired gas and by warming of the inspired gas from the temperature in the chamber to that in the alveoli. Conversely, pressure diminishes during expiration because of condensation of water vapor and cooling of expired gas. Whole-body plethysmography has been validated against pneumotachography in adult mice [41] but not in newborn mice. In newborn mice, whole-body plethysmography provides semiquantitative measurements of tidal volume and ventilation while allowing valid measurements of breathing frequency and apnea. Despite its limitations, whole-body flow barometric plethysmography remains the only noninvasive method and is therefore the method of choice for studying unrestrained newborn mice. Because body temperature can strongly influence the breathing pattern, breathing variables are measured under thermoneutrality. Thermoneutrality has been found to be around $33°C$ in rat pups [9], a value which corresponds to the temperature measured inside the litter of newborn mice.

Mouse models of CCHS (or any neonatal respiratory disorder) require relatively long recording sessions to take into account the dependence upon sleep-wake states. Prolonged recording is also needed to study the time-course of ventilatory responses to hypoxia or hyperoxia (detailed below) and the effects of repeated stimulations, most notably intermittent hypoxia, which is the current experimental model for recurrent apneas. Continuous measurements can be obtained by using a bias flow through the plethysmograph to prevent CO_2 accumulation and ambient temperature drifts over time.

New methods have been developed to classify sleep-wake states in newborn mice [21] and to assess arousal responses to chemical challenges [15]. The difficulty here is that it is not possible to determine sleep states by electroencephalogram or electro-oculogram in newborn rodents [28]. In newborn rats, nuchal muscle tone, coordinated movements, and motor twitches have proved reliable for identifying sleep-wake states between 2 and 8 days of age [29], and in newborn mice [21]. Behavioral states (wakefulness, active sleep, quiet sleep, or undetermined sleep) were determined recently using these criteria in newborn mice [21].

Apneas, sighs, and gasps can be identified by visual examination of the plethysmographic signal. However, automatic detection methods based on spectral analysis for apnea detection facilitate phenotype determination of large numbers of mutant animals [35]. As a rule, automatic processing of respiratory signals is required to achieve reasonably high-throughput physiological screening of newborn mutant pups and improved access to the genetic determinants of early respiratory control disorders, such as CCHS.

Studies of mouse models of CCHS were conducted shortly after birth. The control of breathing in newborn pups undergoes rapid developmental changes, in particular because resetting of peripheral chemoreceptors occurs within 6-12 hours after birth [46]. A practical implication is that sequential testing of very young pups over several hours may capture large changes due to development over the testing period. Careful control of postnatal age is mandatory.

14.4 Sleep-disordered breathing in *Phox2b+/-* newborn mice

Sleep apnea time was increased about 6-fold in $Phox2b^{+/-}$ mutant pups and ventilation during active sleep was decreased by about 20%, compared to wild-type pups on P5 (Fig. 1, [21]). A possible explanation to these abnormalities is a decrease in the tonic drive to breathe provided by chemosensitive sites that act predominantly during sleep [39; 40]. In rats, the CO_2-sensitive neurons of the caudal medullar raphe are active only during sleep [39] and those of the rostral nucleus tractus solitarius are more effective during sleep than during wakefulness [40].

14.5 Sensitivity to CO_2

The ventilatory response to 8% CO_2 was tested on 2, 6, and 10 days of postnatal age in $Phox2b^{+/+}$ and $Phox2b^{+/-}$ mouse pups [18]. Mutant $Phox2b^{+/-}$ mice 2 days after birth exhibited blunting of the ventilatory response to hypercapnia, a feature reminiscent of CCHS (Fig. 2). The ventilatory increase during hypercapnia was about 40% smaller in $Phox2b^{+/-}$ than $Phox2b^{+/+}$ pups, due to depression of the breath duration (TTOT) response. Both the number and the total duration of apneas were significantly higher in $Phox2b^{+/-}$ than in $Phox2b^{+/+}$ pups during hypercapnia.CO_2-sensitive sites (locus ceruleus and area postrema [38; 49]), and afferent pathways from the carotid bodies (which contribute to CO_2 sensitivity) also depend on

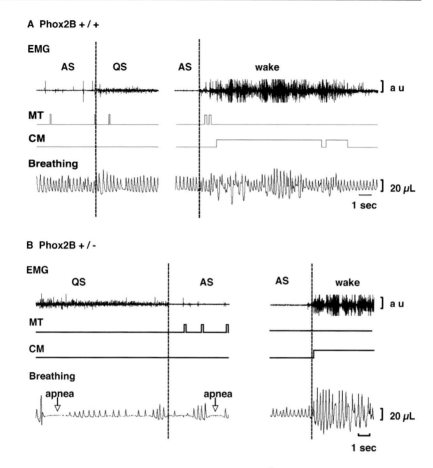

Fig. 1. Sleep-disordered breathing in 5-day-old *Phox2b*[+/-] mutant pups. Examples of nuchal EMG recordings, behavioral scoring (motor twitches, MT; and coordinated movements, CM) and breathing as assessed by plethysmography in a 5-day old *Phox2b*[+/+] mouse (panel A) and in a 5-day-old *Phox2b*[+/-] mouse (panel B) during active sleep (AS), quiet sleep (QS), and wakefulness. Behavioral scores were determined visually by the experimenter during recordings. AS was defined as silent EMG with MTs and QS as low-voltage EMG without MTs. Note that some MTs are visible on both the EMG and the behavioral scoring signal, with a delay due to experimenter response latency. Wakefulness was characterized by high EMG voltage with CMs. During sleep, the *Phox2b*[+/+] pup breathed regularly, without apneas, whereas the *Phox2b*[+/-] mouse pup had irregular breathing interrupted by apneas ([21] with permission).

Phox2b for their development [18]. However, no differences in ventilatory responses to CO_2 were detected between *Phox2b*[+/+] and *Phox2b*[+/-] pups aged 10 days, whereas 6-day-old pups exhibited an intermediate phenotype between 2- and 10-day-old pups. Thus, the postnatal impairment of the hypercapnic ventilatory response seen in *Phox2b*[+/-] pups was short-lived. The mechanisms of this rapid functional recovery are unknown. Such mechanisms obviously do not operate in patients with CCHS, in whom the symptoms are typically severe and irreversible.

Thus, heterozygous null mutation in mice yielded a less severe phenotype than the human CCHS phenotype.

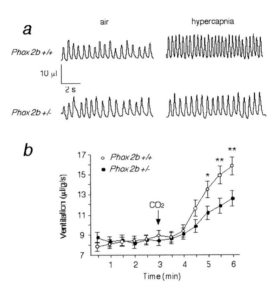

Fig. 2. Abnormal ventilatory response to hypercapnia in 2-day-old Phox2b$^{+/-}$ mutant mice. a: Ventilatory tracings in one Phox2b$^{+/+}$ pup (top) and one Phox2b$^{+/-}$ pup (bottom). Both pups had similar baseline ventilation but the Phox2b$^{+/-}$ pup showed a weaker ventilatory response to hypercapnia. b: Values are means ± SEM ([18], reproduced with permission of the Company of Biologists).

14.6 Sensitivity to hypoxia

In 2-day-old *Phox2b*$^{+/-}$ pups, a biphasic pattern of V_E changes occurred in response to hypoxia (5% O_2, Fig. 3 [18]). This pattern is characteristic of the ventilatory response to hypoxia in newborn mammals. The immediate hyperpneic response to hypoxia (i.e., the ascending limb of the biphasic ventilatory response) was normal in mutant pups. In CCHS patients, sensitivity to hypoxia varied across patients [42]. In CCHS patients who were able to ventilate adequately during wakefulness, peripheral chemosensitivity to oxygen was intact [26], suggesting that defects in peripheral chemosensitivity may exist only in the most severe cases. Thus, the normal hyperpneic response to hypoxia in *Phox2b*$^{+/-}$ pups further confirms their relatively mild phenotype, compared to patients with CCHS.

In contrast, the hypoxic decline was markedly increased in the *Phox2b*$^{+/-}$ pups (Fig. 3), mainly due to abnormal T$_{TOT}$ control. Apneas occurred chiefly during the post-hypoxic decline. Their numbers were similar in *Phox2b*$^{+/+}$ and *Phox2b*$^{+/-}$ pups, but their total duration was considerably longer during the post-hypoxic decline in *Phox2b*$^{+/-}$ than in *Phox2b*$^{+/+}$ pups (Fig. 3).

Post-hypoxic ventilatory depression has not been examined in CCHS patients. The two components of the ventilatory response to hypoxia (i.e., hyperpnea and hypoxic ventilatory decline) involve different neuronal systems, which may be differently affected by *PHOX2B* mutations.

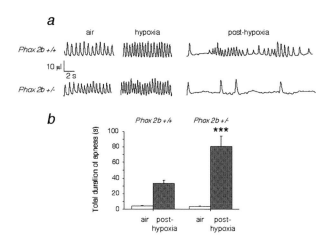

Fig. 3. Abnormal post-hypoxic depression in 2-day-old *Phox2b^+/-* mutant mice. a: Ventilatory tracings in one *Phox2b^+/+* pup (top) and one *Phox2b^+/-* pup (bottom). Both pups had similar baseline ventilation and initial increases in ventilation, but the *Phox2b^+/-* pup showed long post-hypoxic apneas. b: Total duration of apneas (defined as respiratory pauses longer than twice the duration of the preceding breathing cycle) while breathing air (3 min) and after hypoxia (6 minutes). Although the number of apneas was not different between the two groups (0.7 and 2.8 apneas/minute with air and hypoxia in *Phox2b^+/-* pups and 0.8 and 2.8 apneas/minute in *Phox2b^+/+* pups, respectively), apneas were considerably longer in *Phox2b^+/-* mice (***$p < 0.001$). Values are group means ± SEM ([18] reproduced with permission of the Company of Biologists).

14.7 Arousal response to hypoxia

Arousal from sleep is an important protective response that prevents hypoxemia during apneas or impaired gas exchange. Behavioral arousal was first defined as a stereotyped motor response characterized by sudden neck and forepaw extension followed by rising. This coordinated response required cortically mediated activation of the neck, limbs, and thoracic muscles and, therefore, reflected cortical arousal. Behavioral assessment of the arousal response and nuchal EMG criteria yielded practically identical findings [22]. Arousal responses to hypoxia as assessed by behavioral criteria were not significantly different between *Phox2b^+/+* and *Phox2b^+/-* pups.

14.8 Sensitivity to hyperoxia

Chemosensitivity to oxygen can be assessed by inducing "physiological chemodenervation", which consists in inhibiting ventilation by having the subject inhale a bolus of oxygen. The extent of ventilation inhibition in response to hyperoxia provides a functional estimate of the tonic chemoreceptor drive [30; 37; 55]. Hyperoxic and hypoxic tests are complementary and may show divergent results in newborn mammals [8; 36], raising the possibility that a decrease in tonic chemoreceptor drive might coexist with normal ventilatory responses to hypoxia.

The ventilatory decrease caused by hyperoxia was larger in newborn $Phox2b^{+/-}$ mutant mice than in their wild-type littermates and was magnified by longer apnea durations (Fig. 4, [45]). Furthermore, compared to wild-type pups mutant pups showed a more sustained ventilatory decrease, which outlasted the return to normoxia (Fig. 4). These results suggest stronger tonic activity of oxygen-sensitive peripheral chemoreceptors in mutant pups. This augmented peripheral tonic input may be ascribable to low arterial PO_2 levels, which, unfortunately, cannot be measured in newborn mice using currently available techniques. A plausible hypothesis is that the low CO_2 chemosensitivity [18] and sleep-related apneas [21] of mutant pups were associated with low PaO_2 values. In human full-term or preterm infants, periodic breathing was associated with low PaO_2 values and with greater decreases in ventilation and longer apnea times in response to 100% O_2 [2; 47; 48]. $Phox2b^{+/-}$ pups may display a similar pattern of breathing disorders.

14.9 Effects of ambient temperature

CCHS is frequently associated with thermoregulatory disorders (sporadic profuse sweating, decreased basal body temperature with cool extremities, and absence of fever with infections) [56]. Parents of children with CCHS frequently report worse breathing discomfort at high ambient temperatures [31]. These observations suggest that warmer ambient temperatures may aggravate the disruption of breathing control in CCHS. We investigated possible interactions between thermoregulatory and respiratory disorders in $Phox2b^{+/-}$ mice.

Our preliminary data indicated clearly that the impairment in the ventilatory response to CO_2 in 2-day-old pups, as assessed by the difference in this response between $Phox2b^{+/+}$ and $Phox2b^{+/-}$ pups, was larger at 35°C ambient temperature than at 29°C. Interestingly, this amplifying effect of higher ambient temperature on the ventilatory impairment was not related to differences in body temperature changes (measured subcutaneously in the interscapular region) between $Phox2b^{+/+}$ and $Phox2b^{+/-}$ pups. In fact, body temperatures were very similar in these two groups. The hypothalamus, which is the primary locus for thermoregulation and exerts direct control over respiratory brainstem structures, does not express $PHOX2B$ [18]. On the other hand, recent magnetic resonance imaging studies using T2 relaxation time procedures in patients with CCHS revealed morphological abnormalities in the anterior hypothalamus (see Chapter 5). Thus, temperature-related influences on ventilatory control in mutant pups may be caused by hypothalamic abnormalities secondary to structural and neurological abnormalities in *Phox2b*-expressing brain regions (see Chapter 5).

14.10 Cognitive evaluation of *Phox2b$^{+/-}$* mutant mice

As noted above, a relatively large proportion of children with CCHS exhibit speech impairments or learning disabilities [56]. It is difficult to determine whether these developmental disorders are due to the genetic defect or to the severity of the medical condition. Cognitive development can be examined in newborn mice by using odor preference paradigms [5; 10; 11]. Odor preference conditioning consists in assessing preference between two odors, one of which is first paired with a tactile stimulus that replicates maternal care. *Phox2b$^{+/-}$* mutant mice and their wild-type littermates displayed similar learning abilities, suggesting that invalidation of one *Phox2b* allele had no direct effects on early learning abilities (V. Vaubourg et al., unpublished data).

14.11 Comparison between *Phox2b$^{+/-}$* and CCHS phenotypes

Features in *Phox2b$^{+/-}$* mice that replicate CCHS in humans include increased sleep apneas, smaller V_E during sleep, lower body weight (45% of patients with CCHS experience delayed growth [56]), lower body temperature (one of the symptoms often associated with CCHS [1]), and diminished ventilatory response to hypercapnia [18]. On the other hand, the respiratory phenotype of *Phox2b$^{+/-}$* pups differs from CCHS in several ways. First, this phenotype is considerably less severe than CCHS. *Phox2b$^{+/-}$* mutant mice survive normally, whereas CCHS patients die unless they receive lifelong ventilatory support during sleep. The postnatal impairment in the hypercapnic ventilatory response is short-lived in *Phox2b$^{+/-}$* pups and irreversible in patients with CCHS, whose symptoms show no decrease in severity over time. The ventilatory impairments in mutant pups are more pronounced during active than quiet sleep, whereas the opposite occurs in patients with CCHS. Conceivably, the small proportion of quiet sleep in 5-day-old pups, due to sleep organization immaturity, may have led to inaccuracies in ventilation estimates. In patients with CCHS, polysomnography studies were conducted at later stages of sleep devel opment, compared to studies in newborn mice [25; 27], a fact that hinders comparisons. Finally, mutant mice have dilated pupils [14], whereas pupil constriction is the most common ocular abnormality in patients with CCHS [56].

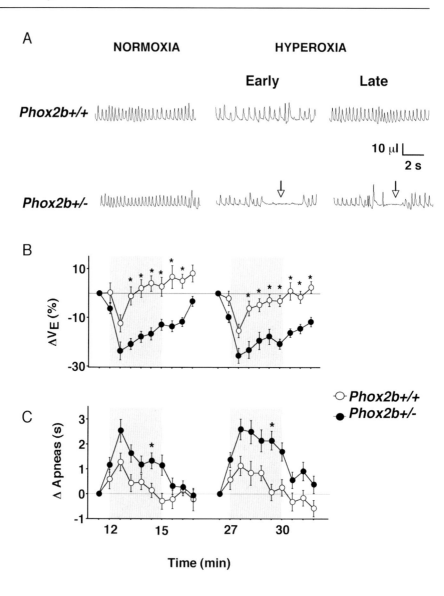

Fig. 4. Abnormal oxygen-induced ventilatory depression in 2-day-old *Phox2b$^{+/-}$* pups. The initial VE decrease caused by hyperoxia was chiefly ascribable to a TTOT increase, whereas VT changes were small and nonsignificant (from [45] with permission).

The phenotype differences between $Phox2b^{+/-}$ mice and patients with CCHS may be due to functional differences between the $Phox2b$-targeted mutation in mice (a null mutation) and the alanine expansion generally found in the $PHOX2B$ gene of CCHS patients. Alanine expansion may result in a protein that can bind to the correct targets in the genome but is unable to carry out its normal regulatory function and therefore competes with the product of the wild-type allele [6]. In contrast, a single functional $Phox2b$ allele may ensure correct protein function, leading to a less severe phenotype [14]. Furthermore, alanine-expanded proteins may be toxic to vulnerable cells that express them, due to aggregate formation [12; 54]. Clearly, knock-in mutant mouse strains with alanine expansions similar to those found in most of the patients with CCHS need to be studied. Preliminary results in such knock-in mice revealed far more severe blunting in CO_2 sensitivity, compared to null mutant $Phox2b^{+/-}$ mice (J. Gallego et al., unpublished data).

The more severe phenotype in CCHS patients may also be due to genetic variations that are not present in the $Phox2b^{+/-}$ model. Heterozygous mutations affecting seven genes involved in neural crest cell development (RET, $GDNF$, $PHOX2A$, $HASH1$, $EDN1$, $GFRA1$, and $EDN3$) and one gene coding for a neurotrophic factor ($BDNF$) have been found in several CCHS patients [23; 50; 58] (see Chapter 4), suggesting that modifier genes may be involved. Associated $PHOX2B$ mutations have been identified in patients with RET, $GDNF$, $PHOX2A$, $BDNF$, $HASH1$, and $GFRA1$ mutations [58]. The possible coexistence of multiple mutations may account for the considerable interindividual variability in symptom severity in patients with CCHS. Abnormal ventilatory responses to hypercapnia have also been reported in newborn mice lacking genes involved in the development of the autonomic nervous system, most notably via the endothelin-1 and the $Mash1$-Ret-$Phox2a$-$Phox2b$ signaling pathways [13; 18; 19; 32]. The phenotype in mice heterozygous for mutations in $Phox2b$ and another gene previously found to be mutated in CCHS may resemble CCHS more closely than the phenotype produced by a single $Phox2b$ mutation.

14.12 Conclusion

Over the last decade, both genetic screening of patients with CCHS and studies of the ventilatory effects of neural-crest gene mutations in mice have shed light on the genetic basis of CCHS [23; 24; 58]. For example, impaired sensitivity to CO_2 in mouse pups heterozygous for a null allele of $Mash1^{+/-}$, which encodes a tissue-specific basic helix–loop–helix transcription factor [16; 19], pointed to the human $Mash1$ ortholog $HASH1$ as a potential candidate gene for CCHS. Heterozygous nucleotide substitution of the $HASH1$ gene was found in three patients [20]. On the other hand, mutations associated with early hypoventilation or respiratory failure in mice, but without the chemosensitivity impairments that constitute the hallmark of CCHS (e.g., Rnx [51]) were not good candidates for CCHS [4].

Phenotype determination in $Phox2b^{+/-}$ mutant pups demonstrates clearly that heterozygous $Phox2b$ null mutation is causally related to CO_2 chemosensitivity and apneas early after birth. Furthermore, $Phox2b^{+/-}$ newborn mice show sleep-related respiratory disorders. The respiratory phenotype of $Phox2b^{+/-}$ newborn mice is reminiscent of CCHS, although it is milder and transient. The polyalanine expansions found in patients with CCHS probably induce different cellular processes from those induced by $Phox2b$ allele inactivation. However, the $Phox2b^{+/-}$ model provides an interesting basis for investigating the involvement of $Phox2b$ in respiratory control development and the potential mechanisms of recovery from respiratory control impairments.

References

1. (1999) Idiopathic congenital central hypoventilation syndrome: diagnosis and management. American Thoracic Society. Am J Respir Crit Care Med 160: 368-373

2. Al-Matary A, Kutbi I, Qurashi M, Khalil M, Alvaro R, Kwiatkowski K, Cates D, Rigatto H (2004) Increased peripheral chemoreceptor activity may be critical in destabilizing breathing in neonates. Semin Perinatol 28: 264-272

3. Amiel J, Laudier B, Attie-Bitach T, Trang H, de Pontual L, Gener B, Trochet D, Etchevers H, Ray P, Simonneau M, Vekemans M, Munnich A, Gaultier C, Lyonnet S (2003) Polyalanine expansion and frameshift mutations of the paired-like homeobox gene PHOX2B in congenital central hypoventilation syndrome. Nat Genet 33: 459-461

4. Amiel J, Pelet A, Trang H, de Pontual L, Simonneau M, Munnich A, Gaultier C, Lyonnet S (2003) Exclusion of RNX as a major gene in congenital central hypoventilation syndrome (CCHS, Ondine's curse). Am J Med Genet A 117: 18-20

5. Armstrong CM, DeVito LM, Cleland TA (2006) One-trial associative odor learning in neonatal mice. Chem Senses 31: 343-349

6. Bachetti T, Matera I, Borghini S, Duca MD, Ravazzolo R, Ceccherini I (2005) Distinct pathogenetic mechanisms for PHOX2B associated polyalanine expansions and frameshift mutations in congenital central hypoventilation syndrome. Hum Mol Genet 14: 1815-1824

7. Bachetti T, Robbiano A, Parodi S, Matera I, Merello E, Capra V, Baglietto MP, Rossi A, Ceccherini I, Ottonello G (2006) Brainstem anomalies in two patients affected by congenital central hypoventilation syndrome. Am J Respir Crit Care Med 174: 706-709

8. Bamford OS, Carroll JL (1999) Dynamic ventilatory responses in rats: normal development and effects of prenatal nicotine exposure. Respir Physiol 117: 29-40

9. Blumberg MS, Sokoloff G (1998) Thermoregulatory competence and behavioral expression in the young of altricial species--revisited. Dev Psychobiol 33: 107-123

10. Bouslama M, Chauviere L, Fontaine RH, Matrot B, Gressens P, Gallego J (2006) Treatment-induced prevention of learning deficits in newborn mice with brain lesions. Neuroscience 141: 795-801

11. Bouslama M, Durand E, Van den Bergh O, Gallego J (2005) Olfactory classical conditioning in newborn mice. Behav Brain Res 161: 102-106

12. Brown LY, Brown SA (2004) Alanine tracts: the expanding story of human ill-
     ness and trinucleotide repeats. Trends Genet 20: 51-58
13. Burton MD, Kawashima A, Brayer JA, Kazemi H, Shannon DC, Schuchardt A,
     Costantini F, Pachnis V, Kinane TB (1997) RET proto-oncogene is important for
     the development of respiratory CO_2 sensitivity. J Auton Nerv Syst 63: 137-143
14. Cross SH, Morgan JE, Pattyn A, West K, McKie L, Hart A, Thaung C, Brunet JF,
     Jackson IJ (2004) Haploinsufficiency for Phox2b in mice causes dilated pupils
     and atrophy of the ciliary ganglion: mechanistic insights into human congenital
     central hypoventilation syndrome. Hum Mol Genet 13: 1433-1439
15. Dauger S, Aizenfisz S, Renolleau S, Durand E, Vardon G, Gaultier C, Gallego J
     (2001) Arousal response to hypoxia in newborn mice. Respir Physiol 128: 235-
     240
16. Dauger S, Guimiot F, Renolleau S, Levacher B, Boda B, Mas C, Nepote V, Si-
     monneau M, Gaultier C, Gallego J (2001) MASH-1/RET pathway involvement in
     development of brain stem control of respiratory frequency in newborn mice.
     Physiol Genomics 7: 149-157
17. Dauger S, Nsegbe E, Vardon G, Gaultier C, Gallego J (1998) The effects of re-
     straint on ventilatory responses to hypercapnia and hypoxia in adult mice. Respir
     Physiol 112: 215-225
18. Dauger S, Pattyn A, Lofaso F, Gaultier C, Goridis C, Gallego J, Brunet JF (2003)
     Phox2b controls the development of peripheral chemoreceptors and afferent vis-
     ceral pathways. Development 130: 6635-6642
19. Dauger S, Renolleau S, Vardon G, Nepote V, Mas C, Simonneau M, Gaultier C,
     Gallego J (1999) Ventilatory responses to hypercapnia and hypoxia in Mash-1
     heterozygous newborn and adult mice. Pediatr Res 46: 535-542
20. De Pontual L, Nepote V, Attie-Bitach T, Al Halabiah H, Trang H, Elghouzzi V,
     Levacher B, Benihoud K, Auge J, Faure C, Laudier B, Vekemans M, Munnich A,
     Perricaudet M, Guillemot F, Gaultier C, Lyonnet S, Simonneau M, Amiel J
     (2003) Noradrenergic neuronal development is impaired by mutation of the
     proneural HASH-1 gene in congenital central hypoventilation syndrome
     (Ondine's curse). Hum Mol Genet 12: 3173-3180
21. Durand E, Dauger S, Pattyn A, Gaultier C, Goridis C, Gallego J (2005) Sleep-
     disordered breathing in newborn mice heterozygous for the transcription factor
     Phox2b. Am J Respir Crit Care Med 172: 238-243
22. Durand E, Lofaso F, Dauger S, Vardon G, Gaultier C, Gallego J (2004) Intermit-
     tent hypoxia induces transient arousal delay in newborn mice. J Appl Physiol 96:
     1216-1222
23. Gaultier C, Amiel J, Dauger S, Trang H, Lyonnet S, Gallego J, Simonneau M
     (2004) Genetics and early disturbances of breathing control. Pediatr Res 55: 729-
     733
24. Gaultier C, Dauger S, Simonneau M, Gallego J (2003) Genes modulating chemi-
     cal breathing control: lessons from mutant animals. Respir Physiol Neurobiol
     136: 105-114
25. Gaultier C, Trang H, Praud JP, Gallego J (1997) Cardiorespiratory control during
     sleep in the Congenital Central Hypoventilation Syndrome. Pediatr Pulmonol 23:
     140-142
26. Gozal D, Marcus CL, Shoseyov D, Keens TG (1993) Peripheral chemoreceptor
     function in children with the congenital central hypoventilation syndrome. J Appl
     Physiol 74: 379-387

27.   Guilleminault C, McQuitty J, Ariagno RL, Challamel MJ, Korobkin R, McClead RE, Jr. (1982) Congenital central alveolar hypoventilation syndrome in six infants. Pediatrics 70: 684-694

28.   Karlsson KA, Blumberg MS (2002) The union of the state: myoclonic twitching is coupled with nuchal muscle atonia in infant rats. Behav Neurosci 116: 912-917

29.   Karlsson KA, Kreider JC, Blumberg MS (2004) Hypothalamic contribution to sleep-wake cycle development. Neuroscience 123: 575-582

30.   Kline DD, Yang T, Huang PL, Prabhakar NR (1998) Altered respiratory responses to hypoxia in mutant mice deficient in neuronal nitric oxide synthase. J Physiol 511: 273-287

31.   Kumar R, Macey PM, Woo MA, Alger JR, Keens TG, Harper RM (2005) Neuroanatomic deficits in congenital central hypoventilation syndrome. J Comp Neurol 487: 361-371

32.   Kuwaki T, Cao WH, Kurihara Y, Kurihara H, Ling GY, Onodera M, Ju KH, Yazaki Y, Kumada M (1996) Impaired ventilatory responses to hypoxia and hypercapnia in mutant mice deficient in endothelin-1. Am J Physiol 270: R1279-1286

33.   Marcus CL, Livingston FR, Wood SE, Keens TG (1991) Hypercapnic and hypoxic ventilatory responses in parents and siblings of children with congenital central hypoventilation syndrome. Am Rev Respir Dis 144: 136-140

34.   Matera I, Bachetti T, Puppo F, Di Duca M, Morandi F, Casiraghi G, Cilio M, Hennekam R, Hofstra R, Schöber J, Ravazzolo R, Ottonello G, Ceccherini I (2004) PHOX2B mutations and polyalanine expansions corellate with the severity of the respiratory phenotype and associated symptoms in both congenital and late onset central hypoventilation syndrome. J Med Genet 41: 373-380

35.   Matrot B, Durand E, Dauger S, Vardon G, Gaultier C, Gallego J (2005) Automatic classification of activity and apneas using whole body plethysmography in newborn mice. J Appl Physiol 98: 365-370

36.   Milerad J, Larsson H, Lin J, Sundell HW (1995) Nicotine attenuates the ventilatory response to hypoxia in the developing lamb. Pediatr Res 37: 652-660

37.   Mortola JP, Tenney SM (1986) Effects of hyperoxia on ventilatory and metabolic rates of newborn mice. Respir Physiol 63: 267-274

38.   Nattie EE (2001) Central chemosensitivity, sleep, and wakefulness. Respir Physiol 129: 257-268

39.   Nattie EE, Li A (2001) CO_2 dialysis in the medullary raphe of the rat increases ventilation in sleep. J Appl Physiol 90: 1247-1257

40.   Nattie EE, Li A (2002) CO_2 dialysis in nucleus tractus solitarius region of rat increases ventilation in sleep and wakefulness. J Appl Physiol 92: 2119-2130

41.   Onodera M, Kuwaki T, Kumada M, Masuda Y (1997) Determination of ventilatory volume in mice by whole body plethysmography. Jpn J Physiol 47: 317-326

42.   Paton JY, Swaminathan S, Sargent CW, Keens TG (1989) Hypoxic and hypercapnic ventilatory responses in awake children with congenital central hypoventilation syndrome. Am Rev Respir Dis 140: 368-372

43.   Pattyn A, Hirsch M, Goridis C, Brunet JF (2000) Control of hindbrain motor neuron differentiation by the homeobox gene Phox2b. Development 127: 1349-1358

44.   Pattyn A, Morin X, Cremer H, Goridis C, Brunet JF (1997) Expression and interactions of the two closely related homeobox genes Phox2a and Phox2b during neurogenesis. Development 124: 4065-4075

45.   Ramanantsoa N, Vaubourg V, Dauger S, Matrot B, Vardon G, Chettouh Z, Gaultier C, Goridis C, Gallego J (2006) Ventilatory response to hyperoxia in newborn mice heterozygous for the transcription factor Phox2b. Am J Physiol Regul Integr Comp Physiol 290: R1691-1696

46.   Renolleau S, Dauger S, Autret F, Vardon G, Gaultier C, Gallego J (2001) Maturation of baseline breathing and of hypercapnic and hypoxic ventilatory responses in newborn mice. Am J Physiol Regul Integr Comp Physiol 281: R1746-1753

47.   Rigatto H, Brady JP (1972) Periodic breathing and apnea in preterm infants. I. Evidence for hypoventilation possibly due to central respiratory depression. Pediatrics 50: 202-218

48.   Rigatto H, Brady JP (1972) Periodic breathing and apnea in preterm infants. II. Hypoxia as a primary event. Pediatrics 50: 219-228

49.   Ruggiero DA, Gootman PM, Ingenito S, Wong C, Gootman N, Sica AL (1999) The area postrema of newborn swine is activated by hypercapnia: relevance to sudden infant death syndrome? J Auton Nerv Syst 76: 167-175

50.   Sasaki A, Kanai M, Kijima K, Akaba K, Hashimoto M, Hasegawa H, Otaki S, Koizumi T, Kusuda S, Ogawa Y, Tuchiya K, Yamamoto W, Nakamura T, Hayasaka K (2003) Molecular analysis of congenital central hypoventilation syndrome. Hum Genet 114: 22-26

51.   Shirasawa S, Arata A, Onimaru H, Roth KA, Brown GA, Horning S, Arata S, Okumura K, Sasazuki T, Korsmeyer SJ (2000) Rnx deficiency results in congenital central hypoventilation. Nat Genet 24: 287-290

52.   Spengler CM, Gozal D, Shea SA (2001) Chemoreceptive mechanisms elucidated by studies of congenital central hypoventilation syndrome. Respir Physiol 129: 247-255

53.   Stornetta RL, Moreira TS, Takakura AC, Kang BJ, Chang DA, West GH, Brunet JF, Mulkey DK, Bayliss DA, Guyenet PG (2006) Expression of Phox2b by brainstem neurons involved in chemosensory integration in the adult rat. J Neurosci 26: 10305-10314

54.   Trochet D, Jong Hong S, Lim JK, Brunet JF, Munnich A, Kim KS, Lyonnet S, Goridis C, Amiel J (2005) Molecular consequences of PHOX2B missense, frameshift and alanine expansion mutations leading to autonomic dysfunction. Hum Mol Genet 14: 3697-3708

55.   Ungar A, Bouverot P (1980) The ventilatory responses of conscious dogs to isocapnic oxygen tests. A method of exploring the central component of respiratory drive and its dependence on O_2 and CO_2. Respir Physiol 39: 183-197

56.   Vanderlaan M, Holbrook CR, Wang M, Tuell A, Gozal D (2004) Epidemiologic survey of 196 patients with congenital central hypoventilation syndrome. Pediatr Pulmonol 37: 217-229

57.   Weese-Mayer D, Berry-Kravis E (2004) Genetics of congenital central hypoventilation syndrome. Am J Respir Crit Care Med 170: 16-21

58.   Weese-Mayer DE, Berry-Kravis EM, Marazita ML (2005) In pursuit (and discovery) of a genetic basis for congenital central hypoventilation syndrome. Respir Physiol Neurobiol 149: 73-82

59.   Weese-Mayer DE, Berry-Kravis EM, Zhou L, Maher BS, Silvestri JM, Curran ME, Marazita ML (2003) Idiopathic congenital central hypoventilation syndrome: analysis of genes pertinent to early autonomic nervous system embryologic development and identification of mutations in PHOX2b. Am J Med Genet 123: 267-278

15. Respiratory control abnormalities in necdin-null mice: implications for the pathogenesis of Prader-Willi syndrome

John J. GREER[1] and Rachel WEVRICK[2]

[1]Department of Physiology, [2]Department of Medical Genetics, University of Alberta, Edmonton, Alberta, Canada

15.1 Introduction

Prader-Willi syndrome (PWS) is a sporadic contiguous gene deletion syndrome that occurs at a frequency of approximately 1:15,000 births. Symptoms are variable and include neonatal hypotonia, hypogonadism, failure to thrive in infancy, childhood-onset hyperphagia leading to severe obesity, somatosensory deficits, behavioral problems and mild to moderate mental retardation [10; 42]. Further, PWS is associated with respiratory instability in the newborn period that is manifest as apneas and blunted chemosensitivity [1; 6; 11; 26; 28; 33; 39; 42; 47]. Here we will review data from animal studies that are beginning to provide insights into the developmental origins of the pathophysiology underlying respiratory disorders manifesting in people with PWS.

15.2 Transgenic mouse models

Individuals with PWS carry a congenital inactivation of a set of paternally expressed, imprinted genes on chromosome 15q11-q13, subsequent to microdeletion on the paternally inherited chromosome, maternal uniparental disomy, or a genetic or epigenetic imprinting mutation (reviewed in [10]). Mice with mutations targeted to individual PWS genes enable investigations of the functions of these genes independently. Indeed, mouse models with a deficiency of paternal expression of PWS orthologs in the murine 7C chromosomal region have been generated. This has included mice with gene-targeted loss of function of *Ndn* (encoding necdin), one of four known protein coding genes that are deficient in PWS [19; 25; 43]. Four necdin-null mouse strains were independently generated, with two of the strains demonstrating neonatal lethality of variable penetrance [9; 18; 30; 46]. When present, morbidity in necdin-null mice occurred during the immediate neonatal period and resulted from severe hypoventilation [9; 30]. Two strains exhibit other adult phenotypes that include hypothalamic and behavioral deficits [30; 48] and maldevelopment of the dorsal root ganglia with subsequent deficits in pain sensitivity [18].

15.3 Central Respiratory Deficit

15.3.1 Electrophysiological studies

The respiratory phenotype observed in some necdin-deficient mice could be due to functional defects of the lungs, respiratory musculature, chemoreception or central neural control mechanisms. *In vitro* preparations consisting of the brainstem-spinal cord or medullary slices isolated from $Ndn^{tm2Stw+mat/-pat}$ necdin-null [9], hereafter referred to as necdin-null) mice were used to test the hypothesis that the hypoventilation results from a defective central respiratory drive. Fig. 1A shows recordings of respiratory motor discharge from a necdin-null brainstem-spinal cord preparation. The respiratory rhythms were consistently irregular with prominent bouts of respiratory depression characterized by burst frequencies of 1-3 bursts per 10-minute period and central apneas persisting for up to several minutes. The bouts of suppressed respiratory rhythmic discharge were interspersed with periods of inspiratory motor bursts close to frequencies observed in wild-type preparations.

To further evaluate the central origins of the abnormal respiratory discharge, recordings were made from medullary slice preparations. The medullary slice preparation is a derivative of the brainstem-spinal cord preparation [41]. It contains the minimum component of neuronal populations within the ventrolateral medulla necessary for generating a respiratory rhythm, the pre-Bötzinger complex (pre-BötC, reviewed in [7]). The medullary slice also contains a significant portion of the rostral ventral respiratory group, hypoglossal (XII) nucleus and XII cranial nerve rootlets from which inspiratory motor discharge is recorded. As illustrated in Fig. 1B, the rhythmic neuronal discharge was irregular in necdin-null mice, while robust and regular in all wild-type preparations.

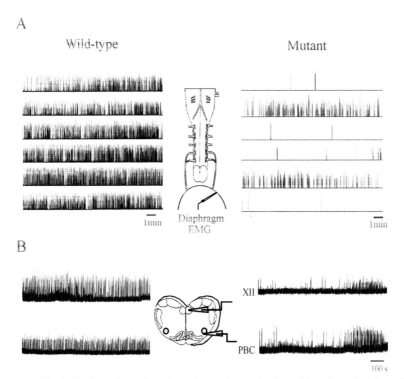

Fig. 1. Necdin-null mice have irregular respiratory rhythms with prolonged periods of central apnea. A) Sample rectified and integrated suction electrode recordings of diaphragm EMG were made from brainstem-spinal cord-diaphragm preparations isolated from E18.5 wild type (left panel) and mutant (right panel) mice. Recordings of 80-minute duration demonstrate the regularity of respiratory discharge frequency (~4-5 sec interspike interval) in wild-type preparations. In contrast, the respiratory frequency is very unstable in mutant preparations over time. B) Defects in respiratory rhythm are observed within the putative respiratory rhythm-generating center. Sample rectified and integrated suction electrode recordings were made from inspiratory neurons located in the pre-BötC (PBC) and neurons within the hypoglossal (XII) nucleus in medullary slice preparations isolated from E18.5 wild-type (left panel) and mutant (right panel) mice.

We next determined whether or not the abnormal respiratory rhythm was present within the pre-BötC *per se*. The rhythmic discharge of neurons within the pre-BötC of mutant preparations had the same abnormal characteristics as the XII motor discharge (Fig. 1B). Further, whole-cell patch clamp recordings of inspiratory neurons within the pre-BötC demonstrated that inspiratory neurons fired with an irregular rhythm with prolonged periods of suppressed rhythmogenesis (Fig. 2). The resting membrane potential of inspiratory neurons became more depolarized during epochs of increased respiratory rhythmic frequency. There were also bouts of longer duration bursting activity that is not of respiratory origin [13].

Fitting with these data, Zanella et al. [48] report recurrent apneas in independently derived [30] necdin-null mice, both *in vitro* and *in vivo*. In addition, those necdin-null mice had blunted respiratory responses to hypoxia and hyper-capnia.

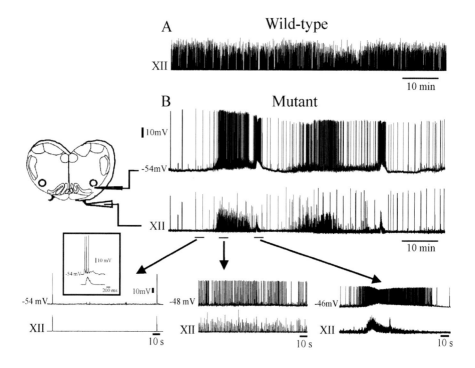

Fig. 2. Abnormal rhythmogenesis is apparent from whole-cell patch-clamp recordings from an inspiratory neuron within the pre-BötC. A) Rectified and integrated suction electrode recordings were made from the XII nerve roots of a wild-type E18.5 medullary slice preparation. B) Top panel shows whole-cell patch clamp recording from an inspiratory neuron located within the region of the pre-BötC. The rhythmic discharge fluctuates between periods of very slow rhythms (left bottom panel) to those where the respiratory rhythm is similar in frequency to wild-type preparations (middle bottom panel). There are also occurrences of high frequency non-respiratory bursts (right bottom panel). Insert shows whole-cell and integrated nerve recordings during a single inspiratory burst.

15.3.2 Anatomical studies

Subsequent work was performed to extend upon the electrophysiological study to determine if there were anatomical abnormalities within neuronal structures of the developing medulla, spinal cord and diaphragm in necdin-null mice that could account for respiratory dysfunction [34]. Anatomical defects in the medulla from necdin-null mice included defasciculation and irregular projections of axonal tracts, aberrant neuronal migration and a major defect in the cytoarchitecture of the cuneate/gracile nuclei which included dystrophic axons. Further, the presence of disarrangement of the radial glia during the development of the brain-stem may suggest a larger role for necdin in axonal patterning and orientation. Defects in axonal fasciculation and extension were also present in a subset of spinal neurons and within phrenic nerve intramuscular branches.

Contrary to the initial hypothesis, the gross structure of the pre-BötC was normal in necdin-null mice, as shown by the presence of neurokinin1 receptors (NK1R) and somatostatin (SST) immunopositive neurons within the putative region of the pre-BötC. Rather, there were clear anatomical abnormalities in surrounding medullary structures that provide conditioning synaptic input to respiratory rhythmogenic neurons. Several tyrosine hydroxylase (TH) immunopositive neurons within the medulla were swollen and irregularly distributed in ectopic areas. Further, TH immunopositive fibers in the ventral bundles and in the ventral medulla were enlarged. Abnormal morphology and orientation of fibers within the ventrolateral medulla was also observed within incoming axons labeled for Substance P (SubP) and serotonin (5HT). The abnormal morphology of neuronal fibers expressing various neurotransmitters suggests that the absence of necdin determines a defect in the formation and the morphology of axonal tracts and fibers within the medulla in different neuronal phenotypes that regulate pre-BötC function.

Biochemical and immunohistochemical studies revealed abnormal brainstem contents of noradrenaline and 5HT in neonatal $Ndn^{-/-}$ survivors with normal counts of bioaminergic neurons in the Muscatelli strain of necdin-null mice [48].

15.3.3 Pharmacological studies

Data from the Ren et al. [35] study is consistent with respiratory defect in necdin-null mice being explained by abnormal respiratory rhythmogenesis emanating from pre-BötC. Further pharmacological experiments were performed to determine if the abnormal breathing pattern in necdin-null mice were due to intrinsic defects within pre-BötC or accounted for by abnormalities of modulatory neuronal inputs that regulate respiratory rhythmogenesis. The primary conditioning excitatory drive that maintains the oscillatory state within the pre-BötC arises from activation of glutamatergic receptors [8; 12]. Further conditioning is provided by a diverse group of neuromodulators including GABA, 5HT, noradrenaline, opioids, prostaglandins, SubP and acetylcholine [14; 20; 29]. Thus, absence of necdin expression could result in the loss, or perturbation of function, of rhythmogenic neurons in the pre-BötC. Alternatively, as suggested by the anatomical data, necdin expression may be necessary for the proper functioning of neurons providing appropriate conditioning drive impinging on rhythmogenic neurons within the pre-BötC. Toward addressing this question, neurotransmitter receptor agonists were administered to determine if the abnormal respiratory neural discharge could be alleviated (i.e. to assess whether the pre-BötC functioned normally in the presence of sufficient neuromodulatory drive).

The frequency of the respiratory rhythm generated by brainstem-spinal cord and medullary slice preparations from necdin-null could be increased and periods of apnea alleviated by the administration of SubP, thyrotrophin releasing hormone (TRH), 5HT and noradrenaline. However, the fluctuations in the respiratory frequency continued in the necdin-null mice. Our interpretation is that the endogenous application of neuromodulators overcame much of the deficit resulting from the abnormalities in medullary structures that normally provide conditioning drive to the pre-BötC. A residual abnormality in the function of the pre-BötC *per se* remains, however. The functional defect could reflect changes in neuronal properties or abnormalities in the pre-BötC network connectivity due to problems with axon guidance and fasciculation.

15.4 General anatomical abnormalities within the CNS

Further extensive analyses of CNS structure in necdin-null mice revealed morphological abnormalities in several regions of the nervous system. Defects were observed in the migration or axonal outgrowth of sympathetic, retinal ganglion cell, serotonergic, and catecholaminergic neurons [21]. Necdin contains a protein-protein interaction domain called the MAGE homology domain (MHD). This MHD can interact with the intracellular domain of the p75 low affinity neurotrophin receptor to facilitate nerve growth factor (NGF)-mediated apoptosis [3; 4; 38]. During embryonic development, necdin normally increases in abundance during the differentiation of neuronal precursor cells into neurons [19; 25; 27; 43]. Ectopic expression of necdin, for example by transfection of sympathoadrenal PC-12 cells with a cDNA encoding necdin, increases differentiation and accelerates neurite extension in response to NGF in these cells. Furthermore, expression of necdin enhances signaling through the tyrosine kinase A (TrkA) neurotrophin receptor [17; 45]. Ectopic necdin expression also induces neurite outgrowth in neuroblastoma cells [16]. Thus necdin promotes terminal differentiation and neurite outgrowth in neurons during embryonic development, in part through its interactions with neurotrophin receptors [17; 18].

We recently demonstrated that necdin interacts with centrosome (microtubule organizing center) proteins and other proteins, such as Jun kinase interacting proteins, that transduce differentiation signals upon cell cycle exit [21]. This suggests that necdin facilitates cytoskeletal rearrangements that occur downstream of neurotrophin signaling. Furthermore, the defects in migration and neurite outgrowth observed in the brainstem of necdin-null mice may result from delayed or disrupted cytoskeletal rearrangement in neuronal precursor cells.

The loss of necdin action in neurotrophin signaling affects many different neuronal populations. For example, repression of necdin in embryonic dorsal root ganglion cells suppresses differentiation and affects communication of sensory information from the periphery to the spinal cord [44]. The loss of necdin in sensory neurons diminishes the association between p75NTR and TrkA, thereby diminishing NGF-induced phosphorylation of TrkA and MAP kinase, ultimately reducing TrkA signaling [17]. In necdin-null mice *in vivo*, the sensory neurons of the dorsal root ganglia form normally, but have defects in neurite outgrowth and survival because of defective NGF signaling [17]. Loss of the association between necdin and neurotrophin receptors likely accounts for the observation of increased cell death in the developing dorsal root ganglia sensory neurons and high tolerance to thermal pain in their necdin-null mice, similar to the pain insensitivity documented in PWS [17]. We have observed similar consequences of reduced neurotrophin receptor function in cortical and sympathetic neurons, implying that necdin deficiency affects diverse neuronal subtypes. In summary, necdin is an intracellular protein that associates with transmembrane receptor tyrosine kinases to promote microtubule rearrangements necessary for neuronal differentiation and neurite outgrowth.

15.5 Respiratory dysfunction in PWS

Respiratory dysfunction in PWS is a major cause of morbidity and mortality in children and adults with PWS (Table 1). Young children are susceptible to sudden, unexpected death due to aspiration or after upper airway infections, and adults are susceptible to respiratory compromise at lower body weights than non-PWS of similar body weights [31; 40]. Daytime breathing is impaired in PWS, and of particular note are abnormal ventilatory responses to hypoxia and hypercapnia [1; 2; 11; 24]. The blunted response is observed in normal weight individuals with PWS, but is exacerbated by obesity, restricted lung volume consequent to scoliosis, and thoracic muscle weakness. Abnormal ventilatory responses are proposed to be due to a primary abnormality in the peripheral chemoreceptor pathways. Sleep disturbances, sleep apneas, and sleep-disordered breathing are relatively common in PWS. Abnormal arousal thresholds during hypercapnic and hypoxic episodes are observed in PWS. The sleep apneas are determined to be of mixed central and obstructive origins. Compounding this primary respiratory dysfunction, individuals with PWS are at additional risk for obstructive sleep apnea because of hypotonia and narrow larygopharangeal space, and in many cases severe obesity. A sleep study to monitor sleep apneas is recommended in PWS to assess the extent of sleep apneas and nighttime hypoxia, and in some cases, adenoid-tonsillectomy or use of night-time Continuous Positive Airway Pressure (CPAP) improves obstructive sleep apneas. A primary hypothalamic abnormality is associated with excessive daytime sleepiness and disorganized REM sleep, as well as hypogonadism, appetite dysregulation, and thermal instability. In summary, it is likely that a primary defect in arousal mechanisms and central respiratory drive is compounded by narrowed airways, hypotonia, and obesity which contribute to obstructive apneas, and by an independent primary abnormality of circadian rhythm that causes excessive daytime sleepiness with abnormalities in the organization of REM sleep in PWS.

Table 1. Abnormalities of respiration in PWS

Clinical findings	Pathophysiology
Obstructive sleep apnea	Facial dysmorphism with narrow naso-pharynx Hypotonia, scoliosis Obesity Adenotonsillar hypertrophy can be exacerbated by initiation of GH treatment
Central Sleep apnea	Primary defect in central respiratory drive
Abnormal ventilatory responses	Primary defect in peripheral chemosensitivity
Excessive daytime sleepiness	Primary hypothalamic defect with abnormal circadian rhythm
Sleep-related hypoventilation	Abnormal REM sleep

Treatment of primary growth hormone deficiency in PWS with recombinant human growth hormone (GH) is now standard of care. Growing evidence suggests that GH treatment improves linear growth, improves muscle tone and endurance, partially normalizes facial dysmorphism, and improves the ratio of muscle to fat mass [5]. GH would therefore be predicted to decrease the incidence of obstructive apneas and indeed a small study suggests that GH treatment increases CO_2 response, ventilation and central respiratory drive in PWS, by as yet undetermined mechanisms [23].

15.6 Conclusion

It is clear that there are wide-spread defects within the developing medulla and other CNS structures in necdin null-mice. Imaging studies are ongoing to determine if a similar scope of defects occur in PWS patients which would explain the vast array of abnormalities. Necdin-deficient mice have marked breathing symptoms. The respiratory phenotype can be traced to abnormal rhythmogenic respiratory drive within the pre-BötC. The gross anatomical structure of the pre-BötC appears normal, however there are defects in neuronal structures that synapse upon and modulate pre-BötC function. Further, there may be defects in the cellular properties, axonal projections and circuitry within the pre-BötC. Those aspects have not been adequately examined. As with many of the mutant mouse models with central respiratory defects, it may be a multitude of defects that results in abnormal rhythmogenesis. In general, our understanding of the neuronal mechanisms acting within the pre-BötC and the relative impact of distributed defects in mutant models is not sufficient to attribute abnormal rhythmogenesis to any one specific neuronal deficit. However, the mutant mouse models, particularly strains that have neonatal survivors with apneas, do provide an opportunity to design strategies that might compensate for loss of necdin. This could include examining the efficacy of pharmacological interventions, such as the recently proposed ampakine administration [36] to counter respiratory depression and instabilities.

References

1. Arens R, Gozal D, Omlin K, Livingston FR, Liu J, Keens TG, Ward SL (1994) Hypoxic and hypercapnic ventilatory responses in Prader-Willi syndrome. J Appl Physiol 77: 2224-2230
2. Arens R, Gozal D, Burrell BC, Bailey SL, Bautista DB, Keens TG, Ward SL (1996) Arousal and cardiorespiratory responses to hypoxia in Prader-Willi syndrome. Am J Respir Crit Care 153: 283-287
3. Barker PA, Salehi A (2002) The MAGE proteins: Emerging roles in cell cycle progression, apoptosis, and neurogenetic disease. J Neurosci Res 67: 705-712
4. Bragason BT, Palsdottir A (2005) Interaction of PrP with NRAGE, a protein involved in neuronal apoptosis. Mol Cell Neurosci 29: 232-244
5. Carrel AL, Myers SE, Whitman BY, Allen DB (2001) Sustained benefits of growth hormone on body composition, fat utilization, physical strength and agility, and growth in Prader-Willi syndrome are dose-dependent. J Pediatr Endocrinol Metab 14: 1097-1105
6. Clift S, Dahlitz M, Parkes JD (1994) Sleep apnoea in the Prader-Willi syndrome. J Sleep Res 3: 121-126
7. Feldman JL, Del Negro CA (2006) Looking for inspiration: new perspectives on respiratory rhythm. Nat Rev Neurosci 7: 232-241
8. Funk GD, Smith JC, Feldman JL (1993) Generation and transmission of respiratory oscillations in medullary slices: role of excitatory amino acids. J Neurophysiol 70: 1497-1515
9. Gerard M, Hernandez L, Wevrick R, Stewart C (1999) Disruption of the mouse necdin gene results in early postnatal lethality: a model for neonatal distress in Prader-Willi syndrome. Nat Genet 23: 199-202
10. Goldstone AP (2004) Prader-Willi syndrome: advances in genetics, pathophysiology and treatment. Trends Endocrinol Metab 15: 12-20
11. Gozal D, Arens R, Omlin KJ, Ward SL, Keens TG (1994) Absent peripheral chemosensitivity in Prader-Willi syndrome. J Appl Physiol 77: 2231-2236
12. Greer JJ, Smith JC, Feldman JL (1991) Role of excitatory amino acids in the generation and transmission of respiratory drive in neonatal rat. J Physiol 437: 727-749
13. Greer JJ, Smith JC, Feldman JL (1992) Respiratory and locomotor patterns generated in the fetal rat brain stem-spinal cord in vitro. J Neurophysiol 67: 996-999
14. Greer JJ, Funk GD, Ballanyi K (2006) Preparing for the first breath: Prenatal maturation of respiratory neural control. J Physiol 570: 437-444
15. Hayashi M, Itoh M, Kabasawa Y, Hayashi H, Satoh J, Morimatsu Y (1992) A neuropathological study of a case of the Prader-Willi syndrome with an interstitial deletion of the proximal long arm of chromosome 15. Brain Dev 14: 58-62
16. Kobayashi M, Taniura H, Yoshikawa K (2002) Ectopic expression of necdin induces differentiation of mouse neuroblastoma cells. J Biol Chem 277: 42128-42135

17.   Kuwako K, Hosokawa A, Nishimura I, Uetsuki T, Yamada M, Nada S, Okada M, Yoshikawa K (2005) Disruption of the paternal necdin gene diminishes TrkA signaling for sensory neuron survival. J Neurosci 25: 7090-7099

18.   Kuwako K, Taniura H, Yoshikawa K (2004) Necdin-related MAGE proteins differentially interact with the E2F1 transcription factor and the p75 neurotrophin receptor. J Biol Chem 279: 1703-1712

19.   Jay P, Rougeulle C, Massacrier A, Moncla A, Mattei MG, Malzac P, Roeckel N, Taviaux S, Lefranc JL, Cau P, Berta P, Lalande M, Muscatelli F (1997) The human necdin gene, NDN, is maternally imprinted and located in the Prader-Willi syndrome chromosomal region. Nat Genet 17: 357-361

20.   Lagercrantz H (1987) Neuromodulators and respiratory control during development. Trends Neurosci 10: 368-372

21.   Lee S, Walker CL, Karten B, Kuny SL, Tennese AA, O'Neill MA, Wevrick R (2005) Essential role for the Prader-Willi syndrome protein necdin in axonal outgrowth. Hum Mol Genet 14: 627-637

22.   L'Hermine AC, Aboura A, Brisset S, Cuisset L, Castaigne V, Labrune P, Frydman R, Tachdjian G (2003) Fetal phenotype of Prader-Willi syndrome due to maternal disomy for chromosome 15. Prenat Diagn 23: 938-943

23.   Lindgren AC, Hellstrom LG, Ritzen EM, Milerad J (1999) Growth hormone treatment increases CO_2 response, ventilation and central inspiratory drive in children with Prader-Willi syndrome. Eur J Pediatr 158: 936-940

24.   Livingston FR, Arens R, Bailey SL, Keens TG, Ward SL (1995) Hypercapnic arousal responses in Prader-Willi syndrome. Chest 108: 1627-1631

25.   MacDonald HR, Wevrick R (1997) The necdin gene is deleted in Prader-Willi syndrome and is imprinted in human and mouse. Hum Mol Genet 6: 1873-1878

26.   Manni R, Politini L, Nobili L, Ferrillo F, Livieri C, Veneselli E, Biancheri R, Martinetti M, Tartara A (2001) Hypersomnia in the Prader-Willi syndrome: clinical-electrophysiological features and underlying factors. Clin Neurophysiol 112: 800-805

27.   Maruyama K, Usami M, Aizawa T, Yoshikawa K (1991) A novel brain-specific mRNA encoding nuclear protein (necdin) expressed in neurally differentiated embryonal carcinoma cells. Biochem Biophys Res Commun 178: 291-296

28.   Menendez AA (1999) Abnormal ventilatory responses in patients with Prader-Willi syndrome. Eur J Pediatr 158: 941-942

29.   Moss IR, Inman JG (1989) Neurochemicals and respiratory control during development. J App Physiol 67: 1-13

30.   Muscatelli F, Abrous DN, Massacrier A, Boccaccio I, Moal ML, Cau P, Cremer H (2000) Disruption of the mouse necdin gene results in hypothalamic and behavioral alterations reminiscent of the human Prader-Willi syndrome. Hum Mol Genet 9: 3101-3110

31.   Nagai T, Obata K, Tonoki H, Temma S, Murakami N, Katada Y, Yoshino A, Sakazume S, Takahashi E, Sakuta R, Niikawa N (2005) Cause of sudden, unexpected death of Prader-Willi syndrome patients with or without growth hormone treatment. Am J Med Genet 136: 45-48

32.   Nicholls RD, Knepper JL (2001) Genome organization, function, and imprinting in Prader-Willi and Angelman syndromes. Annu Rev Genomics Hum Genet 2: 153-175

33.   Nixon GM, Brouillette RT (2002) Sleep and breathing in Prader-Willi syndrome. Pediatr Pulmonol 34: 209-217

34. Pagliardini S, Ren J, Wevrick R, Greer JJ (2005) Developmental abnormalities of neuronal structure and function in prenatal mice lacking the Prader-Willi gene necdin. Am J Pathol 167: 175-191

35. Ren J, Lee S, Pagliardini S, Gerard C L, Stewart CL, Greer JJ, Wevrick R (2003) Absence of Ndn, encoding the Prader-Willi syndrome-deleted gene necdin, results in congenital deficiency of central respiratory drive in neonatal mice. J Neurosci 23: 1569-1573

36. Ren J, Poon BY, Tang Y, Funk GD, Greer JJ (2006) Ampakines alleviate respiratory depression in rats. Am J Resp Crit Care 174: 1384-1391

37. Salehi AH, Roux PP, Kubu CJ, Zeindler C, Bhakar A, Tannis LL, Verdi JM, Barker PA (2000) NRAGE, a novel MAGE protein, interacts with the p75 neurotrophin receptor and facilitates nerve growth factor-dependent apoptosis. Neuron 27: 279-288

38. Salehi AH, Xanthoudakis S, Barker PA (2002) NRAGE, a p75 neurotrophin receptor-interacting protein, induces caspase activation and cell death through a JNK-dependent mitochondrial pathway. J Biol Chem 277: 48043-48050.

39. Schluter B, Buschatz D, Trowitzsch E, Aksu F, Andler W (1997) Respiratory control in children with Prader-Willi syndrome. Eur J Pediatr 156: 65-68

40. Schrander-Stumpel CT, Curfs LM, Sastrowijoto P, Cassidy SB, Schrander JJ, Fryns JP (2004) Prader-Willi syndrome: causes of death in an international series of 27 cases. Am J Med Genet 124: 333-338

41. Smith JC, Ellenberger HH, Ballanyi K, Richter DW, Feldman JL (1991) Pre-Bötzinger Complex: a brainstem region that may generate respiratory rhythm in mammals. Science 254: 726-728

42. Stevenson DA, Anaya TM, Clayton-Smith J, Hall BD, Van Allen MI, Zori RT, Zackai EH, Frank G, Clericuzio CL (2004) Unexpected death and critical illness in Prader-Willi syndrome: report of ten individuals. Am J Med Genet 124: 158-164

43. Sutcliffe JS, Han M, Christian SL, Ledbetter DH (1997) Neuronally-expressed necdin gene: an imprinted candidate gene in Prader-Willi syndrome. Lancet 350: 1520-1521

44. Takazaki R, Nishimura I, Yoshikawa K (2002) Necdin is required for terminal differentiation and survival of primary dorsal root ganglion neurons. Exp Cell Res 277: 220-232

45. Tcherpakov M, Bronfman FC, Conticello SG, Vaskovsky A, Levy Z, Niinobe M, Yoshikawa K, Arenas E, Fainzilber M (2002) The p75 neurotrophin receptor interacts with multiple MAGE proteins. J Biol Chem 277: 49101-49104

46. Tsai TF, Armstrong D, Beaudet AL (1999) Necdin-deficient mice do not show lethality or the obesity and infertility of Prader-Willi syndrome. Nat Genet 22: 15-16

47. Wharton RH, Loechner KJ (1996) Genetic and clinical advances in Prader-Willi syndrome. Curr Opin Pediatr 8: 618-624

48. Zanella S, Roux JC, Muscatelli F, Hilaire G (2006) Necdin gene, respiratory disturbances and Prader-Willi Syndrome. X[th] Oxford Conference Lake Louise Canada 17-24 September: p62

16. Possible role of bioaminergic systems in the respiratory disorders of Rett syndrome

John BISSONNETTE[1] and Gerard HILAIRE[2]

[1] Oregon Health and Science University, 3181 SW Sam Jackson Park Road, Portland, Oregon, 97239, U.S.A.
[2] CNRS, Formation de Recherche en Fermeture, FRE 2722, 280 Boulevard Sainte Marguerite, 13009 Marseille, France

16.1 Introduction

Rett syndrome (RTT) is a neurodevelopmental disorder caused by mutations in the X-linked gene that encodes methyl-CpG-binding protein 2 (MeCP2) [2]. MECP2 is a DNA binding protein that is involved in gene silencing [59]. MECP2 binds to methylated CpG dinucleotides and recruits corepressors. One of these, the Sin3A complex, contains histone deacetylases that remove acetyl groups from histones resulting in a compact form of chromatin, that in turn suppresses gene expression. Over 200 mutations have been reported in RTT patients. More than half of these involve cytosine to thymidine transitions that arise de novo during spermatogenesis [23]. The specific genes that are regulated by MECP2 have not been completely determined. The clinical RTT is characterized by normal development until the onset of symptoms between 6 and 18 months of age [31]. Included in the RTT phenotype are a number of abnormal respiratory patterns. In this review we discuss these patterns that have been observed in patients and the progress that has recently been made in understanding their underlying mechanisms from studies in mouse models. These new data suggest the working hypothesis that the respiratory disorders of mouse models and probably RTT patients originate at least in part from central bioaminergic deficits, and therefore that compensating these bioaminergic deficits by pharmacological treatments might alleviate RTT respiratory symptoms.

16.2 Clinical manifestations of respiratory disorders in RTT

Neither of the two initial descriptions of RTT gave detailed accounts of the respiratory disorders. Rett [75] mentioned that "all the children (21) tended to hyperventilate, sometimes mildly but in a few cases with very forced respirations, almost synchronized with their other movements." Hagberg et al. [31] stated that "bouts of hyperpnea" were often associated with the exaggerated jerking movements of the trunk and limbs. Hyperpnea was seen in 23 of 35 patients examined before age 5 [31]. Lugaresi et al. [49] appear to have been the first to examine RTT subjects in detail. EEG, electrooculogram, chin muscle EMG combined with oral, nasal and thoracic spirograms were used together with ear oximetry. Three of the four subjects described showed a respiratory pattern characterized by clusters of irregular amplitude and rate followed by apnea. Oxygen saturation fell during the apneas. This pattern was confined to wakefulness and was not observed during an all night sleep study. The subjects were between 6 and 18 years old when first studied. In two the respiratory abnormality greatly lessened or disappeared by age 11 and 20. Glase et al. [25] reported similar findings in 11 of 11 patients studied (4 between 2 and 5 years old and 7 between 5 and 15). One subject had 4 episodes of obstructive apnea during the overnight sleep recording. In the remainder respiratory disturbances were confined to the 90-120 min recordings that were made while patients were awake. As with Lugaresi et al. [49], these authors did not observe breath-holding or Valsalva maneuvers. Southall et al. [88] added end tidal carbon dioxide measurements made with a face mask to the parameters used in earlier studies. They compared 18 RTT subjects (age 6-17 years) to 23 normal children (4-15 years). 10 RTT patients, age 6-16, showed periodic hyperventilation followed by apnea as had been previously described. Four, age 13-17, did not hyperventilate when studied but had histories consistent with hyperventilation at earlier ages. The remaining four, age 6-14, had neither a history of hyperventilation nor demonstrated it when studied. In 9 of the 10 subjects that hyperventilated end tidal CO_2 fell progressively to reach levels between 1.4 and 2.4 volumes % (control range 4.6-5.3). Arterial blood gas samples in 9 showed CO_2 tensions between 13 and 30 mm Hg (control 35-45) and pH between 7.48 and 7.60 (control 7.35-7.45). The role of hypocapnia in generating the post hyperventilation apnea has subsequently been strengthened by a case report in which 5% CO_2 reduced their incidence [85]. In contrast to the two earlier reports Southall et al. [88] described the post-hyperventilation apneas as Valsalva maneuvers in which the inspiratory position of the respitrace was held and full expiration occurred only at the termination of the apnea. This pattern has also been reported in 26 of 47 RTT subjects [40]. Southall et al. [88] expanded the patterns of abnormal breathing in RTT by noting the high incidence of periodic breathing, defined as at least two cycles of a group of 1-19 breaths followed by a pause of ≥ 4.0 sec. This pattern was seen in 17 of the 18 subjects studied and occupied a median duration of 11.3 min/hr. Interestingly, periodic breathing was also present in 15 of 23 normal children. It occupied, however, only 0.13 min/hr. Periodic breathing has not been described in all subsequent studies. A report involving 30 RTT patients and an equal number of controls failed to demonstrate a greater duration of periodic breathing in RTT subjects [50], while it was seen in 12 of 47 patients studied by Julu et al. [40].

As noted above there is the suggestion that abnormal breathing in RTT declines with advancing age. This has been confirmed in a largely crossectional study involving 47 patients [40]. Normal breathing patterns were seen in 38% of subjects over age 18, but only 21% of those under 9 years. In addition the incidence of hyperventilation/apnea fell with advancing age while that of the Valsalva maneuver rose. Recently, Weese-Mayer et al. [102] studied 47 girls aged 2-7 years with *MECP2* mutation-confirmed RTT. Recording respiratory activity and ECG during daytime wakefulness confirmed irregular breathing, with increased breathing frequency, airflow, and heart rate in RTT. The authors concluded that RTT girls have cardiorespiratory dysregulation during breathholds as well as during "normal" breaths and during breaths preceding and following breathholds. In RTT patients, Stettner et al. [89] noted alternated episodes of hyperventilation and apneas, the latter being frequent (10 to 60 per hour) and long-lasting (from 24 to 59 s, and occasionally up to 89 s), with oxygen saturation drops. These authors pointed to a post-inspiratory nature of these apneas.

16.3 Mouse models of RTT

16.3.1 Spatial and temporal expression of MeCP2

The tissue and developmental timetable of MECP2 expression may suggest potential hypotheses for its role in respiratory control. In human and mouse protein levels are highest in brain, lung and spleen [44; 82]. Interestingly protein levels did not correlate with RNA expression indicating that the transcripts are under tissue specific post-transcriptional control [82]. Double immunofluorescent labeling with neuronal, glial and microglial markers showed that MeCP2 is expressed exclusively in neurons [55; 82].

MeCP2 protein has been measured from E10.5 days to P40 weeks in mouse brain. It first appears at E11.5 when it is detected in pons, medulla and spinal cord but not in higher centres [82]. It then appears in other brain regions in a pattern that correlates with synapse formation [55; 82]. Developmental expression in the medulla is of special interest in that high levels are seen at E19 (the earliest time point in this study) and P2. These, then, decline before returning at P10 weeks to levels that are similar to those at P40 weeks [8]. Thus MeCP2 may play a role in establishing respiratory neuronal circuitry during the perinatal period.

16.3.2 Mecp2-deficiency in mice: models of RTT

The Cre-*loxP* recombination system has been used to generate two mouse lines with Mecp2 deficiency. Guy et al. [30] deleted exons 3 and 4, while Chen et al. [14] deleted exon 3. The resultant phenotypes appear to be similar. Null males ($^{-/y}$) and females ($^{-/-}$) were normal to 3-8 weeks when a stiff uncoordinated gait, reduced body movements, hindlimb clasping, irregular or hard respiration, body trembling, pila erection and uneven wearing of the teeth developed. Few survived beyond 70-80 days of age. Body weight depended on genetic background. Mutants on a C57BL/6 background were smaller than wild type (WT) while those on 129 [30] or BALB/c backgrounds [14] were heavier. Heterozygous females ($^{+/-}$) were normal to 3 to 4 months when some began to show inertia, ataxia and clasping. By 10 months 70% of $^{+/-}$ females were symptomatic including breathing irregularities [30]. The observation that the onset of symptoms in mice corresponds in real time not in relative developmental time to that in the human was unanticipated. As pointed out by the authors this result argues against a developmental role for Mecp2 and is consistent with the hypothesis that the protein is necessary to maintain neuronal function.

Both groups extended their observations by using Cre-recombinase under control of the nestin transgene to generate mice with Mecp2 deficiency confined to neurons. These animals had phenotypes similar to those with ubiquitous deletion. In addition Chen et al. [14] used a Cre transgene controlled by the CamKII promoter. In these mice deletion was largely confined to the forebrain, hippocampus and brainstem. Importantly deletion started at the perinatal stage and was maximal at P21. These mice had an onset of symptoms (weight gain, ataxic gait and reduced activity) at three months, that is later than the onset at one month in $^{-/y}$ males. The result adds additional support to the hypothesis that Mecp2 is essential for ongoing neuronal function but not early development.

A third model of RTT in mice was generated by homologous recombination in which a stop codon was inserted after codon 308 [83]. This engineered truncation of Mecp2 retained the methyl binding and transcriptional regulating domains of the protein. Males were normal until 6 weeks when a subtle tremor developed. The tremor was visible by 4 months and 40% of the mice had kyphosis at 5 months. 90% were alive at one year of age. Body weight was not affected. Interestingly the Mecp2308/^{y} mice showed rapid and repetitive movements of their forelimbs when suspended by the tail, reminiscent of the hand-wringing seen in RTT subjects. There is no mention of respiratory pattern in these mice whose truncated protein resembles mutations that have been described in humans.

Recently Pelka et al. [69] introduced a nonfunctioning splicing site at the 5' end of the transcriptional repression domain of Mecp2. This prevented expression of sequences downstream. In males brain tissue did not show any Mecp2 transcripts and antibodies to both the N and C terminals of the protein failed to detect it with western blots. On a pure 129 background 50% of the males had died at 5 weeks while those on a mixed 129/C57BL/6 background survived to 8 weeks. These males developed symptoms similar to earlier described null males [14; 30]. The phenotype included labored breathing.

16.4 Respiratory studies in mouse models of RTT

To date detailed studies of respiratory patterns and mechanisms underlying abnormalities have been confined to mice with deletion of the 3rd and 4th exons of Mecp2 [30]. Different approaches have been used with *in vivo* recordings of breathing in restrained awake mice [4; 5], unrestrained animals [96], *in vitro* recordings of respiratory-like activity produced in brainstem slices [96] and *in situ* recordings of phrenic and vagal nerves in perfused heart-brainstem preparations [89].

16.4.1 In vivo studies of respiratory pattern of Mecp2$^{-/y}$ mice

Mecp2 null males (*Mecp2$^{-/y}$*) were examined in a whole body plethysmograph from P4-5 until death at between P32 and P60 [96]. Recordings were made while animals were quiet without limb, body or head movements. At ~ 6 weeks of age respiratory rate tended to be slower (175 ± 30 breaths/min) than WT (195 ± 16). Preliminary studies have found a significant difference in respiratory frequency at P21 [5]. Viemari et al. [96] showed that respiratory patterns in *Mecp2$^{-/y}$* did not differ from WT littermates up to P28. Thereafter *Mecp2$^{-/y}$* showed a pattern that alternated between rapid and slow frequency. Both cycle period and tidal volume histograms showed wide distributions compared to WT. At 15 days prior to death (~ 6 weeks) *Mecp2$^{-/y}$* mice developed apneas (Fig. 1A) which were at times preceded by a large amplitude inspiration. These apneas (> 1.0 sec) occurred a median of 6 per 15 min (range 3-25). By 7 days before death apneas had increased to 10 per 15 min (range 5-75) and respiratory rate had declined to 93 ± 23 (WT = 200 ± 14 breaths per min). While behavioral state could not be defined in these experiments it is interesting that the respiratory disturbances occurred while the animals may have been asleep or awake. These authors made the interesting observation that after light anesthesia (30 mg/kg i.p. pentobarbitone) apneas disappeared and both respiratory cycle and tidal volume variability were reduced to levels comparable to WT.

Heterozygous females (*Mecp2$^{+/-}$*) have been studied using body plethysmography [4]. In this method the unanesthetized animals are restrained by a close fitting hole in Parafilm®, through which their head emerges. These studies concentrated on the ventilatory response to hypoxia. At 5 months of age *Mecp2$^{+/-}$* mice had slower respiratory rates (182 ± 17 breaths/min) than WT (227 ± 17). Tidal volume was larger in *Mecp2$^{+/-}$* so that there was no difference in minute ventilation (VE). In the first min of exposure to 8% oxygen VE in *Mecp2$^{+/-}$* exceeded that in WT. This augmented hypoxic ventilatory response was also seen in null males. This finding has been confirmed by Hilaire et al. [unpublished data] who compared the ventilatory response to hypoxia of WT and Mecp2 null males at one month of age. Using constant flow plethysmography in unrestrained, unanesthetized mice, preliminary results have been obtained revealing that the mean VE increase in response to 10 % O_2 for 5 min was significantly larger in *Mecp2$^{-/y}$* than WT. The exaggerated response of the *Mecp2$^{-/y}$* was due to stronger responses in

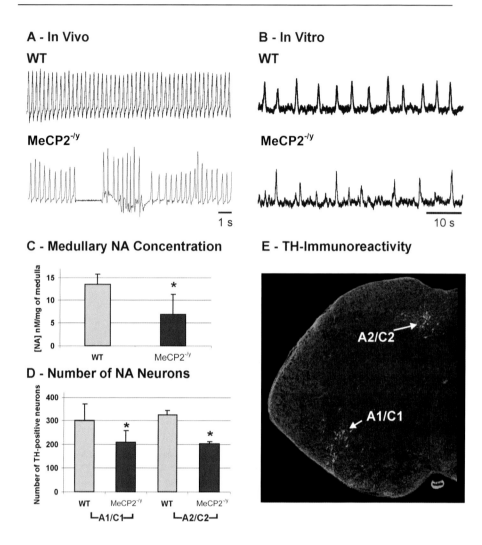

Fig. 1. Respiratory and noradrenergic deficits of *Mecp2⁻/y* mice. *A – In vivo*: Plethysmography in WT and *Mecp2⁻/y* adult mice (upper and lower traces, respectively) reveals long lasting apneas in mutants. *B – In vitro*: Recording of the multi-unitary activity (integrate) produced by neurons of ventral respiratory group in transverse brainstem slices from WT and *Mecp2⁻/y* mice at P14 (upper and lower traces, respectively) reveals that the duration and amplitude of individual respiratory-like cycles are highly variable in mutants; stabilization is observed after adding NA to the superfusate (not shown). *C – Medullary NA Concentration*: bars show median (and interquartile range) concentrations of NA measured in the medulla of WT and *Mecp2⁻/y* adult mice (age about 2 months) by HPLC. Note the decreased NA concentration in *Mecp2⁻/y* mice; *, significant statistical difference at $p < 0.05$. *D – Number of NA Neurons*: bars show mean (and SEM) numbers of NA neurons in the ventral A1/C1 and dorsal A2/C2 groups in the medulla of WT and *Mecp2⁻/y* adult mice (age about 2 months) as revealed by TH-immunoreactivity. Note the decreased number of NA neurons in A1/C1 and A2/C2 areas of *Mecp2⁻/y* mice; *, significant statistical difference at $p < 0.05$. *E – TH-immunoreactivity*: location of ventral A1/C1 and dorsal A2/C2 neurons in slice from the mouse medulla.

both frequency and VT. No data are available about the post-hypoxia decline. This augmented hypoxic ventilatory response is not part of a generalized enhanced respiratory phenotype as carbon dioxide sensitivity was the same in $Mecp2$ heterozygous females as that in WT. By the 5th min of hypoxia VE in $Mecp2^{+/-}$ females was not different from WT [4]. In the initial 30 sec of recovery in 100% oxygen (to remove the influence of peripheral chemoreceptors) the decline in VE for $Mecp2^{+/-}$ animals was greater than that in WT. This respiratory depression was characterized by apneas. When post-hypoxia minute ventilation was examined as a function of tidal volume it demonstrated that larger tidal volumes were associated with greater respiratory depression. This is consistent with hypocapnia being the primary cause of this depression. The hypocapnia suggestion was strengthened by studies in which 4% carbon dioxide was added to the 8% oxygen exposure. In these experiments VE remained at control levels on return to 100% oxygen for both $Mecp2^{+/-}$ and WT. In addition there was no longer a difference between $Mecp2$ deficient females and normal mice. Direct measurement of static lung volumes and anatomical dead space in these animals showed that $Mecp2^{+/-}$ had larger lung volumes but similar dead space volumes to WT. This structural difference would favor carbon dioxide elimination during hyperventilation in $Mecp2^{+/-}$ mice.

Bissonnette and Knopp extended their study of the hypoxic ventilatory response by examining animals with selective deficiency of Mecp2 in neurons [30]. With the protein present in the lung, tidal volumes were not larger than the appropriate WT (one floxed allele but without the nestinCre transgene). As with $Mecp2^{+/-}$ the initial response to hypoxia was greater in females with selective deficiency. In contrast their post-hypoxia decline in VE did not exceed that of WT. This demonstration of distinct respiratory phenotypes depending on the tissues in which Mecp2 is deficient has implications beyond respiratory control. As summarized above when phenotype was defined by body weight and motor behavior mice with conditional deletion confined to neurons were the same as those with ubiquitous deficiency.

16.4.2 In vitro and in situ studies of Mecp2$^{-/y}$ mice

Although breathing of $Mecp2^{-/y}$ has been reported to be normal *in vivo* from birth until P28, *in vitro* experiments performed in transverse brainstem slices from WT and $Mecp2^{-/y}$ mice revealed slight alterations of respiratory rhythmogenic mechanisms at the age P14-P21 [96]. These experiments were performed in slices containing neurons from the ventral respiratory group (VRG) that continued to generate a basic respiratory-like rhythm although totally deafferented. The VRG respiratory-like bursts were similar in shape, duration, and mean frequency in WT and $Mecp2^{-/y}$ but the variability of individual cycle duration was significantly larger in $Mecp2^{-/y}$ (Fig. 1B). Thus the respiratory network of $Mecp2^{-/y}$ produces an irregular respiratory command at P14-P21, well before the first breathing disturbances are detectable *in vivo*. Adding noradrenaline (NA) to the aCSF superfusing the P14-P21 slices significantly increased the VRG respiratory-like rhythm in both WT and $Mecp2^{-/y}$ preparations but stabilized the cycle duration of $Mecp2^{-/y}$

preparations, an observation suggestive of a role of NA in respiratory deficits of $Mecp2^{-/y}$.

An intra-arterially perfused working heart-brainstem preparation has been developed in mice that allows studies of respiratory neurons in an *in vitro* milieu [63]. In these *in situ* preparations from WT and $Mecp2^{-/y}$ aged P40, recordings of eupneic-like activity in both phrenic and vagal motor nerves revealed frequent apneas in $Mecp2^{-/y}$ that were accompanied by a tonic post-inspiratory activity of vagal motorneurons [89]. These results fully confirm those obtained *in vivo*, i.e. breathing of adult $Mecp2^{-/y}$ is frequently interrupted by drastic apneas [96]. In addition, they reveal tonic post-inspiratory discharges occurring during apneas that may be reminiscent of breath-holding of RTT.

16.5 Bioaminergic systems, RTT and *Mecp2^{/y}* mice

Bioaminergic systems are known to modulate numerous functions and it has been suggested that bioaminergic dysfunctions might contribute to some of the different symptoms described in RTT [61; 70]. Indeed results from postmortem studies argued for bioaminergic deficits in RTT, a conclusion that was confirmed by some biochemical analyses performed in live RTT patients and by biochemical and immunohistochemical studies performed in $Mecp2^{-/y}$ mice.

16.5.1 Bioaminergic systems of RTT patients

Postmortem brain analyses have been performed to evaluate pathogenetic aspects of the RTT. Bioaminergic alterations were consistently reported in brains of autopsied RTT patients with reduction of serotonin (5HT) and its metabolite 5-hydroxylindol acetic acid (5HIAA) levels [77] and also reduction of dopamine (DA), NA, 5HT and increased 5HIAA/5HT ratio [9]. In four autopsied RTT patients, taking into account their age, concentrations of DA, NA, 5HT and 5HIAA were found significantly reduced in two adults but nearly normal in two young patients [45]. No significant differences were found in 5HT cell number between autopsied RTT patients and controls but binding studies suggested differences of the serotonin transporter in the dorsal motor nucleus of the vagus [62].

In live RTT patients, biochemical studies were performed using high pressure liquid chromatography (HPLC) to measure the concentrations of bioamines and metabolites in cerebrospinal fluid (CSF). However comparisons between RTT patients and control subjects gave conflicting results. The first studies were performed in small samples of RTT patients and concluded that metabolites for DA and NA were significantly reduced but not the 5HT metabolite 5HIAA (n = 6, [105]) and in separate studies that the DA, NA and 5HT concentrations were found in the normal range (n = 5, [71]). In larger numbers of patients, however, DA, NA and 5HT concentrations were found either significantly reduced (n = 32, [105]) or normal (n = 38 patients, [46]). Low values of 5HIAA levels were reported in the CSF of four RTT patients with mutations of the *MECP2* gene [74]).

Discrepancies between recent and older results might be concerned with clinical classification of RTT patients prior to the recently developed *MECP2* genotyping. In addition, variability exists within RTT since in a sample of 199 RTT children that were all severely functionally dependent, some variations in ability exists, even in children with identified *MECP2* mutations [18].

Several pharmacological treatments targeting the bioaminergic systems have been attempted in RTT patients. Treatments with bioamine precursors, tyrosine and tryptophane for NA and 5HT, respectively, significantly increased CSF metabolite concentrations by 30-40 % but did not result in clinical improvement in RTT girls [60]. However in 28 RTT girls aged 1-14 years, heart rate variability and plasma 5HT levels were significantly correlated, suggesting that treatment with a 5HT analogue could be useful in improving the sympathovagal balance [28]. In addition, pharmacological treatment of one RTT patient with buspirone, a well-known 5HT1A receptor agonist with anxiolytic, non-sedative properties, has been found efficient in treating respiratory dysfunction [3].

16.5.2 Bioaminergic systems of Mecp2$^{-/y}$ mice

Two different research groups have simultaneously conducted biochemical studies with HPLC to determine whether *Mecp2*-deficiency in mice altered bioaminergic systems. They both compared the bioamine contents in the CNS of WT and *Mecp2*$^{-/y}$ males at different developmental ages, measuring the bioamine concentrations within either the whole brain [39] or three dissected regions of the CNS, the forebrain, pons and medulla [96]. Both studies revealed that *Mecp2* deficiency altered NA and 5HT systems. Ide et al. [39] reported that NA and 5HT levels were not statistically different in WT and *Mecp2*$^{-/y}$ brains from birth up to postnatal day 14. However a significant difference in NA concentration was found at P28, with lower mean values in *Mecp2*$^{-/y}$ than WT, while 5HT values were not altered. By P42, the difference in NA level was still present and a significant difference in 5HT level appeared, with again lower mean values in *Mecp2*$^{-/y}$ than WT. Viemari et al. [96] reported that bioaminergic concentrations were normal in the forebrain and pons of *Mecp2*$^{-/y}$ at all ages but observed significant differences in the medulla. At one month of age, the mean NA concentration in the medulla was 40 % lower in *Mecp2*$^{-/y}$ than WT while the mean 5HT concentration was similar in both genotypes. The decreased NA concentration was observed in *Mecp2*$^{-/y}$ who presented breathing alterations as well as in *Mecp2*$^{-/y}$ apparently retaining normal breathing. At two months of age, when all *Mecp2*$^{-/y}$ had severe breathing deficits, the NA concentration in the medulla was ~50 % depressed in *Mecp2*$^{-/y}$ (Fig. 1C) and the 5HT concentration was also significantly lower (40 %) in *Mecp2*$^{-/y}$ than WT. As the medullary level of the 5HT metabolite 5HIAA was not depressed, the 5HIAA/5HT ratio was significantly increased in *Mecp2*$^{-/y}$ mice, suggesting an altered 5HT metabolism. These studies suggested that *Mecp2* deficiency does not impair the early development of bioaminergic neurons but causes succeeding impairment of those neuronal systems, probably from postnatal day 14 until drastic alterations of the NA system developed at one month of age, followed by alteration of the 5HT system at two months of age.

To further analyze the bioaminergic alterations resulting from *Mecp2* deficiency in mice, immunohistochemical studies were conducted in mice to label the medullary neurons expressing either 5HT or the NA synthesis enzyme tyrosine hydroxylase, TH, and therefore to allow quantification in WT and *Mecp2*$^{-/y}$ mice [96]. In the medulla of WT and *Mecp2*$^{-/y}$ neonates, the number of TH neurons was similar at birth. At one month of age, however, the number of TH neurons was significantly lower in the dorsal A2/C2 area of *Mecp2*$^{-/y}$ when compared to WT and it became significantly lower in both dorsal A2/C2 and ventral A1/C1 areas at two months of age (Fig. 1D-E). Interestingly, the number of 5HT neurons was not altered in *Mecp2*$^{-/y}$ mice, even at 2 months of age, a result fully consistent with the 5HT neuron number in autopsied RTT patients [62]. It is worth noticing that repeated injections of the 5HT re-uptake blocker fluoxetine in adult rat induced expression of Mecp2 protein [12]. Thus it is highly likely that the decreased NA content in the *Mecp2*$^{-/y}$ medulla resulted from a loss of TH-expressing neurons. Whether these neurons were definitively lost or whether they are still present but have lost (transiently or definitively) the ability to express TH and to synthesize NA remains a crucial open question. The answer could be obtained by analyzing other markers of TH-expressing neurons.

16.6 Possible link between bioaminergic and respiratory alterations in mice and RTT patients

In rodents, the respiratory rhythm generator (RRG) is a complex, diffuse neural network in the ventrolateral medulla, with pacemaker neurons in the Pre-Bötzinger Complex area that possibly play a primary role [34; 35; 86]. The *Mecp2* respiratory deficits are unlikely to result from impairment of the rhythmogenic mechanisms, as shown in *MafB*$^{-/-}$ fetal mice where respiratory rhythmicity is drastically impaired, both *in vivo* and *in vitro*, without any bioaminergic alterations [6]. In *Mecp2*$^{-/y}$, the RRG is able to generate a normal activity under some circumstances: *in vivo* during the first days of life and even during adulthood under slight anesthesia, *in vitro* under NA facilitation. As compelling evidence exist that endogenous bioamines, both 5HT and NA, affect the RRG maturation and function during the perinatal period, the hypothesis has been put forward that respiratory deficits of *Mecp2*$^{-/y}$, and probably RTT patients, originate at least in part from alterations of bioaminergic regulations impinging onto the RRG [96]. The NA and the 5HT phenotypes develop in early gestation, under the control of several key genes (for 5HT, see [16; 64; 81]; for NA see [26; 37; 48; 66; 67; 72; 73; 92]) and share at least one common transcription factor [64]. They both play crucial roles in CNS development [24; 91] and modulate numerous functions after birth such as locomotion, sleep, food intake, etc., and respiration.

16.6.1 Bioaminergic systems and central respiratory disorders

Pontine and medullary NA neurons contribute to RRG maturation and function and their alterations disrupt neonatal respiration [21; 22; 33; 36; 95; 97-99; 103]. Thus, alterations of NA system may contribute to respiratory disturbances in $Mecp2^{-/y}$ mice and probably RTT patients, especially alteration of the medullary ventral A1/C1 and dorsal A2/C2 neurons that have been shown to facilitate and stabilize the RRG, respectively [103]. In $Mecp2^{-/y}$ mice, the decreased NA concentration at P14 with a significant loss of A2/C2 neurons at one month of age may contribute to the appearance of a variable rhythm both *in vitro* and *in vivo*. The later loss of A1/C1 neurons between one and two months of age may be responsible for the drastic rhythm reduction.

Endogenous 5HT also exerts a pivotal influence on the RRG maturation and function. In fetuses and neonates, activation and blockade of medullary 5HT1A receptors facilitates and depresses, respectively, the RRG [20; 52; 53]. Although still under debate, 5HT may also contribute to respiratory regulation by CO_2 and pH [54; 58; 76]. Raphe neurons fire with a respiratory modulation [47], and respond to CO_2 [94; 100; 101]. Inhibition or lesions of raphe decreases the respiratory responses to CO_2 [58; 87; 90] while raphe acidification increases ventilation [57]. Furthermore, 5HT may also participate to respiratory responses to hypoxia, probably via 5HT2A receptors [93]. Excess of 5HT in monoamine oxidase A-deficient mice [11; 43] alters RRG maturation, rhythm stability and regulation by 5HT, hypoxia and lung afferents [7; 10]. The late reduction of 5HT content in the $Mecp2^{-/y}$ medulla may therefore worsen their respiratory deficits, either directly by affecting the 5HT modulation of the RRG or indirectly via A2/C2 neurons since raphe 5HT neurons send synaptic projections to A2/C2 neurons [68]. Indeed NA and 5HT neurons constitute two complex, interconnected networks [42; 68] since both affect RRG maturation and function, one cannot exclude the possibility that alteration of one network may affect the other and that unbalanced NA and 5HT changes may have drastic consequences.

16.6.2 BDNF, bioamines and central respiratory disorders

Recently, it has been suggested that BDNF plays a role in RTT since Mecp2 binds selectively to BDNF promoter and functions to repress expression of the *BDNF* gene [15]. The possibility of a functional interaction between *Mecp2* and *BDNF* is supported by the observation that the progression of the disease in $Mecp2^{-/y}$ is affected by the level of BDNF expression [13]. Complex interactions exist between BDNF and bioaminergic systems [51]. BDNF is crucial for survival and plasticity of NA neurons [1; 19], for the development of A5 neurons [29], and for functional changes in presynaptic axon terminals of A6 neurons in the aging brain [56]. In addition, BDNF mRNA is expressed in A1/C1 and A2/C2 neurons [17]. Null mutants for TrkB, the tyrosine kinase receptor for BDNF, display a loss of NA neurons whereas TrkB ligands increase their number [38]. BDNF is also crucial for 5HT neurons since BDNF upregulates 5HT neuronal phenotype [79; 80] and increases 5HT synthesis and/or turnover in the mature rat forebrain [84].

Thus, BNDF, NA and 5HT, via complex interactions, may contribute to the development of respiratory disturbances that accompanied MeCP2 deficiency and RTT.

16.6.3 Bioamines and peripheral respiratory disorders

Although the above reviewed data support a causal link between *Mecp2*-deficiency, alterations of bioaminergic systems and central alterations of respiratory rhythmogenesis, alterations of peripheral respiratory regulations cannot be excluded and might also participate in RTT respiratory disturbances. As briefly reviewed [35], two main processes regulate breathing, namely the chemoreflex and the Hering Breuer reflex, that are sensitive to bioamines. Both chemoreflex and Hering Breuer reflex are depressed in MAOA-deficient mice by 5HT excess [10] and the central respiratory response to hypoxia is almost abolished in *Phox2a*- and *Ret*-deficient neonates by NA deficits [97; 98].

The chemoreflex is a facilitatory loop that implicates three steps, first detection of peripheral hypoxia by carotid body chemoreceptors, second integration of the carotid body inputs in the dorsal medulla and third activation of the respiratory network to compensate the peripheral hypoxia and restore normoxia. In the absence of peripheral stimulation central respiratory neurons are depressed, resulting in a slowing of the respiratory rate. One cannot exclude the possibility that the bioaminergic alterations resulting from Mecp2 deficiency affect the sensitivity of the peripheral chemoceptors, the integrative processes of the hypoxic message in the dorsal medulla (where there was about a 30% decrease in TH positive A2/C2 neurons) and/or the responsiveness of the respiratory network to hypoxic challenge. In $Mecp2^{-/y}$ and $Mecp2^{+/-}$ mice, the responses to hypoxia is augmented which may contribute to post-hyperventilation hypocapnia and apnea (G. Hilaire, unpublished results, [4]). In addition in RTT children, imposing transient episodes of hypercapnic hyperoxemia (80% O_2 and 20% CO_2) improved clinical symptoms [70; 85].

The Hering-Breuer reflex is an inhibitory loop with three steps, first detection of lung inflation by pulmonary stretch receptors, second integration of their pulmonary vagal inputs in the dorsal medulla and third inhibition of the respiratory drive to prevent over-inflation of the lungs. Here again, one cannot exclude alterations of this inhibitory loop by the bioaminergic deficits of RTT and *Mecp2* mice. In $Mecp2^{-/y}$ mice, long-lasting apneas are occasionally preceded by a large amplitude inspiration [96]. In $Mecp2^{+/-}$ mice, tidal volume and lung volume are larger than in WT [4]. In RTT patients, forced inspirations with exaggerated movements of the trunk occur [32; 75] with apneas [89].

16.7 Conclusions: bioamines, respiration and RTT

Whether of central and/or peripheral origin, respiratory disturbances in RTT and *Mecp2* mice likely originate from bioaminergic alterations that result from *Mecp2* deficiency. These respiratory disturbances may play a crucial role in shortening life span and may also facilitate and/or induce neurobehavioral deficits during adulthood. In maturing rats, exposure to intermittent hypoxia induces, during adulthood, spatial learning deficits, altered working memory, decreased dendritic branching in the frontal cortex, cortical and hippocampal vulnerability, increased NA concentrations and DA turnover [27; 41; 78]. In RTT patients, respiratory disorders with intermittent apneas frequently occur during which oxygen saturation drastically falls [49; 89]). This may contribute to the neurobehavioral deficits since it has been reported that restoring a normal respiratory pattern in RTT patients (and oxygenation) by means of a respirator induces a near normalization of the EEG [70]. Therefore, it may be that intermittent hypoxia during RTT-disordered breathing contributes, at least in part, to some of the non-respiratory RTT phenotype.

To conclude, finding efficient pharmacological treatments to alleviate bioaminergic deficits in *Mecp2* deficient mice, and in turn to alleviate their respiratory disturbances, is a challenge for future investigations aimed at ameliorating RTT symptoms.

Acknowledgements

The support of the Rett Syndrome Research Foundation, the French Ministry of Research (ACI NIC0054) and the CNRS to G. Hilaire's team is acknowledged. Work in the J. Bissonnette's laboratory has been supported by the International Rett Syndrome Association; HD 044453 from the NIH and FY06-314 from the March of Dimes Birth Defects Foundation. The authors gratefully thank Sébastien Zanella for his constructive comments on the review.

References

1. Akbarian S, Rios M, Liu RJ, Gold SJ, Fong HF, Zeiler S, Coppola V, Tessarollo L, Jones KR, Nestler EJ, Aghajanian GK, Jaenisch R (2002) Brain-derived neurotrophic factor is essential for opiate-induced plasticity of noradrenergic neurons. J Neurosci 22: 4153-4162

2. Amir RE, Van den Veyver IB, Wan M, Tran CQ, Francke U, Zoghbi HY (1999) Rett syndrome is caused by mutations in X-linked MECP2, encoding methyl-CpG-binding protein 2. Nat Genet 23: 185-188

3. Andaku DK, Mercadante MT, Schwartzman JS (2005) Buspirone in Rett syndrome respiratory dysfunction. Brain Dev 2: 437-438

4. Bissonnette JM, Knopp SJ (2006) Separate respiratory phenotypes in methyl-CpG-binding protein 2 (Mecp2) deficient mice. Pediatr Res 59: 513-518

5. Bissonnette J, Knopp S, McKinney B, Blue M (2005) Separate roles for central and pulmonary methyl-CpG-binding protein 2 (Mecp2) in establishing respiratory pattern (2005) FASEB J 19: 372

6.   Blanchi B, Kelly LM, Viemari JC, Lafon I, Burnet H, Bevengut M, Tillmanns S, Daniel L, Graf T, Hilaire G, Sieweke MH (2003) MafB deficiency causes defective respiratory rhythmogenesis and fatal central apnea at birth. Nat Neurosci 6: 1091-1100

7.   Bou-Flores C, Lajard AM, Monteau R, De Maeyer E, Seif I, Lanoir J, Hilaire G (2000) Abnormal phrenic motoneuron activity and morphology in neonatal monoamine oxidase A-deficient transgenic mice: possible role of a serotonin excess. J Neurosci 20: 4646-4656

8.   Braunschweig D, Simcox T, Samaco RC, LaSalle JM (2004) X-Chromosome inactivation ratios affect wild-type *Mecp2* expression within mosaic Rett syndrome and *Mecp2^-/+* mouse brain. Hum Mol Genet 13: 1275-1286

9.   Brucke T, Sofic E, Killian W, Rett A, Riederer P (1987) Reduced concentrations and increased metabolism of biogenic amines in a single case of Rett-syndrome: a postmortem brain study. J Neural Transm 68: 315-324

10.   Burnet H, Bevengut M, Chakri F, Bou-Flores C, Coulon P, Gaytan S, Pasaro R, Hilaire G (2001) Altered respiratory activity and respiratory regulations in adult monoamine oxidase A-deficient mice. J Neurosci 21: 5212-5221

11.   Cases O, Seif I, Grimsby J, Gaspar P, Chen K, Pournin S, Muller U, Aguet M, Babinet C, Shih JC (1995) Aggressive behavior and altered amounts of brain serotonin and norepinephrine in mice lacking MAOA. Science 268: 1763-1766

12.   Cassel S, Carouge D, Gensburger C, Anglard P, Burgun C, Dietrich JB, Aunis D, Zwiller J (2006) Fluoxetine and cocaine induce the epigenetic factors *Mecp2* and MBD1 in adult rat brain. Mol Pharmacol 70: 487-492

13.   Chang Q, Khare G, Dani V, Nelson S, Jaenisch R (2006) The disease progression of Mecp2 mutant mice is affected by the level of BDNF expression. Neuron 49: 341-348

14.   Chen RZ, Akbarian S, Tudor M, Jaenisch R (2001) Deficiency of methyl-CpG binding protein-2 in CNS neurons results in a Rett-like phenotype in mice. Nat Genet 27: 327-331

15.   Chen WG, Chang Q, Lin Y, Meissner A, West AE, Griffith EC, Jaenisch R, Greenberg ME (2003) Derepression of BDNF transcription involves calcium-dependent phosphorylation of Mecp2. Science 302: 885-889

16.   Cheng L, Chen CL, Luo P, Tan M, Qiu M, Johnson R, Ma Q (2003) Lmx1b, Pet-1, and Nkx2.2 coordinately specify serotonergic neurotransmitter phenotype. J Neurosci 23: 9961-9967

17.   Cho HJ, Yoon KT, Kim HS, Lee SJ, Kim JK, Kim DS, Lee WJ (1999) Expression of brain-derived neurotrophic factor in catecholaminergic neurons of the rat lower brainstem after colchicine treatment or hemorrhage. Neuroscience 92: 901-909

18.   Colvin L, Fyfe S, Leonard S, Schiavello T, Ellaway C, De Klerk N, Christodoulou J, Msall M, Leonard H (2003) Describing the phenotype in Rett syndrome using a population database. Arch Dis Child 88: 38-43

19.   Copray JC, Bastiaansen M, Gibbons H, van Roon WM, Comer AM, Lipski J (1999) Neurotrophic requirements of rat embryonic catecholaminergic neurons from the rostral ventrolateral medulla. Brain Res Dev Brain Res 116: 217-222

20.   Di Pasquale E, Monteau R, Hilaire G (1994) Endogenous serotonin modulates the fetal respiratory rhythm: an in vitro study in the rat. Brain Res Dev Brain Res 80: 222-232

21.   Errchidi S, Monteau R, Hilaire G (1991) Noradrenergic modulation of the medullary respiratory rhythm generator in the newborn rat: an in vitro study. J Physiol 443: 477-498

22. Errchidi S, Hilaire G, Monteau R (1990) Permanent release of noradrenaline modulates respiratory frequency in the newborn rat: an in vitro study. J Physiol 429: 497-510

23. Franke U (2006) Mechanisms of disease: neurogenetics of MeCP2 deficiency. Nat Clin Pract Neurol 2: 212-221

24. Gaspar P, Cases O, Maroteaux L (2003) The developmental role of serotonin: news from mouse molecular genetics. Nat Rev Neurosci 4: 1002-1012

25. Glaze DG, Frost JD, Jr., Zoghbi HY, Percy AK (1987) Rett's syndrome: characterization of respiratory patterns and sleep. Ann Neurol 21: 377-382

26. Goridis C, Rohrer H (2002) Specification of catecholaminergic and serotonergic neurons. Nat Rev 3: 531-541

27. Gozal E, Row BW, Schurr A, Gozal D (2001) Developmental differences in cortical and hippocampal vulnerability to intermittent hypoxia in the rat. Neurosci Lett 305: 197–201

28. Guideri F, Acampa M, Blardi P, de Lalla A, Zappella M, Hayek Y (2004) Cardiac dysautonomia and serotonin plasma levels in Rett syndrome. Neuropediatrics 35: 36-38

29. Guo H, Hellard DT, Huang L, Katz DM (2005) Development of pontine noradrenergic A5 neurons requires brain-derived neurotrophic factor. Eur J Neurosci 21: 2019-2023

30. Guy J, Hendrich B, Holmes M, Martin JE, Bird A (2001) A mouse *Mecp2*-null mutation causes neurological symptoms that mimic Rett syndrome. Nat Genet 27: 322-326

31. Hagberg B (2002) Clinical manifestations and stages of Rett syndrome. Ment Retard Dev Disabil Res Rev 8: 61-65

32. Hagberg B, Aicardi J, Dias K, Ramos O (1983) A progressive syndrome of autism, dementia, ataxia, and loss of purposeful hand use in girls: Rett's syndrome: report of 35 cases. Ann Neurol 14: 471-479

33. Hilaire, G (2006) Endogenous noradrenaline affects the maturation and function of the respiratory network: Possible implication for SIDS. Auton Neurosci 127: 320-331

34. Hilaire G, Duron B (1999) Maturation of the mammalian respiratory system. Physiol Rev 79: 325-360

35. Hilaire G, Pasaro R (2003) Genesis and control of the respiratory rhythm in adult mammals. News Physiol Sci 18: 23-28

36. Hilaire G, Viemari JC, Coulon P, Simonneau M, Bévengut M (2004) Modulation of the medullary respiratory rhythm generator by pontine noradrenergic A5 and A6 groups in rodents. Respir Physiol Neurobiol 143: 187-197

37. Hirsch M, Tiveron MC, Guillemot F, Brunet JF, Goridis C (1998) Control of noradrenergic differentiation and Phox2a expression by MASH1 in the central and peripheral nervous system. Development 125: 599-608

38. Holm PC, Rodriguez FJ, Kresse A, Canals JM, Silos-Santiago I, Arenas E (2003) Crucial role of TrkB ligands in the survival and phenotypic differentiation of developing locus coeruleus noradrenergic neurons. Development 130: 3535-3545

39. Ide S, Itoh M, Goto Y (2005) Defect in normal developmental increase of the brain biogenic amine concentrations in the *Mecp2*-null mouse. Neurosci Lett 386: 14-17

40. Julu PO, Kerr AM, Apartopoulos F, Al-Rawas S, Engerstrom IW, Engerstrom L, Jamal GA, Hansen S (2001) Characterisation of breathing and associated central autonomic dysfunction in the Rett disorder. Arch Dis Child 85: 29-37

41. Kheirandish L, Gozal D, Pequignot JM, Pequignot J, Row BW (2005) Intermittent hypoxia during development induces long-term alterations in spatial working memory, monoamines, and dendritic branching in rat frontal cortex. Pediatr Res 58: 594-599

42. Kim MA, Lee HS, Lee BY, Waterhouse BD (2004) Reciprocal connections between subdivisions of the dorsal raphe and the nuclear core of the locus coeruleus in the rat. Brain Res 1026: 56-67

43. Lajard AM, Bou C, Monteau R, Hilaire G (1999) Serotonin levels are abnormally elevated in the fetus of the monoamine oxidase-A-deficient transgenic mouse. Neurosci Lett 261: 41-44

44. LaSalle JM, Goldstine J, Balmer D, Greco CM (2001) Quantitative localization of heterogeneous methyl-CpG-binding protein 2 (MeCP2) expression phenotypes in normal and Rett syndrome brain by laser scanning cytometry. Hum Mol Genet 10: 1729-1740

45. Lekman A, Witt-Engerstrom I, Gottfries J, Hagberg BA, Percy AK, Svennerholm L (1989) Rett syndrome: biogenic amines and metabolites in postmortem brain. Pediatr Neurol 5: 357-362

46. Lekman A, Witt-Engerstrom I, Holmberg B, Percy A, Svennerholm L, Hagberg B (1990) CSF and urine biogenic amine metabolites in Rett syndrome Clin Genet 37: 173-178

47. Lindsey BG, Hernandez YM, Morris KF, Shannon R, Gerstein GL (1992) Respiratory-related neural assemblies in the brain stem midline. J Neurophysiol 67: 905-922

48. Lo L, Tiveron MC, Anderson D (1998) MASH1 activates expression of the paired homeodomain transcription factor Phox2a, and couples pan-neuronal and subtype-specific components of autonomic neuronal identity. Development 125: 609-620

49. Lugaresi E, Cirignotta F, Montagna P (1985) Abnormal breathing in the Rett syndrome. Brain Dev 7: 329-333

50. Marcus CL, Carroll JL, McColley SA, Loughlin GM, Curtis S, Pyzik P, Naidu S (1994) Polysomnographic characteristics of patients with Rett syndrome. J Pediatr 125: 218-224

51. Mojca Juric D, Miklic S, Carman-Krzan M (2006) Monoaminergic neuronal activity up-regulates BDNF synthesis in cultured neonatal rat astrocytes. Brain Res 1108: 54-62

52. Morin D, Monteau R, Hilaire G (1991) 5-Hydroxytryptamine modulates central respiratory activity in the newborn rat: an in vitro study. Eur J Pharmacol 192: 89-95

53. Morin D, Hennequin S, Monteau R, Hilaire G (1990) Serotonergic influences on central respiratory activity: an in vitro study in the newborn rat. Brain Res 535: 281-287

54. Mulkey DK, Stornetta RL, Weston MC, Simmons JR, Parker A, Bayliss DA, Guyenet PG (2004) Respiratory control by ventral surface chemoreceptor neurons in rats. Nat Neurosci 7: 1360-1369

55. Mullaney BC, Johnston MV, Blue ME (2004) Developmental expression of methyl-CpG binding protein 2 is dynamically regulated in the rodent brain. Neuroscience 123: 939-949

56. Nakai S, Matsunaga W, Ishida Y, Isobe K, Shirokawa T (2006) Effects of BDNF infusion on the axon terminals of locus coeruleus neurons of aging rats. Neurosci Res 54: 213-219

57.   Nattie EE, Li A (2001) CO_2 dialysis in the medullary raphe of the rat increases ventilation in sleep. J Appl Physiol 90: 1247-1257

58.   Nattie EE, Li A, Richerson G, Lappi DA (2004) Medullary serotonergic neurones and adjacent neurones that express neurokinin-1 receptors are both involved in chemoreception in vivo. J Physiol 556: 235-253

59.   Neul JL, Zoghbi HY (2004) Rett syndrome: a prototypical neurodevelopmental disorder. Neuroscientist 10: 118-128

60.   Nielsen JB, Lou HC, Andresen J (1990) Biochemical and clinical effects of tyrosine and tryptophan in the Rett syndrome. Brain Dev 12: 143-147

61.   Nomura Y, Segawa M, Higurashi M (1985) Rett syndrome -an early catecholamine and indolamine deficient disorder? Brain Dev 7: 334-341

62.   Paterson DS, Thompson EG, Belliveau RA, Antalffy BA, Trachtenberg FL, Armstrong DD, Kinney HC (2005) Serotonin transporter abnormality in the dorsal motor nucleus of the vagus in Rett syndrome: potential implications for clinical autonomic dysfunction. J Neuropathol Exp Neurol 64: 1018-1027

63.   Paton JF (1996) A working heart-brainstem preparation of the mouse. J Neurosci Meth 65: 63-68

64.   Pattyn A, Simplicio N, van Doorninck JH, Goridis C, Guillemot F, Brunet JF (2004) Ascl1/Mash1 is required for the development of central serotonergic neurons. Nat Neurosci 7: 589-595

65.   Pattyn A, Goridis C, Brunet JF (2000) Specification of the central noradrenergic phenotype by the homeobox gene Phox2b. Mol Cell Neurosci 15: 235-243

66.   Pattyn A, Morin X, Cremer H, Goridis C, Brunet JF (1999) The homeobox gene Phox2b is essential for the development of autonomic neural crest derivatives. Nature 399: 366-370

67.   Pattyn A, Morin X, Cremer H, Goridis C, Brunet JF (1997) Expression and interactions of the two closely related homeobox genes Phox2a and Phox2b during neurogenesis. Development 124: 4065-4075

68.   Pickel VM, Joh TH, Chan J, Beaudet A (1984) Serotoninergic terminals: ultrastructure and synaptic interaction with catecholamine-containing neurons in the medial nuclei of the solitary tracts. J Comp Neurol 225: 291-301

69.   Pelka GJ, Watson CM, Radziewic T, Hayward M, Lahooti H, Christodoulou J, Tam PP (2006) Mecp2 deficiency is associated with learning and cognitive deficits and altered gene activity in the hippocampal region of mice. Brain 129: 887-898

70.   Pelligra R, Norton RD, Wilkinson R, Leon HA, Matson WR (1992) Rett syndrome: stimulation of endogenous biogenic amines. Neuropediatrics 23: 131-137

71.   Perry TL, Dunn HG, Ho HH, Crichton JU (1988) Cerebrospinal fluid values for monoamine metabolites, gamma-aminobutyric acid, and other amino compounds in Rett syndrome. J Pediatr 112: 234-238

72.   QianY, Shirasawa S, Chen C, Cheng L, Ma Q (2002) Proper development of relay somatic sensory neurons and D2/D4 interneurons requires homeobox genes Rnx/Tlx-3 and Tlx-1. Gene Dev 16: 1220-1233

73.   Qian Y, Fritzsch B, Shirasawa S, Chen CL, Choi Y, Ma Q (2001) Formation of brainstem (nor)adrenergic centers and first-order relay visceral sensory neurons is dependent on homeodomain protein Rnx/Tlx3. Gene Dev 15: 2533-2545

74.   Ramaekers VT, Hansen SI, Holm J, Opladen T, Senderek J, Hausler M, Heimann G, Fowler B, Maiwald R, Blau N (2003) Reduced folate transport to the CNS in female Rett patients. Neurology 61: 506-515

75. Rett A (1977) Cerebral atrophy associated with hyperammonaemia. In : Handbook of clinical neurology. Vinken PJ, Bruyn GW (eds): North Holland, Amsterdam; pp 305-329

76. Richerson GB (2004) Serotonergic neurons as carbon dioxide sensors that maintain pH homeostasis. Nat Rev Neurosci 5: 449-461

77. Riederer P, Weiser M, Wichart I, Schmidt B, Killian W, Rett A (1986) Preliminary brain autopsy findings in progredient Rett syndrome. Am J Med Genet Suppl 1: 305-315

78. Row BW, Liu R, Xu W, Kheirandish L, Gozal D (2003) Intermittent hypoxia is associated with oxidative stress and spatial learning deficits in the rat. Am J Respir Crit Care Med 167: 1548–1553

79. Rumajogee P, Verge D, Darmon M, Brisorgueil MJ, Hamon M, Miquel MC (2005) Rapid up-regulation of the neuronal serotoninergic phenotype by brain-derived neurotrophic factor and cyclic adenosine monophosphate: relations with raphe astrocytes. J Neurosci Res 81: 481-487

80. Rumajogee P, Madeira A, Verge D, Hamon M, Miquel MC (2002) Up-regulation of the neuronal serotoninergic phenotype in vitro: BDNF and cAMP share Trk B-dependent mechanisms. J Neurochem 83: 1525-1528

81. Scott MM, Wylie CJ, Lerch JK, Murphy R, Lobur K, Herlitze S, Jiang W, Conlon RA, Strowbridge BW, Deneris ES (2005) A genetic approach to access serotonin neurons for in vivo and in vitro studies. Proc Natl Acad Sci USA 102: 16472-16477

82. Shahbazian M, Young J, Yuva-Paylor L, Spencer C, Antalffy B, Noebels J, Armstrong D, Paylor R, Zoghbi H (2002a) Mice with truncated *Mecp2* recapitulate many Rett syndrome features and display hyperacetylation of histone H3. Neuron 35: 243-254

83. Shahbazian MD, Antalffy B, Armstrong DL, Zoghbi HY (2002b) Insight into Rett syndrome: MeCP2 levels display tissue- and cell-specific differences and correlate with neuronal maturation. Hum Mol Genet 11: 115-124

84. Siuciak JA, Boylan C, Fritsche M, Altar CA, Lindsay RM (1996) BDNF increases monoaminergic activity in rat brain following intracerebroventricular or intraparenchymal administration. Brain Res 710: 11-20

85. Smeets EJ, Julu PO, Van Waardenburg, D, Engerstrom IG, Hansen S, Apartopoulos F, Curfs LM, Schrander-Stumpel C (2006) Management of a severe forceful breather with Rett Syndrome using carbogen. Brain Dev 28: 625-632

86. Smith JC, Ellenberger HH, Ballanyi K, Richter DW, Feldman, JL (1991) Pre-Bötzinger complex: a brainstem region that may generate respiratory rhythm in mammals. Science 254: 726-729

87. Sood S, Raddatz E, Liu X, Liu H, Horner RL (2006) Inhibition of serotonergic medullary raphe obscurus neurons suppresses genioglossus and diaphragm activities in anesthetized but not conscious rats. J Appl Physiol 100: 1807-1821

88. Southall DP, Kerr AM, Tirosh E, Amos P, Lang MH, Stephenson JB (1988) Hyperventilation in the awake state: potentially treatable component of Rett syndrome. Arch Dis Child 63: 1039-1048

89. Stettner GM, Huppke P, Gärtner J, Richter D, Dutschmann M (2007) Disturbances of breathing pattern in Rett Syndrome: results from patients and animals models. Adv Exp Biol Med 579: 863-876

90. Taylor NC, Li A, Nattie EE (2005) Medullary serotonergic neurones modulate the ventilatory response to hypercapnia, but not hypoxia in conscious rats. J Physiol 566: 543-557

91. Thomas SA, Matsumoto AM, Palmiter RD (1995) Noradrenaline is essential for mouse fetal development. Nature 374: 643-646

92. Tiveron MC, Hirsch MR, Brunet JF (1996) The expression pattern of the transcription factor Phox2 delineates synaptic pathways of the autonomic nervous system. J Neurosci 16: 7649-7660

93. Tryba AK, Pena F, Ramirez JM (2006) Gasping activity *in vitro*: A rhythm dependent on 5-HT$_{2A}$ receptors. J Neurosci 26: 2623-2634

94. Veasey SC, Fornal CA, Metzler CW, Jacobs BL (1995) Response of serotonergic caudal raphe neurons in relation to specific motor activities in freely moving cats. J Neurosci 15: 5346-5359

95. Viemari JC, Ramirez JM (2006) Norepinephrine differentially modulates different types of respiratory pacemaker and nonpacemaker neurons. J Neurophysiol 95: 2070-2082

96. Viemari JC, Roux JC, Tryba AK, Saywell V, Burnet H, Pena F, Zanella S, Bevengut M, Barthelemy-Requin M, Herzing LB, Moncla A, Mancini J, Ramirez JM, Villard L, Hilaire G (2005a) Mecp2 deficiency disrupts norepinephrine and respiratory systems in mice. J Neurosci 25: 11521-11530

97. Viemari JC, Maussion G, Bevengut M, Burnet H, Pequignot JM, Nepote V, Pachnis V, Simonneau M, Hilaire G (2005b) Ret deficiency in mice impairs the development of A5 and A6 neurons and the functional maturation of the respiratory rhythm. Eur J Neurosci 22: 2403-2412

98. Viemari JC, Bevengut M, Burnet H, Coulon P, Pequignot JM, Tiveron MC, Hilaire G (2004) Phox2a gene, A6 neurons, and noradrenaline are essential for development of normal respiratory rhythm in mice. J Neurosci 24: 928-937

99. Viemari JC, Burnet H, Bevengut M, Hilaire G (2003) Perinatal maturation of the mouse respiratory rhythm-generator: in vivo and in vitro studies. Eur J Neurosci 17: 1233-1244

100. Wang W, Bradley SR, Richerson GB (2002) Quantification of the response of rat medullary raphe neurones to independent changes in pH(o) and P(CO$_2$). J Physiol 540: 951-970

101. Wang W, Pizzonia JH, Richerson GB (1998) Chemosensitivity of rat medullary raphe neurones in primary tissue culture. J Physiol 511: 433-450

102. Weese-Mayer DE, Lieske SP, Boothby CM, Kenny AS, Bennett HL, Silvestri JM, Ramirez JM (2006) Autonomic nervous system dysregulation: breathing and heart rate perturbation during wakefulness in young girls with Rett syndrome. Pediatr Res 60: 443-449

103. Zanella S, Roux JC, Viemari JC, Hilaire G (2006). Possible modulation of the mouse respiratory rhythm generator by A1/C1 neurons. Respir Physiol Neurobiol 153: 126-138

104. Zoghbi HY, Milstien S, Butler IJ, Smith EO, Kaufman S, Glaze DG, Percy AK (1989) Cerebrospinal fluid biogenic amines and biopterin in Rett syndrome. Ann Neurol 25: 56-60

105. Zoghbi HY, Percy AK, Glaze DG, Butler IJ, Riccardi VM (1985) Reduction of biogenic amine levels in the Rett syndrome. N Engl J Med 313: 921-924

17. Respiratory plasticity following intermittent hypoxia: a guide for novel therapeutic approaches to ventilatory control disorders?

Gordon S. MITCHELL

Department of Comparative Biosciences, University of Wisconsin, Madison, Wisconsin, U.S.A.

17.1 Introduction

The respiratory control system exhibits considerable plasticity, both during development and in adults [67]. In this context, plasticity is defined as a functional change in respiratory control system performance based on prior experience [67]. Indeed, to maintain appropriate respiratory motor output throughout life, neural elements that control breathing must adapt to physiological and/or environmental changes, such as birth, pregnancy, obesity, acute respiratory infection, altitude exposure, neural injury and even normal changes in pulmonary mechanics or gas exchange with aging. Although our appreciation of respiratory plasticity and its significance is still quite recent, we are already making good progress towards an understanding of its underlying mechanisms.

Recently, we have come to appreciate at least some cellular/synaptic mechanisms underlying respiratory long-term facilitation (LTF), a model of respiratory plasticity induced by acute intermittent hypoxia [18; 60; 67]. Acute intermittent hypoxia (AIH) triggers LTF *via* serotonin-dependent synaptic enhancement of inputs to spinal and brainstem respiratory motor neurons [18; 66]. Although we do not yet know the precise physiological role of LTF [60; 66], insights gained from detailed mechanistic studies of this intriguing model may stimulate the development of novel therapeutic approaches in the treatment of multiple ventilatory control disorders with devastating consequences for human health. The fundamental perspective of this chapter is based on the premise that we may be able to "harness" the inherent capacity for respiratory motor neuron plasticity to offset functional deficits caused by three seemingly unrelated disorders: respiratory insufficiency following cervical spinal injury, ventilatory failure during motor neuron disease and obstructive sleep apnea. Although each disorder arises from distinct pathogenic mechanisms, they share common features, leading to the idea that similar therapeutic approaches may be effective.

The first goal of this chapter is to highlight a shared feature of obstructive sleep apnea, respiratory insufficiency following cervical spinal injury and ventilatory failure during motor neuron disease: a need to increase selectively targeted respiratory motor neuron activity. A second goal is to review recent progress in our understanding of the cellular/synaptic mechanisms underlying phrenic LTF following AIH, and enhanced phrenic LTF induced by AIH in animals pretreated with chronic or repetitive intermittent hypoxia. A third goal is to review the severity and human relevance of each ventilatory control disorder, followed by a discussion of normal compensatory mechanisms that mitigate the severity of ventilatory deficits. Finally, four potential therapeutic strategies are discussed based on the concept that we can harness respiratory motor neuron plasticity to further enhance respiratory motor output, thereby reversing respiratory deficits. Although none of these strategies are envisioned as a cure, they may improve respiratory function sufficiently to improve the patient's quality of life. At this point, each approach is completely untested and should be regarded as speculative. Without question, extensive development, testing for efficacy and adverse side effects are necessary before they could be considered for clinical application. Nevertheless, the potential of these approaches, and the lack of adequate alternatives, makes further exploration worthwhile.

17.2 Deficiencies in respiratory motor neuron activity underlie some ventilatory control disorders

Although respiratory disorders caused by cervical spinal injury, motor neuron disease and obstructive sleep apnea are not normally thought of as similar, each involves deficits in respiratory motor neuron output. Furthermore, because they share a common need to increase the neuro-motor output to respiratory pump and/or upper airway muscles, they may be responsive to common therapeutic approaches designed to enhance synaptic inputs to respiratory motor neurons. These similarities, and common therapeutic goals, are featured in Fig. 1. Cervical spinal injury (top) via traumatic injury interrupts descending synaptic pathways linking brainstem pre-motor neurons and spinal inspiratory (cervical and thoracic) or expiratory (thoracic and upper lumbar) motor neurons. However, since most spinal injuries are incomplete, spared synaptic pathways represent an important target for therapeutic interventions. Therapeutic approaches that strengthen remaining synaptic inputs to respiratory motor neurons may increase the contributions of these spared synaptic pathways, thereby restoring at least partial function. In contrast, degenerative *motor neuron diseases* such as Amyotrophic Lateral Sclerosis (ALS), (middle panel), cause progressive death of respiratory motor neurons. To preserve function, surviving motor neurons must carry a greater load, offsetting the loss of their neighbors. Synaptic enhancement of inputs to these surviving motor neurons may enable greater or longer compensation, preserving

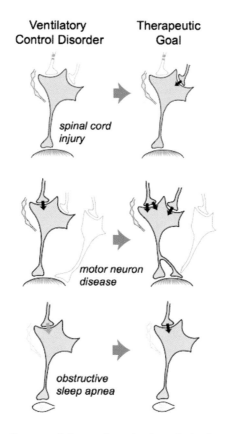

Fig.1. Common features and therapeutic goals of seemingly disparate ventilatory control disorders associated with cervical spinal injury, motor neuron disease and obstructive sleep apnea. On the left, the fundamental disorder is represented as a (relative) deficiency in respiratory motor neuron activation; on the right, a possible therapeutic goal involving compensatory respiratory plasticity is illustrated. In the case of **spinal cord injury** (upper panel), respiratory insufficiency results from traumatic disruption of descending synaptic inputs to spinal respiratory motor neurons (shown as broken axon). However, most spinal injuries are incomplete, leaving spared synaptic pathways that are unable to reliably activate respiratory motor neurons (faded synapse on right side of motor neuron). By harnessing respiratory plasticity to strengthen those spared synaptic pathways (robust synapse on right side of motor neuron), it may be possible to restore at least partial respiratory function below the site of injury. During **motor neuron disease**, loss of synergistic motor units (faded motor neuron on right) requires that surviving motor neurons increase their output to preserve adequate ventilatory function. Respiratory plasticity may strengthen functional, as well as "silent," synapses onto the surviving motor neurons, preserving respiratory motor output to the greatest extent possible. Sprouting at the motor end-plate is another possible means of increasing the contribution of surviving motor neurons, functionally increasing the size of the motor unit and improving respiratory function (illustrated as sprouting terminal onto muscle). **Obstructive sleep apnea** is associated with inadequate neuromodulation and hypo-excitable brainstem respiratory motor neurons (depicted as faded serotonergic input left of the motor neuron). Inadequate motor output from brainstem respiratory motor neurons predisposes to upper airway collapse and repetitive apneas during sleep. By harnessing respiratory plasticity, it may be possible to restore synaptic inputs to brainstem respiratory motor neurons (depicted as more robust synapse), thereby increasing respiratory motor output to upper airway muscles and protecting upper airway patency during sleep.

respiratory function until late in the disease progression. Lastly, obstructive sleep apnea (lower panel) is associated with diminished modulatory influences on upper airway motor neurons. Diminished upper airway tone predisposes the upper airway to collapse and obstructive apneas. Thus, therapeutic approaches that strengthen synaptic inputs to upper airway motor neurons may preserve airway patency and prevent subsequent apneic episodes during sleep. In each disorder, therapeutic strategies that increase the strength of functional or ineffective synaptic inputs onto respiratory motor neurons may provide benefit and restore ventilatory function. Since AIH-induced LTF strengthens synaptic inputs to respiratory motor neurons ([18; 60]; see below), initiating or enhancing its cellular mechanism *via* pharmacological or other interventions represents an interesting but untested strategy to treat the ventilatory control disorders discussed in this chapter.

17.3 Intermittent hypoxia induced respiratory plasticity and metaplasticity

17.3.1 Phrenic Long-Term Facilitation: a model of respiratory motor neuron compensation?

Our conceptual development concerning the induction of respiratory plasticity as a therapeutic approach in the treatment of ventilatory control disorders has been guided by studies of serotonin-dependent synaptic plasticity in phrenic motor neurons following AIH or phrenic long-term facilitation (pLTF) [18; 66; 67]. pLTF is expressed as a prolonged increase in phrenic motor output following AIH (3 to 5 episodes with 5 min interval; Fig. 2) [3; 18; 66]. The magnitude of pLTF is relatively insensitive to the severity or duration of hypoxic episodes—as long as they are intermittent [22, 60]. Similar LTF is expressed in other respiratory nerves, including the inspiratory intercostal nerves [21], or a major regulator of upper airway resistance, the hypoglossal nerve [2; 23].

In our working cellular/synaptic model (Fig. 3), pLTF is initiated by spinal serotonin receptor activation, thereby stimulating new BDNF synthesis within phrenic motor neurons [4]. New BDNF synthesis is necessary for pLTF since RNA interference with small, interfering RNAs (siRNAs) targeting BDNF mRNA abolishes pLTF [4]. BDNF is sufficient to elicit pLTF since intrathecal BDNF administration elicits a long-lasting facilitation of phrenic nerve activity with a magnitude and time course consistent with AIH-induced pLTF [4]. BDNF-induced activation of the high affinity receptor tyrosine kinase (TrkB) strengthens short-latency, synaptic inputs from glutamatergic respiratory pre-motor neurons onto phrenic motor neurons [26; 31].

Fig. 2. Representative phrenic LTF (pLTF) protocol. A 1.5 hour compressed tracing is shown of integrated phrenic activity before, during (arrows) and one hour following three 5 min hypoxic episodes (5 min intervals) in an anesthetized rat (data from [4]). The progressive increase in peak integrated phrenic activity post-intermittent hypoxia is pLTF. pLTF represents a "respiratory memory," induced by synaptic enhancement of respiratory inputs to phrenic motor neurons. The postulated mechanism of pLTF is shown in Fig. 3.

The postulated signaling cascade during pLTF is: 5HT2A-receptor activation of a G protein (Galphaq), phospholipase C and protein kinase C (PKC; [7; 63]. We postulate that PKC activation subsequently initiates new BDNF synthesis *via* (direct or indirect) phosphorylation of relevant translation initiation factors. We propose that new BDNF is released from phrenic motor neuron dendrites, activating post-synaptic TrkB receptors [4]. BDNF activation of TrkB receptors on phrenic motor neurons initiates further signaling cascades that establish pLTF, possibly *via* the activation of extracellular regulated MAP kinases 1/2 (ERK1/2; [87]) and/or protein kinase B (Akt; [31]). ERK1/2 and/or Akt activation are postulated to phosphorylate glutamate receptor subunits, thereby strengthening respiratory synaptic inputs onto phrenic motor neurons [22; 64]. Intermittent serotonin receptor activation elicits phrenic [56] and XII [7] LTF in *in vitro* neonatal rat preparations, largely by increased AMPA currents [7]. ERK1/2 and Akt activation may also initiate gene expression, enhancing future pLTF (ie. metaplasticity) and providing a degree of neuroprotection.

Whereas AIH elicits pLTF, sustained hypoxia of similar cumulative duration does not, demonstrating a profound pattern-sensitivity in the underlying mechanism [3]. The pattern sensitivity of pLTF may relate to differences in the formation of reactive oxygen species (ROS) and protein phosphatase activity between intermittent and sustained hypoxia. The role of phosphatase inhibition is demonstrated by the observation that pLTF is observed following sustained hypoxia when serine/threonine protein phosphatase activity is inhibited in the cervical spinal cord with okadaic acid [89]. Conversely, AIH-induced pLTF requires ROS formation in the cervical spinal cord [58], most likely *via* NADPH oxidase activation [59]. Interestingly, intrathecal okadaic acid restores pLTF following AIH in rats with suppressed ROS formation (P. M. MacFarlane, G. S. Mitchell, unpublished). Thus, a major factor differentiating intermittent *versus* sustained hypoxia

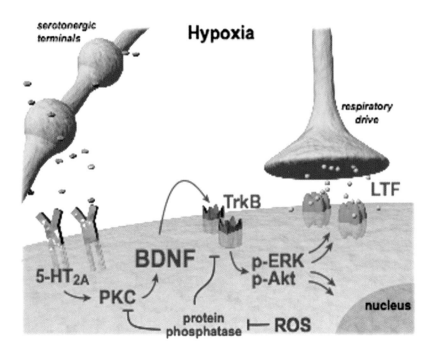

Fig. 3. Working model of cellular/synaptic mechanisms giving rise to pLTF. During intermittent hypoxia (IH), serotonin receptor activation initiates new BDNF synthesis within phrenic motor neurons. BDNF, *via* autocrine actions on the TrkB receptor, activates ERK 1/2 MAP kinases, which in turn establish LTF. Unlike intermittent hypoxia, sustained hypoxia (SH) is associated with greater protein phosphatase activation, halting the mechanism of pLTF by dephosphorylating relevant kinases or their targets. We propose that a major distinguishing feature between IH and SH is greater ROS formation by IH, thereby inhibiting the phosphatases and reducing their constraint on the mechanism of pLTF. To harness the mechanism of pLTF as a therapeutic approach during ventilatory control disorders, repetitive intermittent hypoxia will trigger the mechanism and is easy to administer. To bypass potential complications associated with repetitive hypoxia, other treatments may be devised that manipulate downstream molecular targets, either activating key molecules (e.g. transactivation of TrkB), removing inhibitory constraints (e.g. siRNAs targeting relevant phosphatases), or by providing relevant proteins (e.g. via cell based therapies). For further discussion of these possibilities, see text.

may be differences NADPH oxidase activation, ROS formation and their inhibitory actions on spinal protein phosphatase activity.

Additional factors modulate AIH-induced pLTF expression. For example, pLTF is greatly enhanced by adenosine A2A receptor antagonists [40], an unanticipated effect since A2A receptor activation alone induces phrenic motor facilitation by transactivation of TrkB receptors ([31]; see below). Sex hormones enable AIH-induced phrenic and XII LTF [92-95]. For example, phrenic and (particularly) XII LTF are diminished in middle-aged [92] and neutered male rats [94], largely due to loss of estrogen derived from circulating testosterone [95]. Genetic factors also influence AIH-induced LTF. For example phrenic and hypoglossal LTF differ among inbred rat strains [6], and even among colonies of outbred Sprague Dawley rats [22; 23]. Differential mRNA expression among three inbred rat strains revealed an inverse relationships between BDNF, TrkB and serotonin 5HT7 receptor mRNA and pLTF [10]. Only 5HT2A receptor protein and mRNA were positively correlated with pLTF across inbred strains [10], suggesting that genetic variations in 5HT2A receptor expression influence LTF magnitude among (and possibly within) populations. Regardless, preconditioning with repetitive intermittent hypoxia restores LTF in rats with minimal capacity for LTF following AIH (see below).

17.3.2 Enhanced Long-Term Facilitation following repetitive intermittent hypoxia

Although AIH-induced pLTF normally lasts for approximately two or three hours, pretreatment with repetitive intermittent hypoxia elicits longer-lasting forms of respiratory plasticity, including enhanced AIH-induced pLTF. For example, pretreatment with chronic intermittent hypoxia (CIH; continuous 5 min episodes, 5 min intervals; 12 hrs per day, 7 days) elicits a serotonin-dependent enhancement of phrenic [54; 63; 74] and hypoglossal LTF [96]. CIH-induced plasticity lasts for many days [24]. CIH strengthens ineffective spinal synaptic pathways to phrenic motor neurons, thereby restoring respiratory motor function below a cervical spinal hemisection [27]. In essence, CIH was used to harness respiratory plasticity, thereby reversing functional breathing deficits caused by cervical spinal injury. Unfortunately, severe CIH is associated with morbidity, including systemic hypertension, hippocampal apoptosis and learning deficits [20; 34; 36].

Other, less severe protocols of repetitive intermittent hypoxia also elicit metaplasticity in LTF, but without evidence for pathological consequences. Daily exposure to AIH (10 episodes per day, 7 days) increases ventral spinal BDNF protein levels, ERK1/2 phosphorylation and the magnitude of LTF, but without systemic hypertension [88]. AIH exposures three days per week for 10 weeks upregulate key elements in the mechanism of pLTF, including increased serotonin terminal density, protein expression of 5HT2A receptors, BDNF and TrkB receptors, as well as phosphorylation/activation of ERK 1/2 and Akt within the phrenic motor nucleus [81]. Neither microglia nor astrocytes within the phrenic motor nucleus were activated by thrice weekly AIH, and there was no evidence for apoptosis or cell death. Another potentially important motor neuron protein induced by repetitive intermittent hypoxia is vascular endothelial growth factor (VEGF). Daily 4 hour hypoxic exposures increase VEGF expression in motor neurons, an effect not observed following sustained hypoxia of similar cumulative duration [43]. VEGF is now known to be an important trophic factor for motor neurons, and elicits similar intracellular signaling cascades as BDNF [82; 97]. Since VEGF is expressed in phrenic motor neurons, and is up regulated by thrice-weekly AIH [12], it is a potentially exciting molecule in respiratory motor neuron plasticity and/or survival. Since repetitive AIH up-regulates multiple proteins that play a key

role in respiratory plasticity, it represents an easily administered means of inducing and/or enhancing respiratory plasticity and, thus, is interesting to consider
from a therapeutic context.

17.4 Compensatory plasticity during ventilatory control disorders

17.4.1 Spinal cord injury

Injuries above cervical spinal segment C_3 produce respiratory muscle paralysis and respiratory insufficiency. In fact, the most common cause of death in
spinal cord injury patients is respiratory failure [62]. Many victims of cervical spinal injury die before ventilatory support can be implemented. Other patients survive the initial injury and then become ventilator-dependent, greatly diminishing
their quality of life. In ventilator-dependent spinal injury patients, the 1-year survival rate is only 25% [14]. Clearly, improving respiratory function is of the highest importance in patients with high cervical spinal injuries [76].

The International Spinal Research Trust [78] suggested several priorities
for research and treatment of spinal cord injury: 1) preserve spinal synaptic pathways to the greatest extent possible; 2) reestablish interrupted connections through
axonal regeneration or cell transplantation; and 3) harness mechanisms of spinal
plasticity, thereby enhancing the functional contributions of residual or regener-
ated synaptic connections. Since most spinal injuries are incomplete, many residual axons are spared, bypassing the injury site to innervate their spinal targets [15;
91]. Maximizing function in such spared synaptic pathways is a viable therapeutic
goal.

In most spinal cord injury patients, some degree of functional improvement in respiratory and locomotor function is observed within one year of injury
(reviewed in [62]). Although this spontaneous recovery almost certainly arises
from spared axonal pathways [15; 78], the cellular and synaptic mechanisms underlying spontaneous improvement are not well understood. Several lines of evidence suggest a prominent role for serotonin in spontaneous functional recovery
following cervical spinal injury.

Spinal hemisection rostral to the phrenic motor nucleus interrupts descending projections from brainstem inspiratory pre-motor neurons to phrenic motor neurons, thereby paralyzing the ipsilateral hemidiaphragm. A small degree of
functional recovery is observed in the paralyzed hemidiaphragm during normal
breathing, and this effect is greatly enhanced by chemoreflex activation [32; 33].
Spontaneous functional recovery below cervical hemisection is a long-studied
model of respiratory plasticity known as the "crossed-phrenic phenomenon" [33;
36; 72; 75]. The anatomical basis of the crossed phrenic phenomenon is an existing, but functionally ineffective crossed-spinal synaptic pathway to phrenic motor
neurons [33]. Spinal serotonin receptor activation reveals these crossed-spinal

synaptic pathways to phrenic motor neurons within minutes [53], as well as existing but ineffective ipsilateral synaptic pathways [68].

The crossed phrenic phenomenon following cervical hemisection is serotonin-dependent since it is diminished in rats pretreated with the serotonin synthesis inhibitor, para-chlorophenylalanine [37]. Conversely, the serotonin precursor 5-hydroxytryptophan and the 5HT2 receptor agonist DOI induce the crossed-phrenic phenomenon in hemisected rats [99-102]. Thus, one viable strategy to treat respiratory insufficiency following cervical spinal injury is to harness serotonin-dependent respiratory plasticity, strengthening silent or ineffective synaptic pathways to respiratory motor neurons that bypass the lesion.

Although cervical hemisection initially decreases serotonin terminal density in the ipsilateral phrenic nucleus, it is at least partially restored over weeks [30; 83]. Thus, serotonergic function is a "moving target" in respiratory motor nuclei after cervical injury, suggesting that therapeutic strategies reliant on endogenous activation of the serotonergic system will be most effective in cases of chronic *versus* acute spinal injury.

Experimental treatments that enhance serotonergic function in the phrenic motor nucleus augment phrenic responses to spontaneous and evoked synaptic inputs. For example, chronic sensory denervation of cervical segments associated with the phrenic motor nucleus increases serotonin terminal density near identified phrenic motor neurons [44], increases neurotrophin concentrations in the ventral cervical spinal cord, including BDNF [42], enhances AIH-induced pLTF [44] and strengthens crossed-spinal synaptic pathways to phrenic motor neurons [25]. Thus, spinal sensory denervation, a form of selective spinal injury, induces serotonin-dependent plasticity and metaplasticity of synaptic inputs to phrenic motor neurons.

Intermittent hypoxia elicits plasticity in crossed-spinal synaptic pathways to phrenic motor neurons following chronic (not acute) spinal injury. For example, AIH elicits pLTF in the crossed-spinal synaptic pathway at one and two months, but not at two weeks post-injury [30]. The differences are attributable to progressive recovery serotonergic innervation within the phrenic motor nucleus post-injury [30]. On the other hand, chronic intermittent hypoxia enhances crossed-spinal synaptic pathways to phrenic motor neurons two weeks post-hemisection [27], suggesting that more robust stimulation protocols can overcome deficiencies in serotonergic innervation, possibly due to increased serotonin 2A receptor expression on phrenic motor neurons at that time [28]. Daily acute intermittent hypoxia (10 episodes per day, 7 days) reverses modest ventilatory deficits characteristic of rats with chronic cervical hemisections [5; 29], suggesting that repetitive hypoxic exposures have potential for meaningful functional benefit in intact animals and, possibly, in human patients.

Other transmitter systems tap into the mechanism of pLTF downstream from intermittent hypoxia or serotonin-dependent BDNF synthesis. For example, repetitive activation of the Gs-protein coupled adenosine 2A receptor trans-activates an immature TrkB isoform within phrenic motor neurons, increasing TrkB signaling and inducing phrenic motor facilitation [31]. While similar effects were observed in intact and spinally injured rats, there were some differences in the detailed mechanisms of TrkB trans-activation between intact and spinally-injured rats [31]. Other Gs protein coupled receptors, such as the PACAP receptor, are likely to induce similar phrenic motor facilitation since they also trans-activate TrkB receptors [51]. Small molecules that activate TrkB receptors may have powerful advantages (such as ease of delivery) in the treatment of ventilatory control disorders that require increased respiratory motor output (see below).

17.4.2 Motor neuron disease

Patients with ALS invariably develop respiratory muscle weakness, and the most common cause of death is ventilatory failure [8; 49; 57]. Respiratory muscle weakness decreases the capacity to generate inspiratory force [41; 57]. Deficiencies in the capacity to generate inspiratory force suggest inspiratory motor neuron degeneration in ALS patients, although respiratory motor neuron degeneration has not been demonstrated directly. On the other hand, the extreme inspiratory maneuvers (eg. maximal force during inspiratory sniffing; [57]) necessary to demonstrate functional respiratory deficits prior to late stages of the disease suggest that respiratory motor function is somehow protected long after symptoms of somatic motor weakness have been observed. Regardless, at some point, inspiratory motor neuron death will exceed the capacity for compensation, resulting in ventilator dependence or death.

Although ventilatory failure and sleep-disordered breathing are severe consequences, there are few investigations concerning respiratory motor neurons during ALS. Rodent models of ALS have become available in recent years [90], such as transgenic mice and rats that over-express a mutated form of superoxide dismutase found in familial ALS. $SOD\text{-}1$ mutants exhibit motor neuron disease mimicking important aspects of familial ALS in humans. In the $SOD1^{G93A}$ rat, Llado et al. [55] reported profound degeneration of phrenic (but not hypoglossal) motor neurons, progressive reduction of compound diaphragm action potentials, phrenic nerve fiber loss, and diaphragm atrophy. Thus, there is clear evidence of the potential for respiratory compromise in this rat model of ALS. Tankersley et al. [84] report that $SOD1^{G93A}$ mice preserve ventilation and the ability to increase ventilation during chemoreceptor stimulation until late in disease progression, when ventilatory insufficiency develops rapidly. $SOD1^{G93A}$ rats develop profound losses of phrenic and thoracic (but not hypoglossal) motor neurons at a disease end-stage defined by 20% loss in body mass [73]. Despite more than 80% losses in phrenic motor neurons [73], phrenic motor output decreases only slightly more than 40% [73], and the ability to increase tidal volume during maximal chemoreceptor stimulation is fully preserved [11]. We hypothesize that compensatory spinal neuroplasticity preserves respiratory function despite severe phrenic (>80% loss at end-stage) and thoracic (>60%) motor neuron degeneration in $SOD1^{G93A}$ rats. Possible mechanisms of compensatory plasticity include: 1) synaptic enhancement of descending inputs to phrenic (and other respiratory) motor neurons, thus enabling surviving respiratory motor neurons to compensate for progressive motor neuron death during the disease progression (Fig. 1) neuromuscular junction plasticity, enabling surviving motor neurons to activate larger motor units. We rule out increases in descending respiratory drive from medullary pre-motor neurons since the apneic CO_2 threshold was unchanged in SOD1^{G93A} rats, and because chemoreflex responses in the XII nerve (100% motor neuron survival) are normal [73]. Thus, pre-motor drive was most likely normal and was unlikely to increase output in individual phrenic motor neurons.

In motor neuron disease, we postulate that cellular mechanisms similar to pLTF allow surviving motor neurons to at least partially compensate for the loss of others (Fig. 1), thereby sustaining phrenic (and other respiratory) motor output and breathing despite severe motor neuron death. Indeed, the few surviving phrenic motor neurons at disease end-stage exhibit several important hallmarks suggestive of this mechanism, including increased BDNF and TrkB protein levels [80]. True disease end-stage may be the point where compensation can no longer be achieved, leading to catastrophic ventilatory failure [84]. Our working hypothesis is that increased BDNF and/or VEGF expression initiates synaptic plasticity during disease progression, increasing the strength of glutamatergic synaptic inputs from descending respiratory pre-motor neurons onto spinal, respiratory motor neurons. Increasing synaptic strength may preserve phrenic motor output (and diaphragm function) until motor neuron losses become overwhelming.

17.4.3 Obstructive Sleep Apnea

Obstructive sleep apnea (OSA) is the periodic cessation of breathing due to collapsed upper airways during sleep. OSA causes severe morbidity including systemic hypertension, learning deficits and hypersomnolence [20; 34]. Since intermittent hypoxia is a prominent feature of repetitive apneas during sleep, LTF may represent a form of physiological compensation, promoting breathing stability during sleep in the face of challenges that otherwise lead to apneas and/or hypopneas. This possibility is consistent with observations that ventilatory LTF is most prominent during slow-wave sleep (*versus* wakefulness) in rats [69; 70]. LTF is also most prominent during sleep in human subjects, particularly in the upper airway muscles, consistent with a role in maintaining upper airway patency. We recently reviewed evidence that LTF [1; 39] offsets apneas in subjects that do not yet reach the clinical criteria defining OSA [60]. If LTF is indeed a stabilizing, compensatory mechanism for sporadic apneas, then factors that impair LTF are expected to accelerate the appearance of apneas and the clinical manifestation of OSA. Although there is no direct evidence for a causal, inverse relationship between LTF and OSA, the prevalence of apneas is increased by serotonin receptor antagonists in both sleeping bulldogs [86] and obese Zucker rats [71].

17.5 Possible therapeutic approaches for ventilatory control disorders: lessons from LTF?

Although the precise physiological significance of respiratory LTF is not yet clear, a detailed understanding of its cellular and molecular mechanisms may provide useful insights leading to the development of novel therapeutic strategies in the treatment of ventilatory control disorders that require enhanced respiratory motor output (e.g. SCI, ALS and OSA). None of the ideas suggested in this chapter has been tested for efficacy or undesirable side effects in human patients. For now, they should be regarded only as ideas that warrant further exploration. Creative approaches are necessary before new and effective therapeutic strategies can be devised to treat the devastating ventilatory control disorders discussed in this chapter.

Repetitive AIH may be used as a tool to induce respiratory plasticity, thereby enhancing or restoring function to the upper airways and/or respiratory pump muscles without morbidity associated with more severe protocols of chronic intermittent hypoxia. RNA interference may enable selective gene regulation in respiratory motor neurons, thereby modulating their excitability and respiratory motor output. The key is to identify appropriate molecular targets. siRNAs directed at molecules that normally constrain LTF have the greatest potential to convey the necessary "gain of function." Another interesting approach concerns small molecules that simulate properties of key proteins regulating respiratory motor neuron excitability, such as BDNF. Such small molecules already exist, and may have superior access to respiratory motor neurons, thus allowing regulation of downstream signaling mechanisms without side effects attendant to hypoxia, serotonin receptor activation or siRNA administration. Lastly, cell-based therapies have interesting potential. Differentiated stem cells engineered to express and secrete key trophic proteins may survive *in situ* following transplantation, providing an on-site supply of therapeutic proteins in respiratory muscles and/or motor nuclei.

17.5.1 Repetitive intermittent hypoxia

Since repetitive intermittent hypoxia elicits respiratory motor neuron plasticity and metaplasticity, it may be a viable means of restoring respiratory function in ventilatory control disorders. The critical issue is to develop a protocol of intermittent hypoxia that elicits the intended plasticity without unintended consequences associated with more severe chronic intermittent hypoxia [20; 34]. In cases of spinal injury and/or ALS, the risk of hypertension is lower than during OSA since these patients often exhibit hypotension. To date, our studies indicate that even modest protocols of intermittent hypoxia elicit motor neuron plasticity and metaplasticity [60; 61; 87], suggesting the possibility of developing effective and safe protocols for at least some patients. Major advantages of repetitive intermittent hypoxia are that it is simple to administer and increases endogenous production of the relevant trophic molecules, thereby avoiding complications attendant to exogenous protein delivery (see below). Recent experiments indicate success in restoring ventilatory function to rats with a chronic cervical spinal hemisection following one week of daily acute intermittent hypoxia [5], confirming the potential of using repetitive intermittent hypoxia in patients.

The concept of using repetitive intermittent hypoxia as a therapeutic tool in OSA patients is complex, since OSA causes intermittent hypoxia and triggers considerable morbidity. However, if there is a causal relationship between OSA and deficits in the capacity to elicit upper airway LTF, daytime treatments that restore LTF may improve upper airway patency during sleep. As examples, chronic intermittent hypoxia restores hypoglossal LTF in geriatric female rats [96], and daily AIH enables hypoglossal LTF in a rat strain with minimal LTF expression [87]. Since even modest but prolonged protocols of repetitive intermittent hypoxia have considerable impact on key proteins that underlie LTF [81], there is potential to affect the progression of at least pre-clinical OSA (and possibly snoring).

17.5.2 Regulation of gene expression in respiratory motor neurons

RNAi is a newly discovered mechanism based on double-stranded RNA molecules that trigger the sequence-specific suppression of gene expression *via* an endogenous RNA-induced silencing complex [13; 16; 19; 38]. The mechanism of RNAi was first described in 1998 in invertebrates [19], and is thought to be an evolutionarily conserved, endogenous mechanism to defend against invading viral pathogens or aberrant transcriptional products [19]. Although the exploitation of this technique to trigger silencing of a targeted gene is still in its infancy, many groups have now had success in mammalian cell cultures [17; 65; 98]. Although the utility of RNAi *in vivo* remains limited in mammals, it offers a novel means of regulating gene expression *via* exogenous administration of small interfering RNAs (siRNAs). We utilized this technique successfully to regulate BDNF protein expression in the phrenic motor nucleus, the first report of successful RNAi in the adult mammalian spinal cord *in vivo* [4]. However, the utilization of RNAi to treat ventilatory control disorders will most likely require stable transfection with siRNAs or microRNAs targeting key molecules regulating respiratory motor neuron excitability. Recently, viral vectors have been developed to transfect target cells with the siRNAs of interest. For example, targeting the SOD1 gene in a mouse model of ALS using muscle injections (including the diaphragm) of a lentiviral vector successfully delivers the RNAi molecule and knocks down the mutant gene in motor neurons [77; 79]. Since motor neurons are one of the few neuronal cell types that cross the blood brain barrier (to innervate muscle), they are relatively accessible to siRNA delivery via intramuscular injections into respiratory muscles.

To enhance LTF as a treatment strategy for ventilatory control disorders, siRNAs must be designed that target proteins constraining LTF, thereby resulting in a gain of function. For example, serine-threonine protein phosphatases represent an interesting target since their inhibition enables pLTF following sustained hypoxia [89] or following intermittent hypoxia in rats with suppressed ROS formation (P. M. MacFarlane, G. S. Mitchell, unpublished). Hurdles to overcome include the development of suitable delivery techniques, demonstration that the siRNA selectively degrades the target mRNA and protein for a suitable duration, and the ability to enhance LTF without unacceptable side effects. Although substantial hurdles must be overcome before RNAi will be considered a viable therapy for ventilatory control disorders, further investigations are warranted based on the rapid progress and enormous potential in this field.

17.5.3 Small molecules simulating BDNF

BDNF and TrkB signaling are critical elements in the mechanism of pLTF. Thus, small molecules that initiate TrkB signaling may harness the benefits of pLTF, but with practical advantages relative to BDNF protein delivery. Protein delivery to the CNS is notably difficult due to problems with inflammatory responses, access across the blood brain barrier and receptor down-regulation induced by exogenous protein delivery. Small molecules that trans-activate TrkB offer multiple advantages, including superior delivery to the relevant target sites without problems associated with inflammatory responses or receptor down-regulation. Endogenous ligands with Gs protein coupled metabotropic receptors trans-activate receptor tyrosine kinases [50]. For example, adenosine 2A receptors [50] and the PACAP receptor [51] trans-activate TrkB. Consequently, A2A receptor agonists that readily cross the blood brain barrier induce phrenic and hypoglossal motor facilitation, thereby reversing respiratory deficits in rats with cervical spinal injury [31]. In theory, similar effects may restore upper airway patency in patients with OSA. Intrethecal and intraperitoneal A2A receptor agonists induce phrenic motor facilitation via new synthesis of an immature, intracellular TrkB isoform that initiates signaling cascades normally associated with pLTF [31]. Intravenous adenosine 2A receptor agonists also induce hypoglossal motor facilitation (F. J. Golder, G. S. Mitchell, unpublished observation), suggesting that this effect is generalized to motor neurons that innervate both upper airway and respiratory pump muscles and may represent an effective pharmacological approach in the treatment of OSA.

17.5.4 Cell based therapies: protein delivery and cell replacement

Advances in our ability to culture and engineer (embryonic or differentiated) stem cells have raised expectations concerning their possible therapeutic applications in many diseases. There may be important roles for these cells to play in the treatment of multiple ventilatory control disorders, particularly respiratory insufficiency during motor neuron disease. During the progression of ALS, respiratory motor neuron death leads to respiratory failure. Thus, an appropriate goal is to replace those motor neurons with, for example, embryonic stem cells differentiated along a motor neuron lineage. Such cells have been developed [52], although there are major practical limitations concerning their possible use in restoring respiratory motor neuron pools. The dilemma is that once motor neurons are transplanted, the patients would die before new motor neurons could be incorporated and become functional in their new environment. Furthermore, the ALS spinal cord may be a toxic environment for motor neurons, and the new cells may die, just as the cells they were intended to replace. Before motor neuron replacement will be viable, strategies to harness compensatory plasticity and to restore a suitable environment must be developed. Thus, motor neuron transplants may be effective only in conjunction with strategies that preserve respiratory function (and life) sufficiently to enable the incorporation of new motor neurons. Stem cells may be useful in other contexts, for example as *in situ* sources of molecules that promote plasticity and/or restore a hospitable environment for replacement cells. Such complicated procedures will take considerable effort to develop.

A practicable starting point may be to use differentiated neural or muscle progenitor cells, engineered to synthesize and deliver beneficial proteins [46] that enable mechanisms of plasticity and/or neuroprotection, thereby restoring or delaying deterioration of respiratory motor function in ALS, SCI and OSA. The protein secreting cells could be transplanted into the relevant respiratory motor nuclei, or in the target muscle since, in many cases, the proteins will be taken up by motor neuron terminals and transported to the motor neuron cell body. Finding the most beneficial protein(s) is critical, as is the ability to transplant the cells and regulate their synthesis/release of that protein. Selection of those proteins may be guided by studies of respiratory LTF.

17.6 Conclusion

Regardless of its physiological significance, a detailed understanding of respiratory LTF may lead to new insights concerning the role of compensatory plasticity, and possibly new therapeutic approaches in the treatment of ventilatory control disorders. In addition to the three disorders featured in this chapter, other ventilatory control disorders may share common features associated with abnormal serotonin and/or BDNF-function. For example, defects in serotonergic function have been implicated not only in OSA ([48; 86]; see Chapter 8), but also in sudden infant death syndrome ([18; 45]; see Chapter 7). Similar cellular mechanisms operating at distinct points in the respiratory circuit may lead to, or contribute to severe breathing disorders. For example, Rett Syndrome is caused by mutations in *MECP2*, a regulator of BDNF gene expression ([9; 47], and children with Rett Syndrome have severe difficulties with the control of breathing (see Chapter 16). Further refinements in our understanding may help guide forthcoming applications from the rapidly evolving fields of RNAi and stem cell-based therapies. An understanding of the most promising molecular targets may lead to the development of entirely new therapeutic strategies for devastating respiratory control disorders for which we currently have no effective treatments or cures.

Acknowledgements

I thank B. Hodgeman for preparing the figures. Supported by NIH HL69064 and HL80209, and the ALS Association.

References

1. Babcock M, Shkoukani M, Aboubakr SE, Badr MS (2003) Determinants of long-term facilitation in humans during NREM sleep. J Appl Physiol 94: 53-59

2. Bach KB, Mitchell GS (1996) Hypoxia-induced long-term facilitation of respiratory activity is serotonin dependent. Respir Physiol 104: 251-260

3. Baker TL, Mitchell GS (2000) Episodic but not continuous hypoxia elicits long-term facilitation of phrenic motor output in rats. J Physiol 529: 215-219

4. Baker-Herman TL, Fuller DD, Bavis RW, Zabka AG, Golder FJ, Doperalski NJ, Johnson RA, Watters JJ, Mitchell GS (2004) BDNF is necessary and sufficient for spinal respiratory plasticity following intermittent hypoxia. Nat Neurosci 7: 48-55

5. Barr MRL, Sibigtroth CM, Mitchell GS (2007) Daily acute intermittent hypoxia improves respiratory function in rats with chronic cervical spinal hemisection. FASEB J, in press

6. Bavis RW, Baker-Herman TL, Zabka AG, Golder FJ, Fuller DD, Behan M, Mitchell GS (2003) Respiratory long-term facilitation differs among inbred rat strains. FASEB J 17: A824

7. Bocchiaro CM, Feldman JL (2004) Synaptic activity-independent persistent plasticity in endogenously active mammalian motoneurons. Proc Natl Acad Sci USA 101: 4292-4295

8. Bourke SC, Shaw PJ, Gibson GJ (2001) Respiratory function vs sleep-disordered breathing as predictors of QOL in ALS. Neurology 57: 2040-2044

9. Chen WG, Chang Q, Lin Y, Meissner A, West AE, Griffith EC, Jaenisch R, Greenberg ME (2003) Derepression of BDNF transcription involves calcium-dependent phosphorylation of MeCP2. Science 302: 885-889

10. Dahlberg JM, Wilkerson JER, Mitchell GS (2005) Differential $5HT_{2A}$, $5HT_7$, BDNF and TrkB receptor mRNA expression in ventral spinal segments associated with phrenic motor nucleus of four inbred rat strains. FASEB J 19: A1638

11. Dale EA, Nashold LJ, Mahamed S, Svendsen CN, Mitchell GS (2006) Sustained ventilatory capacity in a rat model of amyotrophic lateral sclerosis. FASEB J 20: A1213

12. Dale EA, Satriotomo I, Mitchell GS (2007) Thrice-weekly acute intermittent hypoxia induces Vascular Endothelial Growth Factor (VEGF) in phrenic motor neurons. FASEB J, in press

13. Denli AM, Hannon GJ (2003) RNAi: an ever-growing puzzle. Trends Biochem Sci 28: 196-201

14. DeVivo MJ, Ivie CS (1995) Life expectancy of ventilator-dependent persons with spinal cord injuries. Chest 108: 226-232

15. Dimitrejivic MR (1988) Model for the study of plasticity of the human nervous system: features of residual spinal cord motor activity resulting from established post-traumatic injury. Ciba Found Symp 138: 227-239

16. Dykxhoorn DM, Novina CD, Sharp PA (2003) Killing the messenger: short RNAs that silence gene expression. Nat Rev Mol Cell Biol 4: 457-467

17. Elbashir SM, Harborth J, Lendeckel W, Yalcin A, Weber K, Tuschl T (2001) RNA interference is mediated by 21- and 22-nucleotide RNAs. Genes Dev 15: 188-200

18. Feldman JL, Mitchell GS, Nattie EE (2003) Breathing: rhythmicity, plasticity, chemosensitivity. Annu Rev Neurosci 26: 239-266

19. Fire A, Xu S, Montgomery MK, Kostas SA, Driver SE, Mello CC (1998) Potent
     and specific genetic interference by double-stranded RNA in *Caenorhabditis ele-
     gans*. Nature 391: 806-811
20. Fletcher EC (1981) Invited review: Physiological consequences of intermittent
     hypoxia: systemic blood pressure. J Appl Physiol 90: 1600-1605
21. Fregosi RF, Mitchell GS (1994) Long-term facilitation of inspiratory intercostal
     nerve activity following carotid sinus nerve stimulation in cats. J Physiol 477:
     469-479
22. Fuller DD, Bach KB, Baker TL, Kinkead R, Mitchell GS (2000) Long term fa-
     cilitation of phrenic motor output. Respir Physiol 121: 135-146
23. Fuller DD, Baker TL, Behan M, Mitchell GS (2001a) Expression of hypoglossal
     long-term facilitation differs between substrains of Sprague-Dawley rat. Physiol
     Genomics 4: 175-181
24. Fuller DD, Wang ZY, Ling L, Olson EB, Bisgard GE, Mitchell GS (2001b) In-
     duced recovery of hypoxic phrenic responses in adult rats exposed to hyperoxia
     for the first month of life. J Physiol 536: 917-926
25. Fuller DD, Johnson SM, Johnson RA, GS Mitchell (2002a) Chronic cervical spi-
     nal sensory denervation reveals ineffective spinal pathways to phrenic motoneu-
     rons in the rat. Neurosci Lett 323: 25-28
26. Fuller DD, Johnson SM, Mitchell GS (2002b) Respiratory long-term facilitation
     (LTF) is associated with enhanced spinally evoked phrenic potentials. Program
     N°. 363.1, *Abstract Viewer/Itinerary Planner,* Washington, DC: Society for Neu-
     roscience
27. Fuller DD, Johnson SM, Olson EB, Mitchell GS (2003) Synaptic pathways to
     phrenic motoneurons are enhanced by chronic intermittent hypoxia after cervical
     spinal cord injury. J Neurosci 23: 2993-3000
28. Fuller DD, Baker-Herman TL, Golder FJ, Doperalski NJ, Watters JJ, Mitchell GS
     (2005) Cervical spinal cord injury upregulates ventral spinal 5HT2A receptors. J
     Neurotrama 22:203-213.
29. Fuller DD, Golder FJ, Olson EB, Mitchell GS (2006) Recovery of phrenic activ-
     ity and ventilation after cervical spinal hemisection in rats. J Appl Physiol 100:
     800-806
30. Golder FJ, Mitchell GS (2005) Spinal synaptic enhancement with acute intermit-
     tent hypoxia improves respiratory function after chronic cervical spinal cord in-
     jury. J Neurosci 25: 2925-2932
31. Golder FJ, Ranganathan L, Satriotomo I, Hoffman S, Mahamed S, Baker-Herman
     TL and Mitchell GS (2007) A2a receptor transactivation of cervical TrkB protein
     improves respiratory function after cervical spinal injury. FASEB J, in press.
32. Goshgarian HG (1981) The role of cervical afferent nerve fiber inhibition of the
     crossed phrenic phenomenon. Exp Neurol 72: 211-225
33. Goshgarian HG (2003) The crossed phrenic phenomenon: a model for plasticity
     in the respiratory pathways following spinal cord injury. J Appl Physiol 94: 795-
     810
34. Gozal D (2001) Morbitity of obstructive sleep apnea in children: facts and theory.
     Sleep Breath 5: 35-42
35. Gozal D, Kheirandish-Gozal L (2006) Sleep apnea in children-treatment consid-
     erations. Paediatr Respir Rev 7 Supp 1: S58-61
36. Guth L (1976) Functional plasticity in the respiratory plasticity of the mammalian
     spinal cord. Exp Neurol 51: 414-420

37.   Hadley SD, Walker PD, Goshgarian HG (1999) Effects of the serotonin synthesis inhibitor p-CPA on the expression of the crossed phrenic phenomenon 4 h following C2 spinal cord hemisection. Exp Neurol 160: 479-488

38.   Hannon GJ (2003) RNAi: A guide to gene silencing. Cold Spring Harbor, New York: Cold Spring Harbor Laboratory Press

39.   Harris DP, Balasubramaniam A, Badr MS, Mateika JH (2006) Long-term facilitation of ventilation and genioglossus muscle activity is evident in the presence of elevated levels of carbon dioxide in awake humans. Am J Physiol Regul Integr Comp Physiol 291: R1111-1119

40.   Hoffman MS, Mahamed S, Golder FJ, Mitchell GS (2007) Adenosine A2A receptors constrain phrenic long term facilitation following acute intermittent hypoxia. FASEB J, in press

41.   Ilzecka J, Stelmasiak Z, Balicka G (2003) Respiratory function in amyotrophic lateral sclerosis. Neurol Sci 24: 288-289

42.   Johnson RA, Okragly AJ, Haak-Frendscho M, Mitchell GS (2000) Cervical dorsal rhizotomy increases brain-derived neurotrophic factor and neurotrophin-3 expression in the ventral spinal cord. J Neurosci 20 RC77: 1-5

43.   Kalaria RN, Spoors L, Laude EA, Emery CJ, Thwaites-Bee D, Fairlie J, Oakley AE, Barer DH, Barer GR (2004) Hypoxia of sleep apnoea: cardiopulmonary and cerebran changes after intermittent hypoxia in rats. Respir Physiol Neurobiol 140: 53-62

44.   Kinkead R, Zhan WZ, Prakash YS, Bach KB, Sieck GC, Mitchell GS (1998) Cervical dorsal rhizotomy enhances serotonergic innervation of phrenic motoneurons and serotonin-dependent long-term facilitation of respiratory motor output in rats. J Neurosci 18: 8436-8443

45.   Kinney HC, Filiano JJ, White WF (2001) Medullary serotonergic network deficiency in the sudden infant death syndrome: review of a 15-year study of a single dataset. J Neuropathol Exp Neurol 60: 228-247

46.   Klein SM, Behrstock S, McHugh J, Hoffmann K, Wallace K, Suzuki M, Aebischer P, Svendsen CN (2005) GDNF delivery using human neural progenitor cells in a rat model of ALS. Hum Gene Ther 16: 509-521

47.   Klose R, Bird A (2003) Molecular biology. MeCP2 repression goes nonglobal. Science 302: 793-795

48.   Kraiczi H, Hedner J, Dahlof P, Ejnell H, Carlson J (1999) Effect of serotonin uptake inhibition on breathing during sleep and daytime symptoms in obstructive sleep apnea. Sleep 22: 61-67

49.   Lechtzin N, Rothstein J, Clawson L, Diette GB Wiener CM (2002) Amyotrophic lateral sclerosis: evaluation and treatment of respiratory impairment. Amyotrophic Lateral Scler Other Motor Neuron Disord 3: 5-13

50.   Lee FS, Chao MV (2001) Activation of Trk neurotrophin receptors in the absence of neurotrophins. Proc Natl Acad Sci USA 98: 3555-3560

51.   Lee FS, Rajagopal R, Chao MV (2002) Distinctive features of Trk neurotrophin receptor transactivation by G protein-coupled receptors. Cytokine Growth Factor Rev. 13: 11-17

52.   Li XJ, Du ZW, Zarnowska ED, Pankratz M, Hansen LO, Pearce RA, Zhang SC (2005) Specification of motoneurons from human embryonic stem cells. Nat Biotechnol. 23: 215-221

53.   Ling L, Bach KB, Mitchell GS (1994) Serotonin reveals ineffective spinal pathways to contralateral phrenic motoneurons in spinally hemisected rats. Exp Brain Res 101: 35-43

54. Ling L, Fuller DD, Bach KB, Kinkead R, Olson EB, Mitchell GS (2001) Chronic intermittent hypoxia elicits serotonin-dependent plasticity in the central neural control of breathing. J Neurosci 21: 5381-5388

55. Llado J, Haenggeli C, Pardo A, Wong V, Benson L, Coccia C, Rothstein JD, Shefner JM, Maragakis NJ (2005) Degeneration of respiratory motor neurons in the SOD1 G93A transgenic rat model of ALS. Neurobiol Dis 21: 110-118

56. Lovett-Barr MR, Mitchell GS, Satriotomo I, Johnson SM (2006) Serotonin-induced in vitro long-term facilitation exhibits differential pattern sensitivity in cervical and thoracic inspiratory motor output. Neuroscience 142: 885-892

57. Lyall RA, Donaldson N, Polkey MI, Leigh PN, Moxham J (2001) Respiratory muscle strength and ventilatory failure in amyotrophic lateral sclerosis. Brain 124: 2000-2013

58. MacFarlane P, Mitchell GS (2006) Respiratory long term facilitation evoked by acute intermittent hypoxia is impaired following intravenous injection of a superoxide dismutase mimetic. FASEB J 20: A372

59. MacFarlane PM, Satriotomo I, Mitchell GS (2007) Reactive oxygen species generated by NADPH oxidase are necessary for phrenic long-term facilitation following acute intermittent hypoxia. FASEB J, in press

60. Mahamed S, Mitchell GS (2006) Is there a link between intermittent hypoxia-induced respiratory plasticity and obstructive sleep apnoea? Exp Physiol 92: 27-37

61. Mahamed S, Mitchell GS (2006) Does simulated apnea elicit respiratory long-term facilitation? FASEB J 20: A372

62. Mansel JK, Norman JR (1990) Respiratory complications and management of spinal cord injuries. Chest 97: 1446-1452

63. McGuire M, Zhang Y, White DP, Ling L (2004) Serotonin receptor subtypes required for ventilatory long-term facilitation and its enhancement after chronic intermittent hypoxia in awake rats. Am J Physiol Regul Integr Comp Physiol 286: R334-341

64. McGuire M, Zhang Y, White DP, Ling L (2005) Phrenic long-term facilitation requires NMDA receptors in the phrenic motonucleus in rats. J Physiol 567: 599-611

65. McManus MT, Sharp PA (2002) Gene silencing in mammals by small interfering RNAs. Nat Rev Genet 3: 737-747

66. Mitchell GS, Baker TL, Nanda SA, Fuller DD, Zabka AG, Hodgeman BA, Bavis RW, Mack KJ, Olson EB (2001) Physiological and genomic, consequences of intermittent hypoxia. J Appl Physiol 90: 2466-2475

67. Mitchell GS, Johnson SM (2003) Neuroplasticity in respiratory motor control. J Appl Physiol 94: 358-374

68. Mitchell GS, Sloan HE, Jiang C, Hayashi F, J Lipski (1992) 5-Hydroxytryptophan (5-HTP) augments spontaneous and evoked phrenic motoneuron discharge in spinalized rats. Neurosci Lett 141: 75-78

69. Nakamura A, Wenninger JM, Olson EB, Bisgard GE, Mitchell GS (2006) Ventilatory long-term facilitation following intermittent hypoxia is state-dependent in rats. J Physiol Sci 56 (Suppl): S75

70. Nakamura A, Wenninger JM, Olson EB, Bisgard GE, Mitchell GS (2005) Ventilatory long-term facilitation in sleeping Lewis rats. FASEB J 19: A1284

71. Nakano H Magalang UJ, Lee SD, Krasney JA, Farkas GA (2001) Serotonergic modulation of ventilation and upper airway stability in obese Zucker rats. Am J Respir Crit Care Med 163: 1191-1197

72.   Nantwi KD, Goshgarian HG (1998) Effects of chronic systemic theophylline injections on recovery of hemidiaphragmatic function after cervical spinal cord injury in adult rats. Brain Res 789: 126-129

73.   Nashold LJ, Wilkerson JER, Satriotomo I, Dale EA, Svendsen CN, Mitchell GS (2006) Phrenic, but not hypoglossal, motor output is diminished in a rat model of amyotrophic lateral sclerosis (ALS). FASEB J 20: A1212

74.   Peng YJ, Prabhakar NR (2003) Reactive oxygen species in the plasticity of respiratory behavior elicited by chronic intermittent hypoxia. J Appl Physiol 94: 2342-2349

75.   Porter WT (1895) The path of the respiratory impulse from the bulb to the phrenic nuclei. J Physiol 17: 455-485

76.   Prochazka A, Mushahwar VK (2001) Spinal cord function and rehabilitation-an overview. J Physiol 533: 3-4

77.   Ralph GS, Radcliffe PA, Day DM, Carthy JM, Leroux MA, Lee DC, Wong LF, Bilsland LG, Greensmith L, Kingsman SM, Mitrophanous KA, Mazarakis ND, Azzouz M (2005) Silencing mutant SOD1 using RNAi protects against neurodegeneration and extends survival in an ALS model. Nat Med. 11: 429-433

78.   Ramer MS, Priestly JV, McMahon SB (2000) Functional regeneration of sensory axons into the adult spinal cord. Nature 403: 312-316

79.   Raoul C, Abbas-Terki T, Bensadoun JC, Guillot S, Haase G, Szulc J, Henderson CE, Aebischer P (2005) Lentiviral-mediated silencing of SOD1 through RNA interference retards disease onset and progression in a mouse model of ALS. Nat Med 11: 423-428

80.   Satriotomo I, Nashold LJ, Svendsen CN, Mitchell GS (2006) Enhancement of BDNF and serotonin terminal density in phrenic and hypoglossal motor nuclei in a rat model of Amyotrophic Lateral Sclerosis (ALS). FASEB J 20: A1212

81.   Satriotomo I, Dale EJ, Mitchell GS (2007) Thrice weekly intermittent hypoxia increases expression of key proteins necessary for phrenic long-term facilitation: a possible mechanism of respiratory metaplasticity? FASEB J, in press

82.   Storkebaum E, Lambrechts D, Carmeliet P (2004) VEGF: once regarded as a specific angiogenic factor, now implicated in neuroprotection. Bioessays 26: 943-954

83.   Tai Q, Palazzolo KL, Goshgarian HG (1996) Synaptic plasticity of 5-hydroxytryptamine-immunoreactive terminals in the phrenic nucleus following spinal cord injury: a quantitative electron microscopic analysis. J Comp Neurol 386: 613-624

84.   Tankersley CG, Haengelli C, Rothstein J (2007) Respiratory impairment in a mouse model of Amyotrophic Lateral Sclerosis. J Appl Physiol 102: 926-932

85.   Veasey SC, Panckeri KA, Hoffman EA, Pack AI, Hendricks JC (1996) The effects of serotonin antagonists in an animal model of sleep-disordered breathing. Am J Respir Crit Care Med 153: 776-786

86.   Veasey SC, Chachkes J, Fenik P, Hendricks JC (2001) The effects of ondansetron on sleep-disordered breathing in the English bulldog. Sleep 24: 155-160

87.   Wilkerson JER, Baker-Herman TL, Mitchell GS (2005) BDNF synthesis and ERK1 activation are induced in ventral cervical spinal cord following intermittent hypoxia in Brown Norway rats. Program No. 635.8. *Abstract Viewer/Itinerary Planner,* Washington, DC: Society for Neuroscience

88.   Wilkerson JER, Mitchell GS (2005) Daily acute intermittent hypoxia enhances hypoglossal, but not phrenic Long Term Facilitation (LTF) in Brown Norway rats FASEB J 19: A1639

89.   Wilkerson JER, Baker-Herman TL, Mitchell GS (2006) Okadaic acid-sensitive protein phosphatases constrain phrenic long-term facilitation following sustained hypoxia. FASEB J 20: A372

90.   Wong PC, Cai H, Borchelt DR, Price DL (2002) Genetically engineered mouse models of neurodegenerative diseases. Nat Neurosci 5: 633-639

91.   Young W (1996) Spinal cord regeneration. Science 273: 451

92.   Zabka AG, Behan M and Mitchell GS (2001a) Long term facilitation of respiratory motor output decreases with age in male rats. J Physiol 531: 509-514

93.   Zabka AG, Behan M and Mitchell GS (2001b) Selected contribution: time-dependent hypoxic respiratory responses in female rats are influenced by age and by the estrus cycle. J Appl Physiol 91: 2831-2838

94.   Zabka AG, Mitchell GS and Behan M (2005) Ageing and gonadectomy have similar effects on hypoglossal long-term facilitation in male Fischer rats. J Physiol 563: 557-568

95.   Zabka AG, Mitchell GS and Behan M (2006) Conversion from testosterone to oestradiol is required to modulate respiratory long-term facilitation in male rats. J Physiol 576: 903-912

96.   Zabka AG, Mitchell GS, Olson EB Jr, Behan M (2003) Selected contribution: chronic intermittent hypoxia enhances respiratory long-term facilitation in geriatric female rats. J Appl Physiol 95: 2614-2623

97.   Zachary I (2005) Neuroprotective role of vascular endothelial growth factor: signalling mechanisms, biological function, and therapeutic potential. Neurosignals 14: 207-221

98.   Zeng Y, Yi R, Cullen BR (2003) MicroRNAs and small interfering RNAs can inhibit mRNA expression by similar mechanisms. Proc Natl Acad Sci USA. 100: 9779-9784

99.   Zhou SY, Goshgarian HG (1999) 5-Hydroxytryptophan-induced respiratory recovery after cervical spinal cord hemisection in rats. J Appl Physiol 89: 1528-1536

100.   Zhou SY, Goshgarian HG (1999) Effects of serotonin on crossed phrenic nerve activity in cervical spinal cord hemisected rats. Exp Neurol 160: 446-453

101.   Zhou SY, Basura GJ, Goshgarian HG (2001) Serotonin(2) receptors mediate respiratory recovery after cervical spinal cord hemisection in adult rats. J Appl Physiol 91: 2665-2673

102.   Zimmer MB, Goshgarian HG (2006) Spinal activation of serotonin 1A receptors enhances latent respiratory activity after spinal cord injury. J Spinal Cord Med 29: 147-155

Index

A

A/J mice
    lung mechanics, 154
    posthypoxic ventilatory behaviour, 141-142
    recurrent apneas, 146-148
A2A receptors, 296, 299, 304
Acethylcholine
    acethylcholinesterase, 233
    carotid bodies, 230
    cholinergic transmission, 181
Adenosine, 296, 299, 304
Adrenergic centers
    C1, 26, 33, 35, 196
    C1-3, 26, 33, 35, 196
Aging, 296, 303
Akt (protein kinase B), 95-96
Ambiguous nucleus, 26, 174, 177, 180
AMPA/kainate glutamatergic receptors, 176
Ampakine, 267
Amyotrophic lateral sclerosis (ALS), 292, 300-305
Andean high altitude natives, 9, 11
Angiotensine-converting enzyme, 125
AP (*see* area postrema)
Apnea
    automatic processing, 246
    central, 140, 145
    defining
        adult Humans, 140
        mice, 145
        rodents, 140
    inbred strains of mice, 146-148
    models, 145-148
    mutant newborn mice, 177, 182, 206-207, 227-229, 246-247
    neurodegenerative disease, 177
    postinspiratory, 273
    posthypoxic recurrent, 146

    post-sigh, 138, 141, 145-146
    pre-BötC neurons, 212
    prematurity, 3, 225, 234
    REM-related, 146
    Rett syndrome, 272-273
    spontaneously hypertensive rats, 145
Apoprotein E, 124
Apoptosis, 76, 169
Apparent life-threatening events (ALTE), 3, 123
Area postrema
    development, 198-199, 205
    expression of *Phox2b*, 26
    central chemosensitivity, 193
Arousal
    defining
        newborn rodents, 245-246
    responses to hypoxia
        Phox2b$^{+/-}$ mutant
        newborn mice, 249
Arterial chemoreflex (*see also* carotid bodies)
    Bdnf mutant newborn mice, 180, 230
    development, 198-199
    neurotrophic factors, 230
Asthmatic patients
    hypoxic ventilatory responses, 13
Ataxic respiration, 175, 212
Attention-deficit hyperactivity disorder, 225, 234
Autonomic nervous system dysregulation
    CCHS, 45, 49, 57
    SIDS, 91-95, 99-100
Autophagy, 76
Axonal
    outgrowth in necdin deficient mice, 264

Printed in the United States of America